TExES 114 Mathematics-Science 4-8
Teacher Certification Exam

By: Sharon Wynne, M.S
Southern Connecticut State University

"And, while there's no reason yet to panic, I think it's only prudent that we make preparations to panic."

XAMonline, INC.
Boston

Copyright © 2007 XAMonline, Inc.
All rights reserved. No part of the material protected by this copyright notice may be reproduced or utilized in any form or by any means, electronic or mechanical, including photocopying, recording or by any information storage and retrievable system, without written permission from the copyright holder.

To obtain permission(s) to use the material from this work for any purpose including workshops or seminars, please submit a written request to:

> XAMonline, Inc.
> 21 Orient Ave.
> Melrose, MA 02176
> Toll Free 1-800-509-4128
> Email: info@xamonline.com
> Web www.xamonline.com
> Fax: 1-781-662-9268

Library of Congress Cataloging-in-Publication Data

Wynne, Sharon A.
    Mathematics-Science 4-8 114: Teacher Certification / Sharon A. Wynne. -2nd ed.
    ISBN 978-1-58197-948-0
    1. Mathematics-Science 4-8 114.   2. Study Guides.   3. TExES
    4. Teachers' Certification & Licensure.   5. Careers

Disclaimer:
The opinions expressed in this publication are the sole works of XAMonline and were created independently from the National Education Association, Educational Testing Service, or any State Department of Education, National Evaluation Systems or other testing affiliates.

Between the time of publication and printing, state specific standards as well as testing formats and website information may change that is not included in part or in whole within this product. Sample test questions are developed by XAMonline and reflect similar content as on real tests; however, they are not former tests. XAMonline assembles content that aligns with state standards but makes no claims nor guarantees teacher candidates a passing score. Numerical scores are determined by testing companies such as NES or ETS and then are compared with individual state standards. A passing score varies from state to state.

Printed in the United States of America   œ-1

TExES: Mathematics-Science 4-8 114
ISBN: 978-1-58197-948-0

TEACHER CERTIFICATION STUDY GUIDE

Table of Contents

DOMAIN I. NUMBER CONCEPTS

Competency 01   The teacher understands the structure of number systems, the development of a sense of quantity, and the relationship between quantity and symbolic representations 1

Competency 02   The teacher understands number operations and computational algorithms 12

Competency 03   The teacher understands ideas of number theory and uses numbers to model and solve problems within and outside of mathematics 22

DOMAIN II. PATTERNS AND ALGEBRA

Competency 04   The teacher understands and uses mathematical reasoning to identify, extend, and analyze patterns and understands the relationships among variable, expressions, equations, inequalities, relations, and functions 29

Competency 05   The teacher understands and uses linear functions to model and solve problems 35

Competency 06   The teacher understands and uses nonlinear functions and relations to model and solve problems 47

Competency 07   The teacher uses and understands the conceptual foundations of calculus related to topics in middle school mathematics 71

DOMAIN III. GEOMETRY AND MEASUREMENT

Competency 08   The teacher understands measurement as a process 85

Competency 09   The teacher understands the geometric relationships and axiomatic structure of Euclidian geometry 92

Competency 010   The teacher analyzes the properties of two- and three- dimensional figures 102

MATHEMATICS-SCIENCE 4-8

Competency 011	The teacher understands transformational geometry and related algebra to geometry and trigonometry using the Cartesian coordinate system	108

DOMAIN IV. PROBABILITY AND STATISTICS

Competency 012	The teacher understands how to use graphical and numerical techniques to explore data, characterize patterns, and describe departures from patterns	116
Competency 013	The teacher understands the theory of probability	124
Competency 014	The teacher understands the relationship among probability theory, sampling, and statistical inference, and how statistical inference is used in making and evaluating predictions	133

DOMAIN V. MATHEMATICAL PROCESSES AND PERSPECTIVES

Competency 015	The teacher understands mathematical reasoning and problem solving	138
Competency 016	The teacher understands mathematical connections within and outside of mathematics and how to communicate mathematical ideas and concepts	143

DOMAIN VI. MATHEMATICAL LEARNING, INSTRUCTION, AND ASSESSMENT

Competency 017	The teacher understands how children learn and develop mathematical skills, procedures, and concepts	148
Competency 018	The teacher understands how to plan, organize, and implement instruction using knowledge of students, subject matter, and statewide curriculum (Texas Essential Knowledge and Skills [TEKS]) to teach all students to use mathematics	151
Competency 019	The teacher understands assessment and uses a variety of formal and informal assessment techniques to monitor and guide mathematics instruction and to evaluate student progress	156

TEACHER CERTIFICATION STUDY GUIDE

Answer Key to Practice Problems 159

Sample Test: Mathematics 161

Answer Key: Mathematics 190

Rationales for Sample Questions: Mathematics 191

DOMAIN VII.	SCIENTIFIC INQUIRY AND PROCESS

COMPETENCY 020 UNDERSTANDING HOW TO MANAGE LEARNING ACTIVITIES TO ENSURE THE SAFETY OF ALL STUDENTS 203

Skill 20.1 Understands safety regulations and guidelines for science facilities and science instruction 203

Skill 20.2 Knows procedures for and sources of information regarding the appropriate handling, use, disposal, care, and maintenance of chemicals, materials, specimens, and equipment 204

Skill 20.3 Knows procedures for the safe handling and ethical care and treatment of organisms and specimens 205

COMPETENCY 021 UNDERSTANDING THE CORRECT USE OF TOOLS, MATERIALS, EQUIPMENT, AND TECHNOLOGIES 207

Skill 21.1 Selects and safely uses appropriate tools, technologies, materials, and equipment needed for instructional activities 207

Skill 21.2 Understands concepts of precision, accuracy, and error with regard to reading and recording numerical data from a scientific instrument 209

Skill 21.3	Understands how to gather, organize, display, and communicate data in a variety of ways 210
Skill 21.4	Understand the international system of measurement and performs unit conversions within measurement systems 211

COMPETENCY 022 UNDERSTANDING THE PROCESS OF SCIENTIFIC INQUIRY AND THE HISTORY AND NATURE OF SCIENCE 212

Skill 22.1	Understands the characteristics of various types of scientific investigations 212
Skill 22.2	Understands how to design, conduct, and communicate the results of a variety of scientific investigations 213
Skill 22.3	Understands the historical development of science and the contributions that diverse cultures and individuals of both genders have made to scientific knowledge 213
Skill 22.4	Understands the roles that logical reasoning, verifiable evidence, prediction, and peer review play in the process of generating and evaluating scientific knowledge 216
Skill 22.5	Understands principles of scientific ethics 216
Skill 22.6	Develops, analyzes, and evaluates different explanations for a given scientific result 217
Skill 22.7	Demonstrates an understanding of potential sources of error in inquiry based investigation 218
Skill 22.8	Demonstrates an understanding of how to communicate and defend the results of an inquiry-based investigation 218

COMPETENCY 023 UNDERSTANDING HOW SCIENCE IMPACTS THE DAILYLIVES OF STUDENTS AND INTERACTS WITH AND INFLUENCES PERSONAL AND SOCIETAL DECISIONS 219

Skill 23.1	Understands that decisions about the use of science are based on factors such as ethical standards, economics, and personal and societal needs 219

Skill 23.2	Applies scientific principles and the theory of probability to analyze the advantages of, disadvantages of, or alternatives to a give decision or course of action	220
Skill 23.3	Applies scientific principles and the processes to analyze factors that influence personal choices concerning fitness and health, including physiological and psychological effect and risks associated with the use of substances and substance abuse	220
Skill 23.4	Understand concepts, characteristics, and issues related to changes in populations and human population growth	221
Skill 23.5	Understand the types and uses of natural resources and the effects of human consumption of the renewal and depletion of resources	222
Skill 23.6	Understands the role science can play in helping resolve personal, societal, and global challenges	223

COMPETENCY 024 UNDERSTANDING THE UNIFYING CONCEPTS AND PROCESSES THAT ARE COMMON TO ALL SCIENCES 225

Skill 24.1	Understands how the following concepts and processes provide and unifying explanatory framework across the science disciplines: systems, order, and organization; evidence, models, and explanation; change, constancy, and measurements; evolution and equilibrium; and form and function	225
Skill 24.2	Demonstrates an understanding of how patterns in observations and data can be used to make explanations and predictions	226
Skill 24.3	Analyzes interactions and interrelationships between systems and subsystems	227
Skill 24.4	Analyzes unifying concepts to explore similarities in a variety of natural phenomena	228
Skill 24.5	Understands how properties and patterns of systems can be described in terms of space, time, energy, and matter	229
Skill 24.6	Understands how change and constancy occur in systems	230
Skill 24.7	Understands the complementary nature of form and function in a given system	231

TEACHER CERTIFICATION STUDY GUIDE

Skill 24.8 Understands how models are used to represent the natural world and how to evaluate the strengths and limitations of a variety of scientific models (e.g., physical, conceptual, mathematical) 232

DOMAIN VIII. THE PHYSICAL SCIENCES

COMPETENCY 025 UNDERSTANDING FORCES AND MOTION AND THEIR RELATIONSHIPS 233

Skill 25.1 Demonstrates an understanding of properties of universal forces 233

Skill 25.2 Understand how to measure, graph, and describe changes in motion using concepts of displacement, velocity, and acceleration 233

Skill 25.3 Understands the vector nature of force 235

Skill 25.4 Identifies the forces acting on an object and applies Newton's laws to describe the motion of an object 235

Skill 25.5 Analyzes the relationship between force and motion in a variety of situations 236

COMPETENCY 026 UNDERSTANDING PHYSICAL PROPERTIES OF AND CHANGES IN MATTER 238

Skill 26.1 Describes the physical properties of substances (e.g., density, boiling point, solubility, thermal and electrical conductivity) 238

Skill 26.2 Describes the physical properties and molecular structure of solids, liquids, and gases 240

Skill 26.3 Describes the relationship between the molecular structure of materials and their physical properties 240

Skill 26.4 Relates the physical properties of an element to its placement in the periodic table 242

Skill 26.5 Distinguishes between physical and chemical changes in matter 245

Skill 26.6 Applies knowledge of physical properties of and changes in matter to processes and situations that occur in life and earth/space science 245

TEACHER CERTIFICATION STUDY GUIDE

COMPETENCY 027	UNDERSTANDING CHEMICAL PROPERTIES AND CHANGES IN MATTER	246
Skill 27.1	Describes the structure and components of the atom	246
Skill 27.2	Distinguishes among elements, mixtures, and compounds and describes their properties	248
Skill 27.3	Relates the chemical properties of an element to its placement in the periodic table	250
Skill 27.4	Describes chemical bonds and chemical formulas	251
Skill 27.5	Analyzes chemical reactions and their associated chemical equations	253
Skill 27.6	Explains the importance of a variety of chemical reactions that occur in daily life	254
Skill 27.7	Understands applications of chemical properties of matter in physical, life, and earth/space science and technology	258
COMPETENCY 028	UNDERSTANDING ENERGY AND INTERACTIONS BETWEEN MATTER AND ENERGY	260
Skill 28.1	Describes concepts of work, power, and potential and kinetic energy	260
Skill 28.2	Understands the concept of heat energy and the difference between heat and temperature	261
Skill 28.3	Understands the principles of electricity and magnetism and their applications	263
Skill 28.4	Applies knowledge of properties of light to describe the function of optical systems and phenomena	267
Skill 28.5	Demonstrates an understanding of the properties, production, and transmission of sound	269
Skill 28.6	Applies knowledge of properties and characteristics of waves to describe a variety of waves	270

TEACHER CERTIFICATION STUDY GUIDE

COMPETENCY 029   **UNDERSTANDING ENERGY TRANSFORMATIONS AND THE CONSERVATION OF MATTER AND ENERGY** 273

Skill 29.1   Describes the processes that generate energy in the sun and other stars 273

Skill 29.2   Applies the law of conservation of matter to analyze a variety of situations 273

Skill 29.3   Describes sources of electrical energy and processes of energy transformation for human uses 274

Skill 29.4   Understands exothermic and endothermic chemical reactions and their applications 275

Skill 29.5   Applies knowledge of the transfer of energy in a variety of situations (e.g., the production of heat, light, sound, and magnetic effects by electrical energy; the process of photosynthesis; weather processes; food webs; food/energy pyramids) 275

Skill 29.6   Applies the law of conservation of energy to analyze a variety of physical phenomena 276

Skill 29.7   Understands applications of energy transformations and the conservation of matter and energy in life and earth/space science 277

DOMAIN IX.   THE LIFE SCIENCES

COMPETENCY 030   **UNDERSTANDING THE STRUCTURE AND FUNCTION OF LIVING THINGS** 278

Skill 30.1   Describes characteristics of organisms from the major taxonomic groups 278

Skill 30.2   Analyzes how structure complements function in cells 280

Skill 30.3   Analyzes how structure complements function in tissues, organs, organ systems, and organisms 282

Skill 30.4   Identifies human body systems and describes their functions 283

Skill 30.5   Describes how organisms obtain and use energy and matter 291

TEACHER CERTIFICATION STUDY GUIDE

Skill 30.6 Applies chemical properties to describe the structure and function of the basic chemical components or living things 293

COMPETENCY 031 UNDERSTANDING REPRODUCTION AND THE MECHANISMS OF HEREDITY 297

Skill 31.1 Compares and contrasts sexual and asexual reproduction 297

Skill 31.2 Understands the organization of heredity material 297

Skill 31.3 Describes how an inherited trait can be determined by one or many genes and how more than one trait can be influenced by a single gene 299

Skill 31.4 Distinguishes between dominant and recessive traits and predict the probable outcomes of genetic combinations 300

Skill 31.5 Evaluates the influence of environmental and genetic factors on the traits of an organism 301

Skill 31.6 Describes current applications of genetic research 302

COMPETENCY 032 UNDERSTANDING ADAPTATION OF ORGANISMS AND THE THEORY OF EVOLUTION 303

Skill 32.1 Describes similarities and differences among various types of organisms and methods of classifying organisms 303

Skill 32.2 Describes traits in a population of species that enhance its survival and reproductive success 305

Skill 32.3 Describes how populations and species change through time 306

Skill 32.4 Applies knowledge of the mechanisms and processes of biological evolution (e.g., variation, mutation, environmental factors, natural selection) 307

Skill 32.5 Describes evidence that supports the theory of evolution on life on Earth 308

TEACHER CERTIFICATION STUDY GUIDE

COMPETENCY 033   **UNDERSTANDING REGULATORY MECHANISMS AND BEHAVIOR** 311

Skill 33.1   Describes how organisms respond to internal and external stimuli 311

Skill 33.2   Applies knowledge of structures and physiological processes that maintain stable internal conditions 312

Skill 33.3   Demonstrates an understanding of feedback mechanisms that allow organisms to maintain stable internal conditions 313

Skill 33.4   Understands how evolutionary history affects behavior 314

COMPETENCY 034   **UNDERSTANDING THE RELATIONSHIP BETWEEN ORGANISMS AND THE ENVIRONMENT** 315

Skill 34.1   Identifies the abiotic and biotic components of an ecosystem 315

Skill 34.2   Analyzes the interrelationships among producers, consumers, and decomposers in an ecosystem 315

Skill 34.3   Identifies factors that influence the size and growth of population in an ecosystem 316

Skill 34.4   Analyzes adaptive characteristics that result in a population's or species' unique niche in an ecosystem 317

Skill 34.5   Describes and analyzes energy flow through various types of ecosystems 318

Skill 34.6   Knows how populations and species modify and affect ecosystems 319

DOMAIN X.   **EARTH AND SPACE SCIENCE**

COMPETENCY 035   **UNDERSTANDING THE STRUCTURE AND FUNCTION OF EARTH SYSTEMS** 321

Skill 35.1   Understands the structure of Earth and analyzes constructive and destructive processes that produce geological change 321

Skill 35.2   Understands the form and function of surface and subsurface water 324

Skill 35.3	Applies knowledge of the composition and structure of the atmosphere and its properties	326
Skill 35.4	Demonstrates an understanding of the interactions that occur among the biosphere, geosphere, hydrosphere, and atmosphere	327
Skill 35.5	Applies knowledge of how human activity and natural processes, both gradual and catastrophic, can alter earth systems	328
Skill 35.6	Identifies the sources of energy in earth systems and describes mechanisms of energy transfer	329
COMPETENCY 036	**UNDERSTANDING CYCLES IN THE EARTH'S SYSTEMS**	**330**
Skill 36.1	Understands the rock cycle and how rocks, minerals, and soils are formed	330
Skill 36.2	Understands the water cycle and its relationship to earth systems	332
Skill 36.3	Understand the nutrient cycle and its relationship to earth systems	332
Skill 36.4	Applies knowledge of how human and natural processes affect earth systems	333
Skill 36.5	Understands the dynamic interactions that occur among the various cycles in the biosphere, geosphere, hydrosphere, and atmosphere	333
COMPETENCY 037	**UNDERSTANDING THE ROLE OF ENERGY IN WEATHER AND CLIMATE**	**334**
Skill 37.1	Understand the elements of weather and how they are measured	334
Skill 37.2	Compares and contrasts weather and climate	334
Skill 37.3	Analyzes weather charts and data to make weather predictions	335
Skill 37.4	Applies knowledge of how transfers of energy among earth systems affect weather and climate	336
Skill 37.5	Analyzes how Earth's position, orientation, and surface features affect weather and climate	337

COMPETENCY 038 UNDERSTANDING THE CHARACTERISTICS OF THE SOLAR SYSTEM AND THE UNIVERSE 339

Skill 38.1 Understand the properties and characteristics of celestial objects 339

Skill 38.2 Applies knowledge of the earth-moon-sun system and the interactions among them 340

Skill 38.3 Identifies properties of the components of the solar system 341

Skill 38.4 Recognizes characteristics of stars and galaxies and their distribution in the universe 343

Skill 38.5 Demonstrates an understanding of scientific theories of the origin of the universe 344

COMPETENCY 039 UNDERSTANDING THE HISTORY OF THE EARTH SYSTEM 346

Skill 39.1 Understands the scope of the geologic time scale and its relationship to geologic processes 346

Skill 39.2 Demonstrates an understanding of theories about the earth's origin and geologic history 347

Skill 39.3 Demonstrates an understanding of how tectonic forces have shaped landforms over time 349

Skill 39.4 Understands the formation of fossils and the importance of the fossil record in explaining the earth's history 350

DOMAIN XI. SCIENCE LEARNING, INSTRUCTION, AND ASSESSMENT

COMPETENCY 040 THEORETICAL AND PRACTICAL KNOWLEDGE ABOUT TEACHING SCIENCE AND HOW STUDENTS LEARN SCIENCE 351

Skill 40.1 Understands how the developmental characteristics, prior knowledge, and experience, and attitudes of students influence science learning 351

Skill 40.2 Selects and adapts science curricula, content, instructional materials, and activities to meet the interests, knowledge, understanding, abilities, experiences, and needs of all students, including English Language Learners 353

Skill 40.3	Understands how to use situations from students' daily lives to develop instructional materials that investigate how science can be used to make informed decisions	354
Skill 40.4	Understands common misconceptions in science and effective ways to address these misconceptions	354
Skill 40.5	Understands the rationale for the use of active learning and inquiry processes for students	355
Skill 40.6	Understands questioning strategies designed to elicit higher-level thinking and how to use them to move students from concrete to more abstract understanding	356
Skill 40.7	Understands the importance of planning activities that are inclusive and accommodate the needs of all students	358
Skill 40.8	Understands how to sequence learning activities in a way that allows students to build upon their prior knowledge and challenges them to expand their understanding of science	358
COMPETENCY 041	**UNDERSTANDING THE PROCESS OF SCIENTIFIC INQUIRY AND ITS ROLE IN SCIENCE INSTRUCTION**	360
Skill 41.1	Plans and implement instruction that provides opportunities for all students to engage in non-experimental and experimental inquiry investigations	360
Skill 41.2	Focuses inquiry-based instruction on questions and issues relevant to students and uses strategies to assist students with generating, refining, and focusing scientific questions and hypotheses	361
Skill 41.3	Instructs students in the safe and proper use of a variety of grade-appropriate tools, equipment, resources, technology, and techniques to access, gather, store, retrieve, organize, and analyze data	362
Skill 41.4	Knows how to guide students in making systematic observations and measurements	365
Skill 41.5	Knows how to promote the use of critical-thinking skills, logical reasoning, and scientific problem solving to reach conclusions based on evidence	368

Skill 41.6	Knows how to teach students to develop, analyze, and evaluate different explanations for a given scientific investigation	369
Skill 41.7	Knows how to teach students to demonstrate an understanding of potential sources or error in inquiry-based investigation	370
Skill 41.8	Knows how to teach students to demonstrate an understanding of how to communicate and defend the results of an inquiry-based investigation	370

COMPETENCY 042 VARIED AND APPROPRIATE ASSESSMENTS AND ASSESSMENT PRACTICES TO MONITOR SCIENCE LEARNING IN LABORATORY, FIELD, AND CLASSROOM SETTINGS 371

Skill 42.1	Understands the relationship among science curriculum, assessment, and instruction and bases instruction on information gathered through assessment of students' strengths and needs	371
Skill 42.2	Understands the importance of monitoring and assessing students' understanding of science concepts and skills on an ongoing basis	371
Skill 42.3	Understands the importance of carefully selecting or designing formative and summative assessments for the specific decisions they are intended to inform	372
Skill 42.4	Selects or designs and administers a variety of appropriate assessment methods to monitor students understanding and progress	373
Skill 42.5	Uses formal and informal assessments of student performance and products to evaluate student participation in and understanding of the inquiry process	375
Skill 42.6	Understands the importance of sharing evaluation criteria and assessment results with students	376

Sample Test: Science 377

Answer Key: Science 400

Sample Questions with Rationale: Science 401

TEACHER CERTIFICATION STUDY GUIDE

Great Study and Testing Tips!

What to study in order to prepare for the subject assessments is the focus of this study guide but equally important is *how* you study.

You can increase your chances of truly mastering the information by taking some simple, but effective, steps.

Study Tips:

1. <u>Some foods aid the learning process</u>. Foods such as milk, nuts, seeds, rice, and oats help your study efforts by releasing natural memory enhancers called CCKs (*cholecystokinin*) composed of *tryptophan*, *choline*, and *phenylalanine*. All of these chemicals enhance the neurotransmitters associated with memory. Before studying, try a light, protein-rich meal of eggs, turkey, and fish. All of these foods release the memory enhancing chemicals. The better the connections, the more you comprehend.

Likewise, before you take a test, stick to a light snack of energy boosting and relaxing foods. A glass of milk, a piece of fruit, or some peanuts all release various memory-boosting chemicals and help you to relax and focus on the subject at hand.

2. <u>Learn to take great notes</u>. A by-product of our modern culture is that we have grown accustomed to getting our information in short doses (i.e. TV news sound bites or *USA Today* style newspaper articles.)

Consequently, we have subconsciously trained ourselves to assimilate information better in <u>neat little packages</u>. If your notes are scrawled all over the paper, it fragments the flow of the information. Strive for clarity. Newspapers use a standard format to achieve clarity. Your notes can be much clearer through use of proper formatting. A very effective format is called <u>*"Cornell Method."*</u>

Take a sheet of loose-leaf lined notebook paper and draw a line all the way down the paper about 1-2" from the left-hand edge.

Draw another line across the width of the paper about 1-2" up from the bottom. Repeat this process on the reverse side of the page.

Look at the highly effective result. You have ample room for notes, a left hand margin for special emphasis items or inserting supplementary data from the textbook, a large area at the bottom for a brief summary, and a little rectangular space for just about anything you want.

MATHEMATICS-SCIENCE 4-8

3. Get the concept than the details. Too often, we focus on the details and do not gather an understanding of the concept. However, if you simply memorize only dates, places, or names, you may well miss the whole point of the subject.

A key way to understand things is to put them in your own words. If you are working from a textbook, automatically summarize each paragraph in your mind. If you are outlining text, do not simply copy the author's words.

Rephrase them in your own words. You remember your own thoughts and words much better than someone else's and subconsciously tend to associate the important details with the core concepts.

4. Ask Why? Pull apart written material paragraph by paragraph and do not forget the captions under the illustrations.

Example: If the heading is "Stream Erosion", flip it around to read "Why do streams erode?" Then answer the questions.

If you train your mind to think in a series of questions and answers, not only will you learn more, but it also helps to lessen the test anxiety because you are used to answering questions.

5. Read for reinforcement and future needs. Even if you only have ten minutes, put your notes or a book in your hand. Your mind is similar to a computer; you have to input data in order to have it processed. *By reading, you are creating the neural connections for future retrieval.* The more times you read something, the more you reinforce the learning of ideas.

Even if you do not fully understand something on the first pass, *your mind stores much of the material for later recall.*

6. Relax to learn, so go into exile. Our bodies respond to an inner clock called biorhythms. Burning the midnight oil works well for some people but not everyone.

If possible, set aside a particular place to study that is free of distractions. Shut off the television, cell phone, and pager and exile your friends and family during your study period.

If you really are bothered by silence, try background music. Light classical music at a low volume has been shown over other types to aid in concentration. Music without lyrics that evokes pleasant emotions is highly suggested. Try just about anything by Mozart. It relaxes you.

7. Use arrows not highlighters. At best, it is difficult to read a page full of yellow, pink, blue, and green streaks. Try staring at a neon sign for a while and you will soon see that the horde of colors obscures the message.

A quick note, a brief dash of color, an underline, or an arrow pointing to a particular passage is much clearer than a horde of highlighted words.

8. Budget your study time. Although you should not ignore any of the material, *allocate your available study time in the same ratio that topics may appear on the test.*

TEACHER CERTIFICATION STUDY GUIDE

Testing Tips:

1. <u>Get smart; play dumb. Don't read anything into the question.</u> Do not assume that the test writer is looking for something else than what is asked. Stick to the question as written and do not read extra things into it.

2. <u>Read the question and all the choices *twice* before answering the question.</u> You may miss something by not carefully reading, and then re-reading, both the question and the answers.

If you really do not have a clue as to the right answer, leave it blank on the first time through. Go on to the other questions, as they may provide a clue as to how to answer the skipped questions.

If later on, you still cannot answer the skipped ones . . . ***Guess.*** The only penalty for guessing is that you *might* get it wrong. Only one thing is certain; if you don't put anything down, you will get it wrong!

3. <u>Turn the question into a statement.</u> Look at the way the questions are worded. The syntax of the question usually provides a clue. Does it seem more familiar as a statement rather than as a question? Does it sound strange?

By turning a question into a statement, you may be able to spot if an answer sounds right, and it may also trigger memories of material you have read.

4. <u>Look for hidden clues.</u> It is actually very difficult to compose multiple-foil (choice) questions without giving away part of the answer in the options presented.

In most multiple-choice questions you can often readily eliminate one or two of the potential answers. This leaves you with only two real possibilities and automatically your odds go to fifty-fifty for very little work.

5. <u>Trust your instincts.</u> For every fact that you have read, you subconsciously retain something of that knowledge. On questions that you are not certain about, go with your basic instincts. **Your first impression on how to answer a question is usually correct.**

6. <u>Mark your answers directly on the test booklet.</u> Do not bother trying to fill in the optical scan sheet on the first pass through the test.

Just be very careful not to mis-mark your answers when you eventually transcribe them to the scan sheet.

7. <u>Watch the clock!</u> You have a set amount of time to answer the questions. Do not get bogged down trying to answer a single question at the expense of ten questions you can more readily answer.

DOMAIN I. NUMBER CONCEPTS

Competency 001 The teacher understands the structure of number systems, the development of a sense of quantity, and the relationship between quantity and symbolic representations.

A numeration system is a set of numbers represented a by a set of symbols (numbers, letters, or pictographs). Sets can have different bases of numerals within the set. Instead of our base 10, a system may use any base set from 2 on up. The position of the number in that representation defines its exact value. Thus, the numeral 1 has a value of ten when represented as "10". Early systems, such as the Babylonian used position in relation to other numerals or column position for this purpose since they lacked a zero to represent an empty position.

A base of 2 uses only 0 and 1.

Decimal Binary Conversion		
Decimal	Binary	Place Value
1	1	2^0
2	10	2^1
4	100	2^2
8	1000	2^3

Thus, 9 in Base 10 becomes 1001 in Base 2.

$\quad$ 9+4 = 13 (Base 10) becomes 1001 + 100 = 1101 (Base 2).

Fractions, ratios and other functions alter in the same way.

Computers use a base of 2 but combine it into 4 units called a byte to function in base 16 (hexadecimal). A base of 8 (octal) was also used by older computers.

The real number system includes all rational and irrational **Rational numbers** can be expressed as the ratio of two integers, $\frac{a}{b}$ where b ≠ 0, for example $\frac{2}{3}$, $-\frac{4}{5}$, $5 = \frac{5}{1}$.

The rational numbers include integers, fractions and mixed numbers, terminating and repeating decimals. Every rational number can be expressed as a repeating or terminating decimal and can be shown on a number line.

Integers are positive and negative whole numbers and zero.
...-6, -5, -4, -3, -2, -1, 0, 1, 2, 3, 4, 5, 6, ...

Whole numbers are natural numbers and zero.
0, 1, 2, 3, ,4 ,5 ,6 ...

Natural numbers are the counting numbers.
1, 2, 3, 4, 5, 6, ...

Irrational numbers are real numbers that cannot be written as the ratio of two integers. These are infinite non-repeating decimals.
numbers.

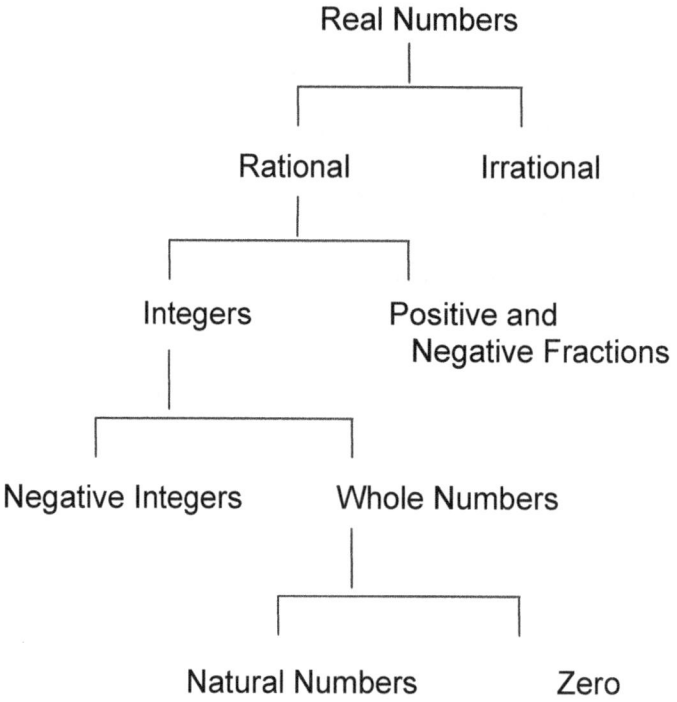

Examples: $\sqrt{5} = 2.2360..$, pi $= \Pi = 3.1415927...$

The shaded region represents 47 out of 100 or 0.47 or $\frac{47}{100}$ or 47%.

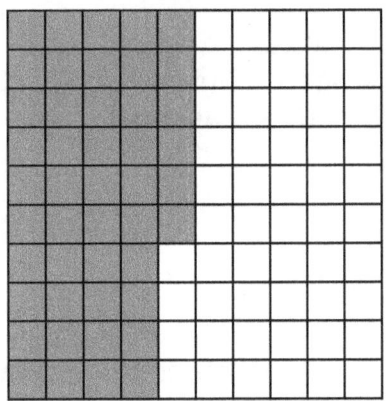

Fraction Strips:

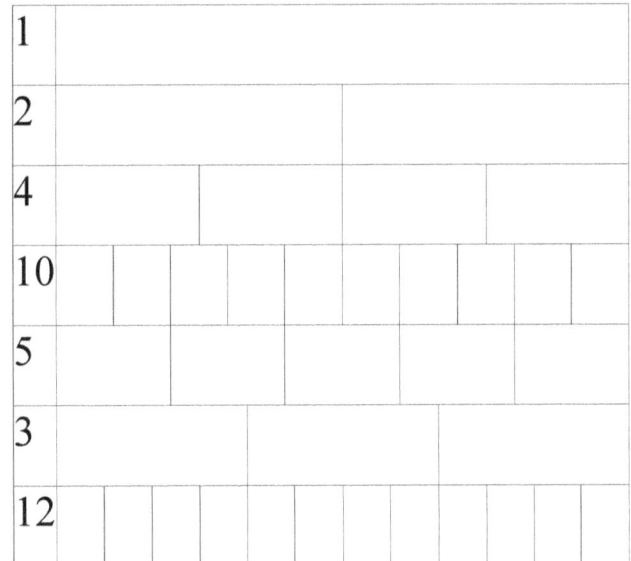

Number Lines:

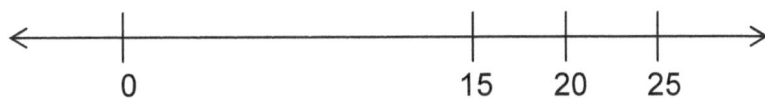

Diagrams:

4   ☐ ☐ ☐ ☐

11 ||||||||||

Percent means parts of one hundred. Fractions, decimals and percents can be interchanged.

$$100\% = 1$$

If a fraction can easily be converted to an equivalent **fraction** whose denominator is a power of 10 (for example, 10, 100, 1000), then it can easily be expressed as a **decimal** or **%**.

Examples: $\frac{1}{10} = 0.10 = 10\%$
$\frac{2}{5} = \frac{4}{10} = 0.40 = 40\%$
$\frac{1}{4} = \frac{25}{100} = 0.25 = 25\%$

Alternately, the **fraction** can be converted to a **decimal** and then a **percent** by dividing the numerator by the denominator, adding a decimal point and zeroes.

Example: $\frac{3}{8} = 8\overline{)3.000}^{\,0.375} = 37.5\%$

COMMON EQUIVALENTS

$\frac{1}{2} = 0.5 = 50\%$
$\frac{1}{3} = 0.33\frac{1}{3} = 33\frac{1}{3}\%$
$\frac{1}{4} = 0.25 = 25\%$
$\frac{1}{5} = 0.2 = 20\%$
$\frac{1}{6} = 0.16\frac{2}{3} = 16\frac{2}{3}\%$
$\frac{1}{8} = 0.12\frac{1}{2} = 12\frac{1}{2}\%$
$\frac{1}{10} = 0.1 = 10\%$
$\frac{2}{3} = 0.66\frac{2}{3} = 66\frac{2}{3}\%$
$\frac{5}{6} = 0.83\frac{1}{3} = 83\frac{1}{3}\%$
$\frac{3}{8} = 0.37\frac{1}{2} = 37\frac{1}{2}\%$
$\frac{5}{8} = 0.62\frac{1}{2} = 62\frac{1}{2}\%$
$\frac{7}{8} = 0.87\frac{1}{2} = 87\frac{1}{2}\%$
$1 = 1.0 = 100\%$

A **decimal** can be converted to a **percent** by multiplying by 100, or merely moving the decimal point two places to the right. A **percent** can be converted to a **decimal** by dividing by 100, or moving the decimal point two places to the left.

Examples: 0.375 = 37.5%
0.7 = 70%
0.04 = 4 %
3.15 = 315 %
84% = 0.84
3 % = 0.03
60% = 0.6
110% = 1.1
$\frac{1}{2}$ % = 0.5% = 0.005

A **percent** can be converted to a **fraction** by placing it over 100 and reducing to simplest terms.

Examples: 32 % = $\frac{32}{100}$ = $\frac{8}{25}$
6 % = $\frac{6}{100}$ = $\frac{3}{50}$
111% = $\frac{111}{100}$ = $1\frac{11}{100}$

The **exponent form** is a shortcut method to write repeated multiplication. The **base** is the factor. The **exponent** tells how many times that number is multiplied by itself.

Example: 3^4 is $3 \times 3 \times 3 \times 3 = 81$
where 3 is the base and 4 is the exponent.

x^2 *is read* "x squared"
y^3 *is read* "y cubed"
a^1 = a for all values of a; thus $17^1 = 17$
b^0 = 1 for all values of b; thus $24^0 = 1$

When 10 is raised to any power, the exponent tells the number of zeroes in the product.

Example: $10^7 = 10,000,000$

Scientific notation is a more convenient method for writing very large and very small numbers. It employs two factors. The **first factor** is a **number between 1 and 10**. The **second factor** is a **power of 10**.

Example 1: Write 372,000 in scientific notation
Move the decimal point to form a number between 1 and 10; thus 3.72. Since the decimal point was moved 5 places, the power of 10 is 10^5. The exponent is positive since the decimal point was moved to the left.

$$372,000 = 3.72 \times 10^5$$

Example 2: Write 0.0000072 in scientific notation.
Move the decimal point 6 places to the right.

$$0.0000072 = 7.2 \times 10^6$$

Example 3: Write 2.19×10^8 in standard form.
Since the exponent is positive, move the decimal point 8 places to the right, and add additional zeroes as needed.

$$2.19 \times 10^8 = 219,000,000$$

Example 4: Write 8.04×10^{-4} in standard form.
Move the decimal point 4 places to the left, writing additional zeroes as needed.

$$8.04 \times 10^{-4} = 0.000804$$

** Note: The first factor **must** be between 1 and 10.

Real numbers exhibit the following addition and multiplication properties, where *a, b,* and *c* are real numbers.

Note: Multiplication is implied when there is no symbol between two variables. Thus, $a \times b$ can be written *ab*. Multiplication can also be indicated by a raised dot • .

Closure
a + *b* is a real number

Example: Since 2 and 5 are both real numbers, 7 is also a real number.

ab is a real number

Example: Since 3 and 4 are both real numbers, 12 is also a real number.

The sum or product of two real numbers is a real number.

Commutative
a + *b* = *b* + *a*

Example: 5 + ⁻8 = ⁻8 + 5 = ⁻3

ab = *ba*

Example: ⁻2 × 6 = 6 × ⁻2 = ⁻12

The order of the addends or factors does not affect the sum or product.

Associative
(*a* + *b*) + *c* = *a* + (*b* + *c*)

Example: (⁻2 + 7) + 5 = ⁻2 + (7 + 5)
       5 + 5 = ⁻2 + 12 = 10

(*ab*) *c* = *a* (*bc*)

Example: (3 × ⁻7) × 5 = 3 × (⁻7 × 5)
       ⁻21 × 5 = 3 × ⁻35 = ⁻105

The grouping of the addends or factors does not affect the sum or product.

Distributive
$a(b + c) = ab + ac$

Example: $6 \times (-4 + 9) = (6 \times -4) + (6 \times 9)$
$6 \times 5 = -24 + 54 = 30$

To multiply a sum by a number, multiply each addend by the number, then add the products.

Additive Identity (Property of Zero)
$a + 0 = a$

Example: $17 + 0 = 17$

The sum of any number and zero is that number.

Multiplicative Identity (Property of One)
$a \cdot 1 = a$

Example: $^-34 \times 1 = {^-34}$

The product of any number and one is that number.

Additive Inverse (Property of Opposites)
$a + -a = 0$

Example: $25 + -25 = 0$

The sum of any number and its opposite is zero.

Multiplicative Inverse (Property of Reciprocals)
$a \times \dfrac{1}{a} = 1$

Example: $5 \times \dfrac{1}{5} = 1$

The product of any number and its reciprocal is one.

PROPERTY OF DENSENESS

Between any pair of rational numbers, there is at least one rational number. The set of natural numbers is <u>not</u> dense because between two consecutive natural numbers there may not exist another natural number.

Example:

Between 7.6 and 7.7, there is the rational number 7.65 in the set of real numbers.

Between 3 and 4 there exists no other natural number.

PROPERTIES SATISFIED BY SUBSETS OF REAL NUMBERS

+	Closure	Commutative	Associative	Distributive	Identity	Inverse
Real	yes	yes	yes	yes	yes	yes
Rational	yes	yes	yes	yes	yes	yes
Irrational	yes	yes	yes	yes	no	yes
Integers	yes	yes	yes	yes	yes	yes
Fractions	yes	yes	yes	yes	yes	yes
Whole	yes	yes	yes	yes	yes	no
Natural	yes	yes	yes	yes	no	no
−	Closure	Commutative	Associative	Distributive	Identity	Inverse
Real	yes	no	no	no	yes	yes
Rational	yes	no	no	no	yes	yes
Irrational	no	no	no	no	no	no
Integers	yes	no	no	no	yes	yes
Fractions	yes	no	no	no	yes	yes
Whole	no	no	no	no	yes	no
Natural	no	no	no	no	no	no
×	Closure	Commutative	Associative	Distributive	Identity	Inverse
Real	yes	yes	yes	yes	yes	yes
Rational	yes	yes	yes	yes	yes	yes
Irrational	yes	yes	yes	yes	no	no
Integers	yes	yes	yes	yes	yes	no
Fractions	yes	yes	yes	yes	yes	yes
Whole	yes	yes	yes	yes	yes	no
Natural	yes	yes	yes	yes	yes	no
÷	Closure	Commutative	Associative	Distributive	Identity	Inverse
Real	yes	no	no	no	yes	yes
Rational	yes	no	no	no	yes	yes
Irrational	no	no	no	no	no	no
Integers	no	no	no	no	yes	no
Fractions	yes	no	no	no	yes	yes
Whole	no	no	no	no	yes	no
Natural	no	no	no	no	yes	no

Complex numbers can be written in the form $a + bi$ where i represents $\sqrt{-1}$ and a and b are real numbers. a is the real part of the complex number and b is the imaginary part.

If $b = 0$, then the number has no imaginary part and it is a real number.

If $b \neq 0$, then the number is imaginary.

Complex numbers are found when trying to solve equations with negative square roots.

Example: If $x^2 + 9 = 0$
then $x^2 = -9$
and $x = \sqrt{-9}$ or $+3i$ and $-3i$

When dividing two complex numbers, you must eliminate the complex number in the denominator. If the complex number in the denominator is of the form bi, multiply both the numerator and denominator by i. Remember to replace i^2 with -1 and then continue simplifying the fraction.

Example:

$$\frac{2+3i}{5i} \quad \text{Multiply this by } \frac{i}{i}$$

$$\frac{2+3i}{5i} \times \frac{i}{i} = \frac{(2+3i)\,i}{5i \cdot i} = \frac{2i+3i^2}{5i^2} = \frac{2i+3(-1)}{-5} = \frac{-3+2i}{-5} = \frac{3-2i}{5}$$

If the complex number in the denominator is of the form $a + bi$, multiply both the numerator and denominator by **the conjugate of the denominator**. **The conjugate of the denominator** is the same two terms with the opposite sign between the 2 terms (the real term does not change signs). The conjugate of $2-3i$ is $2+3i$. The conjugate of $-6+11i$ is $-6-11i$. Multiply together the factors on the top and bottom of the fraction. Remember to replace i^2 with -1, combine like terms, and then continue simplifying the fraction.

Example:

$\dfrac{4+7i}{6-5i}$ Multiply by $\dfrac{6+5i}{6+5i}$, the conjugate.

$\dfrac{(4+7i)}{(6-5i)} \times \dfrac{(6+5i)}{(6+5i)} = \dfrac{24+20i+42i+35i^2}{36+30i-30i-25i^2} = \dfrac{24+62i+35(-1)}{36-25(-1)} = \dfrac{-11+62i}{61}$

Example:

$\dfrac{24}{-3-5i}$ Multiply by $\dfrac{-3+5i}{-3+5i}$, the conjugate.

$\dfrac{24}{-3-5i} \times \dfrac{-3+5i}{-3+5i} = \dfrac{-72+120i}{9-25i^2} = \dfrac{-72+120i}{9+25} = \dfrac{-72+120i}{34} = \dfrac{-36+60i}{17}$

Divided everything by 2.

Competency 002 The teacher understands number operations and computational algorithms

The three basic number properties are distributive, commutative and associative. These number properties are the rules of number operations. The distributive property of multiplication over addition states that $x(y + z) = xy + xz$. The commutative property of multiplication and addition states that the order of numbers does not matter. In other words, $a + b = b + a$ and $ab = ba$. Finally, the associative property of addition and multiplication states that the grouping of numbers does not matter. In other words, $(a + b) + c = a + (b + c)$ and $a(bc) = (ab)c$.

Algorithms are methods or strategies for solving problems. There are several different algorithms for solving addition, subtraction, multiplication and division problems involving integers, rational numbers and real numbers.

In general, algorithms make use of number properties to simplify mathematical operations.

Integer algorithms

Addition –

Three common algorithms for addition of integers are the partial sums method, column addition method and fast method. The partial sums method is a two-stage process. First, we sum the columns from left to right. To complete the operation we add the column values.

```
    125
 +   89
 + 376      Step 1 – column addition
    400
 +  170
 +   20     Step 2 – final sum
    590
```

The column addition method is also a two-stage process. First, we add the digits in each column. To complete the operation we perform the place carries from right to left.

```
    1| 2| 5
 +   | 8| 9            Stage 1 – column addition
 +  3| 7| 6
    4|17|20
    4|19| 0       ←——— First carry
    5| 9| 0 = 590 ←——— Second carry = final answer
```

The fast method of addition is the traditional method of right to left addition. We sum the columns from left to right, performing carries mentally or writing them down.

```
   12   ← Carries
  125
+  89
+ 376
  590
```

All of the integer addition algorithms rely on the commutative and associative properties of addition, allowing re-grouping and re-ordering of numbers.

Subtraction –

Three common algorithms of integer subtraction are left to right subtraction, partial differences and the same change rule. In left to right subtraction, we decompose the second number into smaller values and perform the individual subtractions. For example, to solve 335 – 78, we break 78 down into 70 + 8.

```
  335
-  70
  265
-   8
  257
```

The partial differences method is a two-stage process. First, we operate on each column individually, being careful to record the sign of each result. Then, we sum the results to yield the final answer.

```
  335
-  78
 +300
 - 40
 -  3
  257
```

The same change rule takes advantage of the knowledge that subtraction is easier if the smaller number ends in zero. Thus, we change each number by the same amount to produce a smaller number ending in zero.

```
  335  →  333
-  78  →  - 80
           257
```

Like the addition algorithms, the subtraction algorithms rely on the commutative and associative properties of addition (because subtraction is addition of a negative number).

Multiplication –

Two common multiplication algorithms are the partial products method and the short method. In the partial products method, we decompose each term into base-ten forms and multiply each pair of terms.

```
              84
            x 26
80 x 20  →   1600
80 x 6   →    480
20 x 4   →     80
6 x 4    →     24
             2184
```

The short method is a traditional multiplication algorithm. In the short method, we only decompose the second term.

```
              84
            x 26
84 x 20  →   1680
84 x 6   →    504
             2184
```

The multiplication algorithms rely on the associative and commutative properties of multiplication and the distribution of multiplication over addition.

Division –

A common division algorithm is the partial quotients method. In this method, we make note of two, simple products and estimate our way toward a final answer. For example, to find the quotient of 1440 divided by 18 we first make note that 5 x 18 = 90 and 2 x 18 = 32.

```
18)  1440 |
   -  900 | 50
      540 |
   -  360 | 20
      180 |
   -   90 | 5
       90 |
   -   90 | 5
        0   80  → final quotient = 80 with no remainder
```

Rational and real number algorithms –

Operations involving rational numbers represented as fractions require unique algorithms. For example, when adding or subtracting fractions we use the distributive property of multiplication over division to find common denominators.

When completing operations involving real numbers in decimal form we use similar algorithms to those used with integers. We use the associative, commutative and distributive properties of numbers to generate algorithms.

Concrete and visual representations can help demonstrate the logic behind operational algorithms. Blocks or other objects modeled on the base ten system are useful concrete tools. Base ten blocks represent ones, tens and hundreds. For example, modeling the partial sums algorithm with base ten blocks helps clarify the thought process. Consider the sum of 242 and 193. We represent 242 with two one hundred blocks, four ten blocks and 2 one blocks. We represent 193 with one one hundred block, nine ten blocks and 3 one blocks. In the partial sums algorithm, we manipulate each place value separately and total the results. Thus, we group the hundred blocks, ten blocks and one blocks and derive a total for each place value. We combine the place values to complete the sum.

An example of a visual representation of an operational algorithm is the modeling of a two-term multiplication as the area of a rectangle. For example, consider the product of 24 and 39. We can represent the product in geometric form. Note that the four sections of the rectangle equate to the four products of the partial products method.

	30	9
20	A = 600	A = 180
4	A = 120	A = 36

Thus, the final product is the sum of the areas or 600 + 180 + 120 + 36 = 936.

Teachers must justify the procedures used in operational algorithms to ensure student understanding. Algorithms of the basic operations make use of number properties to simplify addition, subtraction, multiplication and division. The following are examples of operational algorithms, their justifications and common errors in implementation.

Addition –

The partial sums method of integer addition relies on the associative property of addition. Consider the partial sum algorithm of the addition of 125 and 89. We first sum the columns from left to right and then add the results.

```
    125
+    89
    100  →  Hundreds column sum
 + 100  →  Tens column sum
 +  14  →  Ones column sum
    214
```

The associative property of addition shows why this method works. We can rewrite 125 plus 89 as follows:

$(100 + 20 + 5) + (80 + 9)$

Using the associative property to group the terms:

$(100) + (20 + 80) + (5 + 9) = 100 + 100 + 14 = 214$

Note the final form is the same as the second step of the partial sums algorithm. When evaluating addition by partial sums, teachers should look for errors in assigning place values of the partial sums; for example, in the problem above, recording the sum of eight and two in the tens column as 10 instead of 100.

Rational number addition relies on the distributive property of multiplication over addition and the understanding that multiplication by any number by one yields the same number. Consider the addition of 1/4 to 1/3 by means of common denominator.

$$\frac{1}{4}+\frac{1}{3}=\frac{3}{3}(\frac{1}{4})+\frac{4}{4}(\frac{1}{3})=(\frac{3}{12})+(\frac{4}{12})=\frac{7}{12} \longrightarrow \text{Recognize that } \frac{3}{3} \text{ and } \frac{4}{4} \text{ both equal 1.}$$

A common error in rational number addition is the failure to find a common denominator and adding both numerators and denominators.

Subtraction –

The same change rule of substitution takes advantage of the property of addition of zero. The addition of zero does not change the value of a quantity.

289 – 97 = 292 – 100 because
289 – 97 = (289 + 3) – (97 + 3) = (289 – 97) + (3 – 3) = 289 – 97 + 0

Note the use of the distributive property of multiplication over addition, the associative property of addition and the property of addition of zero in proving the accuracy of the same change algorithm. A common mistake when using the same change rule is adding from one number and subtracting from the other. This is an error in reasoning resulting from misapplication of the distributive property (e.g. failing to distribute -1).

The same procedure, justification and error pattern applies to the subtraction of rational and real numbers.

$13 - 2\frac{1}{3} = 13\frac{2}{3} - 3$ because

$13 - 2\frac{1}{3} = (13 + \frac{2}{3}) - (2\frac{1}{3} + \frac{2}{3}) = (13 - 2\frac{1}{3}) + (\frac{2}{3} - \frac{2}{3}) =$

$13 - 2\frac{1}{3} + 0$

or

13 – 2.456 = 13.544 – 3 because

13 – 2.456 = (13 + 0.544) – (2.456 + 0.544) =

(13 – 2.456) + (0.544 – 0.544) = 13 – 2.456 + 0

Multiplication –

The partial products algorithm of multiplication decomposes each term into simpler numbers and sums the products of the simpler terms.

$$84 = 80 + 4$$
$$\times\ 26 = 20 + 6$$

80 x 20 →	1600
80 x 6 →	480
20 x 4 →	80
6 x 4 →	24
	2184

We can justify this algorithm by using the "FOIL" method of binomial multiplication and the distributive property of multiplication over addition.

$$(80 + 4)(20 + 6) = (80)(20) + (4)(20) + (6)(80) + (6)(4)$$

Common errors in partial product multiplication result from mistakes in binomial multiplication and mistakes in pairing terms of the partial products (e.g. multiplying incorrect terms).

Division –

We can justify the partial quotients algorithm for division by using the distributive property of multiplication over division. Because multiplication is the reverse of division, we check the result by multiplying the divisor by the partial sums.

```
   18) 1440 |
     -  900 | 50
        540 |
     -  360 | 20
        180 |
     -   90 | 5
         90 |
     -   90 | 5
          0   80   ⟶  final quotient = 80 with no remainder
```

Check:

$18 (50 + 20 + 5 + 5) = (18)(50) + (18)(20) + (18)(5) + (18)(5) = 1440$

Common errors in division often result from mistakes in translating words to symbols. For example, misinterpreting 10 divided by 5 as 5/10. In addition, when using the partial quotients algorithm errors in subtraction and addition can produce incorrect results.

The main algorithm of rational number division is multiplication by the reciprocal. Thus,

$$\frac{\frac{1}{3}}{\frac{1}{4}} = (\frac{1}{3})(\frac{4}{1}) = \frac{4}{3}.$$

The definition of multiplication and division as inverse operations justifies the use of reciprocal multiplication.

Many **algebraic procedures** are similar to and rely upon number operations and algorithms. Two examples of this similarity are the adding of rational expressions and division of polynomials.

Addition of rational expressions is similar to fraction addition. The basic algorithm of addition for both fractions and rational expressions is the common denominator method. Consider an example of the addition of numerical fractions.

$$\frac{3}{5} + \frac{2}{3} = \frac{3(3)}{3(5)} + \frac{5(2)}{5(3)} = \frac{9}{15} + \frac{10}{15} = \frac{19}{15}$$

To complete the sum, we first find the least common denominator

Now, consider an example of rational expression addition.

$$\frac{(x+5)}{(x+1)} + \frac{2x}{(x+3)} = \frac{(x+3)(x+5)}{(x+3)(x+1)} + \frac{(x+1)2x}{(x+1)(x+3)}$$
$$= \frac{x^2+8x+15}{(x+3)(x+1)} + \frac{2x^2+2x}{(x+3)(x+1)} = \frac{3x^2+10x+15}{(x+3)(x+1)}$$

Note the similarity to fractional addition. The basic algorithm, finding a common denominator and adding numerators, is the same.

Division of polynomials follows the same algorithm as numerical long division. Consider an example of numerical long division.

$$\begin{array}{r} 720 \\ 6\overline{)4321} \\ \underline{42} \\ 12 \\ \underline{12} \\ 01 \end{array}$$ → 720 1/6 = final quotient

Compare the process of numerical long division to polynomial division.

$$x+1 \overline{\smash{)}\, x^2 - 8x - 9}$$

with quotient $x - 9$; subtract $-x^2 - x$ to get $-9x - 9$; subtract $+9x + 9$ to get $0 + 0$.

x – 9 = final quotient

Note that the step-by-step process is identical in both cases.

Real numbers: Include two types of numbers, rational and irrational. Rational means fractional. These numbers, when converted to decimals, are either terminating or repeating.

Examples of rational numbers are:

a) $\dfrac{1}{4} = 0.2500$ (terminating decimal).

b) $\dfrac{10101}{40000} = 0.252525 = 0.2\overline{5}$ (repeating decimal).

Irrational numbers, on the other hand, when converted to decimal values are neither repeating nor terminating. Examples of irrational numbers are:

a) $\pi = 3.14159...$

b) $\sqrt{7} = 2.645751311...$

Notation: Radicals ($\sqrt{}$) are inverse operators of exponents and are represented by the following form:

$\sqrt[n]{a^x}$ where:

n   is called the index or root,
a^x is called to radicand,
x   is the exponent or power of 'a'.

In the notation above, we are finding the *n*th root of a^x. In other words, we want to find the number, when multiplied by itself *n* times, gives a^x.

TEACHER CERTIFICATION STUDY GUIDE

Example: $\sqrt{16} = \sqrt[2]{16}$ (when the index or root is omitted, it is always assumed to be 2).

$\sqrt{16} = +4$ or $^-4$ since $(+4)(+4) = 16$ and $(-4)(-4) = 16$.

So we can write $\sqrt{16} = \pm 4$ to show the two results. $+4$ is called the principal square root of 16. This is because the principal square root is the only one that makes since (example: for measurements).

1) We can only add or subtract radicals that have the same index and the same radicand.

   Example: $2\sqrt{5} + 3\sqrt{5} = 5\sqrt{5}$
   Example: $5\sqrt[3]{2} - 3\sqrt[3]{2} = 2\sqrt[3]{2}$

2) If the radicand is raised to a power that is equal to the index then the root operation will cancel out the power operation.

   Example: Perform the indicated operations.

   a) $\sqrt[3]{2^3} = 2$ The power and the index are equal to each other.

   b) $2\sqrt{32}$
   $32 = 2 \times 16 = 2 \times 4 \times 4 = 2 \times 4^2$ so,
   $2\sqrt{32} = 2\sqrt{2 \times 4^2} = 2 \times 4\sqrt[2]{2}$ or $8\sqrt{2}$

3) If the radicand is raised to a power different from the index, convert the radical to its exponential form and apply laws of exponents.

Review of Law of Exponents: If a and b are real numbers and m and n are rational numbers, then,

1. $a^m \times a^n = a^{(m+n)}$
2. $\dfrac{a^m}{a^n} = a^{(m-n)}$
3. $(a^m)^n = a^{(mn)}$
4. $(ab)^m = a^m b^m$
5. $a^{-n} = \dfrac{1}{a^n} = (1/a)^n$
6. If a is any nonzero number, then $a^0 = 1$.
7. $a^{m/n} = \left(a^{1/n}\right)^m = \left(a^m\right)^{1/n}$
8. $\sqrt[n]{a^m}$ (radical form) $= a^{m/n}$ in exponential form.

MATHEMATICS-SCIENCE 4-8

Competency 003 **The teacher understands ideas of number theory and uses numbers to model and solve problems within and outside of mathematics.**

When two or more numbers are multiplied to give a certain product, each of these numbers is called a **factor** of the product. Factors can most easily be derived in pairs.

Example: The factors of 24 are:
 1 x 24 = 24
 2 x 12 = 24
 3 x 8 = 24
 4 x 6 = 24
The factors of 24 are: 1, 2, 3, 4, 6, 8, 12, 24.

The **Greatest Common Factor** (GCF) of two numbers is the largest factor that they share.

Example 1: Find the GCF of 20 and 30
 20: 1, 2, 4, 5, **10**, 20
 30: 1, 2, 3, 5, 6, **10,** 15, 30
The greatest common factor of 20 and 30 is 10

Example 2: Find the GCF of 12 and 18
 12: 1, 2, 3, 4, **6**, 12
 18; 1, 2, 3, **6**, 9, 18
The greatest common factor of 12 and 18 is 6.

A **multiple** is a number that is the *product* of the given number and another *factor*. Multiples are commonly listed in consecutive order.

Example: The multiples of 5 would be calculated by multiplying
 5 × 1, 5 × 2, 5 × 3, 5 × 4, 5 × 5, 5 × 6, and so on.
 Thus, the multiples of 5 are: 5, 10, 15, 20, 25, 30,...

The **Least Common Multiple** (LCM) of two numbers is the smallest multiple they share.

Example1: Find the LCM of 4 and 6.
 Multiples of 4 are: 4, 8, **12,** 16, 20, 24, ...
 Multiples of 6 are: 6, **12,** 18, 24, 30, 36, ...
 Although 12 and 24 are *both* common multiples of 4 and 6, the LCM is 12 because it is less than 24.

Example 2: Find the LCM of 8 and 10.
 8: 8, 16, 24, 32, 40, 48, ...
 10: 10, 20, 30, 40, ...
 40 is the least common multiple of 8 and 10.

A number, x, is divisible by another number, y, when the second number, y, is a factor of the first number, x. Thus x can be divided by y with no remainder. This can be determined, without actually dividing, by some simple rules.

A number is divisible by **2** when **it ends in 0, 2, 4, 6,** or **8**.

Example: 31<u>2</u>, 777<u>4</u>

A number is divisible by **5** when **it ends in 0** or **5**.

Example: 93<u>5</u>, 87<u>0</u>

A number is divisible by **10** when **it ends in 0**.

Example: 973<u>0</u>, 520<u>0</u>

A number is divisible by **3** when **the sum of its digits is divisible by 3**.

Example: 744, 8511

A number is divisible by **9** when **the sum of its digits is divisible by 9**.

Example: 837, 8928

Additional explanation of examples: 744: 7 + 4 + 4 = 15 which is divisible by 3, therefore 744 is divisible by 3. Similarly, 837: 8 + 3 + 7 = 18 which is divisible by 9, therefore 837 is divisible by 9 (and also by 3).

A number is divisible by **4** when **its last two digits are divisible by 4**.

Example: 3<u>28</u>, 13<u>64</u>

A number is divisible by **8** when **its last three digits are divisible by 8**.

Example: 9<u>248</u>, 13<u>720</u>

Additional explanation of examples: 328: 28 divided by 4 is 7, therefore 328 is divisible by 4. Similarly, 9248: 248 divided by 8 is 31, therefore 9248 is divisible by 8. (Interesting fact: all election years are divisible by 4, thus their last two digits are divisible by 4.)

> A number is divisible by **6** when **it ends in 0, 2, 4, 6, or 8** *and* **the sum of its digits is divisible by 3**. In other words, when **it is divisible by both 2 and 3**

Example: 516, 3852

Additional explanation of example: 516 is divisible by 2 since its last digit is 6 and the sum of its digits is 5 + 1 + 6 = 12 which is divisible by 3. Therefore 516 is divisible by 6.

Example 1: 7314 is divisible by:

- 2 because it ends in 4
- 3 because 7 + 3 + 4 + 1 = 15
- 6 because it is divisible by both 2 and 3

Example 2: 6624 is divisible by

- 2 because is ends in 4
- 3 and 9 because 6 + 6 + 2 + 4 = 18
- 4 because 24 is divisible by 4
- 6 because it is divisible by both 2 and 3

A **prime** number has exactly <u>two</u> factors, itself and one. A **composite** number has more than two factors. 0 and 1 are *neither* prime nor composite.

Example: **Prime:** The only factors of 7 are 1 and 7, therefore 7 is prime.

The number two is the only even prime.

Composite: The factors of 18 are: 1, 2, 3, 6, 9, and 18.

Therefore the number 18 has more than two factors and is considered composite.

To find the amount of sales tax on an item, change the percent of sales tax into an equivalent decimal number. Then multiply the decimal number times the price of the object to find the sales tax. The total cost of an item will be the price of the item plus the sales tax.

Example: A guitar costs $120 plus 7% sales tax. How much are the sales tax and the total bill?

$$7\% = .07 \text{ as a decimal} \quad (.07)(120) = \$8.40 \text{ sales tax}$$
$$\$120 + \$8.40 = \$128.40 \leftarrow \text{total cost}$$

An alternative method to find the total cost is to multiply the price times the factor 1.07 (price + sales tax):

$$\$120 \times 1.07 = \$8.40$$

This gives you the total cost in fewer steps.

Example: A suit costs $450 plus 6½% sales tax. How much are the sales tax and the total bill?

$$6\tfrac{1}{2}\% = .065 \text{ as a decimal}$$
$$(.065)(450) = \$29.25 \text{ sales tax}$$
$$\$450 + \$29.25 = \$479.25 \leftarrow \text{total cost}$$

Estimation and approximation may be used to check the reasonableness of answers.

Example: Estimate the answer.

$$\frac{58 \times 810}{1989}$$

58 becomes 60, 810 becomes 800 and 1989 becomes 2000.

$$\frac{60 \times 800}{2000} = 24$$

Word problems: An estimate may sometimes be all that is needed to solve a problem.

Example: Janet goes into a store to purchase a CD on sale for $13.95. While shopping, she sees two pairs of shoes, prices $19.95 and $14.50. She only has $50. Can she purchase everything? (Assume there is no sales tax.)

Solve by rounding:

$19.95 → $20.00
$14.50 → $15.00
$13.95 → $14.00
$49.00   Yes, she can purchase the CD and the shoes.

The difference between permutations and combinations is that in permutations all possible ways of writing an arrangement of objects are given, while in a combination, a given arrangement of objects is listed only once.

Given the set {1, 2, 3, 4}, list the arrangements of two numbers that can be written as a combination and as a permutation.

Combination	Permutation
12, 13, 14, 23, 24, 34	12, 21, 13, 31, 14, 41, 23, 32, 24, 42, 34, 43,
six ways	twelve ways

Using the formulas given below the same results can be found.

$$_nP_r = \frac{n!}{(n-r)!}$$

The notation $_nP_r$ is read "the number of permutations of n objects taken r at a time."

$$_4P_2 = \frac{4!}{(4-2)!}$$

Substitute known values.

$$_4P_2 = 12$$

Solve.

$$_nC_r = \frac{n!}{(n-r)!r!}$$

The number of combinations when r objects are selected from n objects.

$$_4C_2 = \frac{4!}{(4-2)!2!}$$

Substitute known values.

$$_4C_2 = 6$$

Solve.

Elapsed time problems are usually one of two types. One type of problem is the elapsed time between two times given in hours, minutes, and seconds. The other common type of problem is between two times given in months and years.

For any time of day past noon, change it into military time by adding 12 hours. For instance, 1:15 p.m. would be 13:15. Remember when you borrow a minute or an hour in a subtraction problem that you have borrowed 60 more seconds or minutes.

Example: Find the time from 11:34:22 a.m. until 3:28:40 p.m.

First change 3:28:40 p.m. to   15:28:40 p.m.
Now subtract $-$ 11:34:22 a.m.
                                        :18
Borrow an hour and add 60 more minutes. Subtract.
   14:88:40 p.m.
$-$ 11:34:22 a.m.
    3:54:18 ↔ 3 hours, 54 minutes, 18 seconds

Example: John lived in Arizona from September 1991 until March 1995. How long is that?

```
                    year month
March 1995      =   95    03
September 1991  = − 91    09
```

Borrow a year, change it into 12 more months, and subtract.

```
                    year month
March 1995      =   94    15
September 1991  = − 91    09
                    3 yrs 6 months
```

Example: A race took the winner 1 hr. 58 min. 12 sec. on the first half of the race and 2 hr. 9 min. 57 sec. on the second half of the race. How much time did the entire race take?

```
   1 hr. 58 min. 12 sec.
 + 2 hr.  9 min. 57 sec.    Add these numbers
   3 hr. 67 min. 69 sec.
       + 1 min −60 sec.     Change 60 seconds to 1 min.
   3 hr. 68 min.  9 sec.
 + 1 hr.−60 min.       .    Change 60 minutes to 1 hr.
   4 hr.  8 min.  9 sec. ← Final answer
```

TEACHER CERTIFICATION STUDY GUIDE

DOMAIN II. **PATTERNS AND ALGEBRA**

Competency 004 **The teacher understands and uses mathematical reasoning to identify, extend, and analyze patterns and understands the relationships among variable, expressions, equations, inequalities, relations, and functions.**

Proof by induction states that a statement is true for all numbers if the following two statements can be proven:

1. The statement is true for $n = 1$.
2. If the statement is true for $n = k$, then it is also true for $n = k+1$.

In other words, we must show that the statement is true for a particular value and then we can assume it is true for another, larger value (k). Then, if we can show that the number after the assumed value ($k+1$) also satisfies the statement, we can assume, by induction, that the statement is true for all numbers.

The four basic components of induction proofs are: (1) the statement to be proved, (2) the beginning step ("let $n = 1$"), (3) the assumption step ("let $n = k$ and assume the statement is true for k, and (4) the induction step ("let $n = k+1$").

Example:

Prove that the sum all numbers from 1 to n is equal to $\frac{(n)(n+1)}{2}$.

Let $n = 1$.	Beginning step.
Then the sum of 1 to 1 is 1.	
And $\frac{(n)(n+1)}{2} = 1$.	
Thus, the statement is true for $n = 1$.	Statement is true in a particular instance.

MATHEMATICS-SCIENCE 4-8

Assumption

Let $n = k + 1$
$k = n - 1$

Then $[1 + 2 +...+ k] + (k+1) = \dfrac{(k)(k+1)}{2} + (k+1)$ Substitute the assumption.

$= \dfrac{(k)(k+1)}{2} + \dfrac{2(k+1)}{2}$ Common denominator.

$= \dfrac{(k)(k+1) + 2(k+1)}{2}$ Add fractions.

$= \dfrac{(k+2)(k+1)}{2}$ Simplify.

$= \dfrac{((k+1)+1)(k+1)}{2}$ Write in terms of $k+1$.

For $n = 4$, $k = 3$

$= \dfrac{(4+1)(4)}{2} = \dfrac{20}{2} = 10$

Conclude that the original statement is true for $n = k+1$ if it is assumed that the statement is true for $n = k$.

Conditional: If p, then q   p is the hypothesis. q is the conclusion.

Inverse: If ~p, then ~q. Negate both the hypothesis and the conclusion from the original conditional. (If not p, then not q.)

Converse : If q, then p. Reverse the 2 clauses. The original hypothesis becomes the conclusion. The original conclusion then becomes the new hypothesis.

Contrapositive: If ~q, then ~p. Reverse the 2 clauses. The "If not q, then not p" original hypothesis becomes the conclusion. The original conclusion then becomes the new hypothesis. THEN negate both the new hypothesis and the new conclusion.

Example: Given the **conditional**:

If an angle has 60°, then it is an acute angle.

Its **inverse**, in the form "If ~p, then ~q", would be:

If an angle doesn't have 60°, then it is not an acute angle.

NOTICE that the inverse is not true, even though the conditional statement was true.

Its **converse**, in the form "If q, then p", would be:

If an angle is an acute angle, then it has 60°.

NOTICE that the converse is not true, even though the conditional statement was true.

Its **contrapositive**, in the form "If q, then p", would be:

If an angle isn't an acute angle, then it doesn't have 60°.

NOTICE that the contrapositive is true, assuming the original conditional statement was true.

TIP: If you are asked to pick a statement that is logically equivalent to a given conditional, look for the contrapositive. The inverse and converse are not always logically equivalent to every conditional. The contrapositive is ALWAYS logically equivalent.

Find the inverse, converse and contrapositive of the following conditional statement. Also determine if each of the 4 statements is true or false.

Conditional: If $x = 5$, then $x^2 - 25 = 0$.   TRUE
Inverse: If $x \neq 5$, then $x^2 - 25 \neq 0$.   FALSE, x could be $^-5$
Converse: If $x^2 - 25 = 0$, then $x = 5$.   FALSE, x could be $^-5$
Contrapositive: If $x^2 - 25 \neq 0$, then $x \neq 5$.   TRUE
Conditional: If $x = 5$, then $6x = 30$.   TRUE
Inverse: If $x \neq 5$, then $6x \neq 30$.   TRUE
Converse: If $6x = 30$, then $x = 5$.   TRUE
Contrapositive: If $6x \neq 30$, then $x \neq 5$.   TRUE

Sometimes, as in this example, all 4 statements can be logically equivalent; however, the only statement that will always be logically equivalent to the original conditional is the contrapositive.

Conditional statements are frequently written in "**if-then**" form. The "if" clause of the conditional is known as the **hypothesis**, and the "then" clause is called the **conclusion**. In a proof, the hypothesis is the information that is assumed to be true, while the conclusion is what is to be proven true. A conditional is considered to be of the form:

If p, then q
p is the hypothesis. q is the conclusion.

Conditional statements can be diagrammed using a **Venn diagram**. A diagram can be drawn with one circle inside another circle. The inner circle represents the hypothesis. The outer circle represents the conclusion. If the hypothesis is taken to be true, then you are located inside the inner circle. If you are located in the inner circle then you are also inside the outer circle, so that proves the conclusion is true. Sometimes that conclusion can then be used as the hypothesis for another conditional, which can result in a second conclusion.

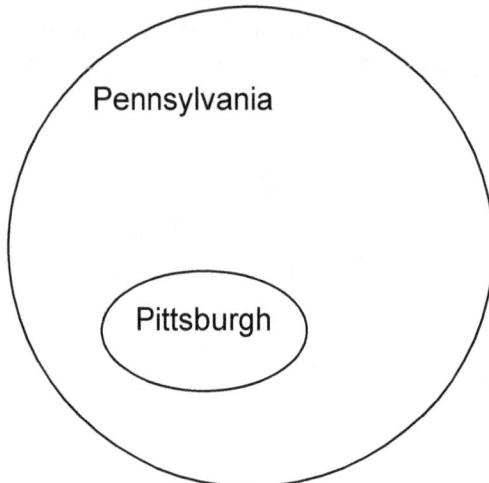

Example:
If an angle has a measure of 90 degrees, then it is a right angle.

> In this statement "an angle has a measure of 90 degrees" is the hypothesis.
> In this statement "it is a right angle" is the conclusion.

Example:
If you are in Pittsburgh, then you are in Pennsylvania.

> In this statement "you are in Pittsburgh" is the hypothesis.
> In this statement "you are in Pennsylvania" is the conclusion.

An **algebraic formula** is an equation that describes a relationship among variables. While it is not often necessary to derive the formula, one must know how to rewrite a given formula in terms of a desired variable.

Example: Given that the relationship of voltage, V, applied across a material with electrical resistance, R, when a current, I, is flowing through the material is given by the formula V = IR. Find the resistance of the material when a current of 10 milliamps is flowing, when the applied voltage is 2 volts.

$V = IR$. Solve for R.

$IR = V$; $R = V/I$ Divide both sides by I.

When $V = 2$ volts; $I = 10 \times 10^{-3}$ amps; find R.

$R = \dfrac{2}{10^1 \times 10^{-3}}$

$R = \dfrac{2}{10^{-2}}$ Substituting in R = V/I, we get,

$R = 2 \times 10^2$

$R = 200$ ohms

Example: Given the formula I = PRT, where I is the simple interest to be paid or realized when an amount P, the principal is to be deposited at simple interest rate of R (in %) and T is the time expressed in years, find what principal must be deposited to yield an interest of $586.00 over a period of 2 years at interest of 23.5%.

$I = PRT$ Solve for P → $P = \dfrac{I}{RT}$ (Divide both sides by RT)

$I = 586$; $R = 23.5\% = 0.235$; $T = 2$

$P = \dfrac{586}{0.235 \times 2} = \dfrac{586}{0.47}$ Substitute.

$P = \dfrac{586}{0.47} = 1246.80 = \$1246.80$

Check; $I = PRT$; $1246.8 \times 0.235 \times 2 = 586$

Loosely speaking, an equation like $y = 3x + 5$ describes a relation between the independent variable x and the dependent variable y. Thus, y is written as $f(x)$ "function of x." But y may not be a "true" function. For a "true" function to exist, there is a relationship between a set of all independent variables (domain) and a set of all outputs or dependent variables (range) such that each element of the domain corresponds to one element of the range. (For any input we get exactly one output).

Example:

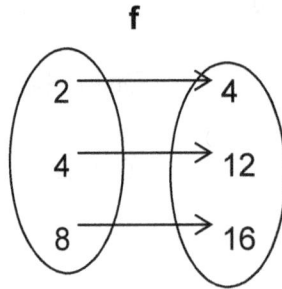   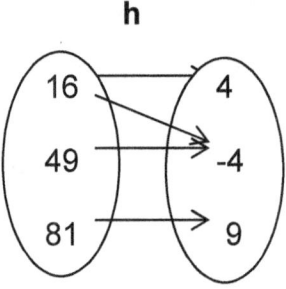

Domain, X   Range, Y   Domain, X   Range, Y

This is a "true" function.   This is not a "true" function.

Example: Given a function $f(x) = 3x + 5$, find $f(2)$; $f(0)$; $f(^-10)$

$f(2)$ means find the value of the function value at $x = 2$.

$$f(2) = 3(2) + 5 = 6 + 5 = 11$$
$$f(0) = 3(0) + 5 = 0 + 5 = 5 \qquad \text{Substitute for } x.$$
$$f(^-10) = 3(^-10) + 5 = ^-30 + 5 = ^-25$$

Example: Given $h(t) = 3t^2 + t - 9$, find $h(^-4)$.

$$h(^-4) = 3(^-4)^2 - 4 - 9$$
$$h(^-4) = 3(16) - 13 \qquad \text{Substitute for } t.$$
$$h(^-4) = 48 - 13$$
$$h(^-4) = 35$$

TEACHER CERTIFICATION STUDY GUIDE

Competency 005 The teacher understands and uses linear functions to model and solve problems.

- A **relation** is any set of ordered pairs.

- The **domain** of a relation is the set made of all the first coordinates of the ordered pairs.

- The **range** of a relation is the set made of all the second coordinates of the ordered pairs.

- A **function** is a relation in which different ordered pairs have different first coordinates. (No x values are repeated.)

- A **mapping** is a diagram with arrows drawn from each element of the domain to the corresponding elements of the range. If two arrows are drawn from the same element of the domain, then it is not a function.

    1. Determine the domain and range of this mapping.

ANSWERS

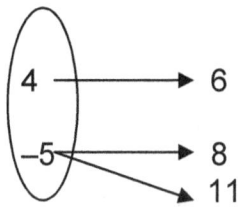

domain: {4, –5}

range : {6, 8, 11 }

MATHEMATICS-SCIENCE 4-8

- If two things vary directly, as one gets larger, the other also gets larger. If one gets smaller, then the other gets smaller too. If x and y vary directly, there should be a constant, c, such that $y = cx$. Something can also vary directly with the square of something else, $y = cx^2$.

- If two things vary inversely, as one gets larger, the other one gets smaller instead. If x and y vary inversely, there should be a constant, c, such that $xy = c$ or $y = c/x$. Something can also vary inversely with the square of something else, $y = c/x^2$.

Example: If $30 is paid for 5 hours work, how much would be paid for 19 hours work?

This is direct variation and $30 = 5c, so the constant is 6 ($6/hour). So $y = 6(19)$ or $y = 114.

This could also be done as a proportion: $\dfrac{\$30}{5} = \dfrac{y}{19}$

$$5y = 570$$
$$y = 114$$

Example: On a 546 mile trip from Miami to Charlotte, one car drove 65 mph while another car drove 70 mph. How does this affect the driving time for the trip?

This is an inverse variation, since increasing your speed should decrease your driving time. Using the equation: rate × time = distance, rt = d.

    65t = 546   and   70t = 546
      t = 8.4   and   t = 7.8

slower speed, more time   faster speed, less time

Example 1:

Consider the average monthly temperatures for a hypothetical location.

Month	Avg. Temp. (F)
Jan	40
March	48
May	65
July	81
Sept	80
Nov	60

Note that the graph of the average temperatures resembles the graph of a trigonometric function with a period of one year. We can use the periodic nature of seasonal temperature fluctuation to predict weather patterns.

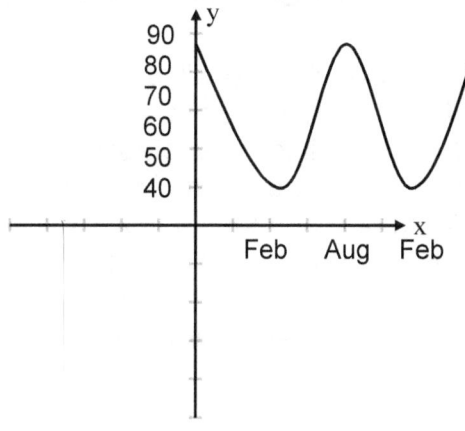

A first degree equation has an equation of the form $ax + by = c$. To graph this equation, find both one point and the slope of the line or find two points. To find a point and slope, solve the equation for y. This gets the equation in **slope intercept form**, $y = mx + b$. The point (0, b) is the y-intercept and m is the line's slope. To find any two points, substitute any two numbers for x and solve for y. To find the intercepts, substitute 0 for x and then 0 for y.

Remember that graphs will go up as they go to the right when the slope is positive. Negative slopes make the lines go down as they go to the right.

If the equation solves to $x =$ **any number**, then the graph is a **vertical line**. It only has an x intercept. Its slope is **undefined**.

If the equation solves to $y =$ **any number**, then the graph is a **horizontal line**. It only has a y intercept. Its slope is 0 (zero).

When graphing a linear inequality, the line will be dotted if the inequality sign is $<$ or $>$. If the inequality signs are either $\geq$ or $\leq$, the line on the graph will be a solid line. Shade above the line when the inequality sign is $\geq$ or $>$. Shade below the line when the inequality sign is $<$ or $\leq$. Inequalities of the form $x >, x \leq, x <,$ or $x \geq$ number, draw a vertical line (solid or dotted). Shade to the right for $>$ or $\geq$. Shade to the left for $<$ or $\leq$. Remember: **Dividing or multiplying by a negative number will reverse the direction of the inequality sign.**

Examples:

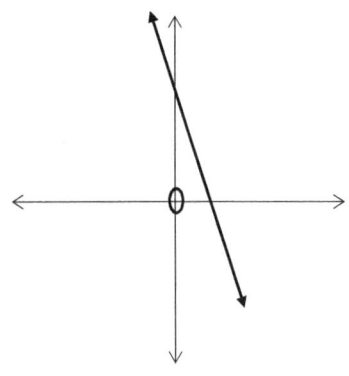

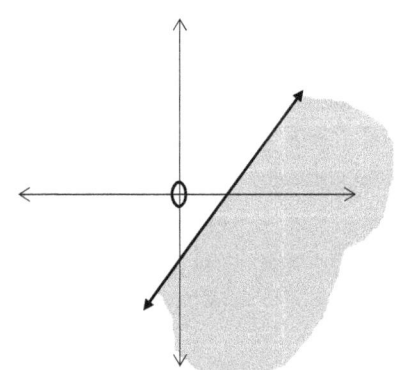

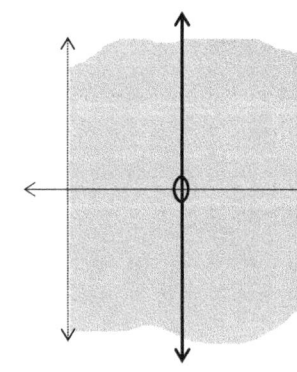

$5x + 2y = 6$ $3x - 2y \geq 6$ $3x + 12 > -3$

$y = -\frac{5}{2}x + 3$ $y \leq \frac{3}{2}x - 3$ $x > -5$

Word problems can sometimes be solved by using a system of two equations in two unknowns. This system can then be solved using substitution, the addition-subtraction method, or graphing.

Example: Mrs. Winters bought 4 dresses and 6 pairs of shoes for $340. Mrs. Summers went to the same store and bought 3 dresses and 8 pairs of shoes for $360. If all the dresses were the same price and all the shoes were the same price, find the price charged for a dress and for a pair of shoes.

Let x = price of a dress
Let y = price of a pair of shoes

Then Mrs. Winters' equation would be: $4x + 6y = 340$
Mrs. Summers' equation would be: $3x + 8y = 360$

To solve by addition-subtraction:

Multiply the first equation by 4: $4(4x + 6y = 340)$
Multiply the other equation by -3: $-3(3x + 8y = 360)$
By doing this, the equations can be added to each other to eliminate one variable and solve for the other variable.

$$16x + 24y = 1360$$
$$-9x - 24y = -1080$$
$$7x = 280$$
$$x = 40 \leftarrow \text{the price of a dress was \$40}$$

solving for y, $y = 30$ ← the price of a pair of shoes, $30

Example: Aardvark Taxi charges $4 initially plus $1 for every mile traveled. Baboon Taxi charges $6 initially plus $.75 for every mile traveled. Determine when it is cheaper to ride with Aardvark Taxi or to ride with Baboon Taxi.

Aardvark Taxi's equation: $y = 1x + 4$
Baboon Taxi's equation: $y = .75x + 6$
Using substitution: $.75x + 6 = x + 4$
Multiplying by 4: $3x + 24 = 4x + 16$
Solving for x : $8 = x$

This tells you that at 8 miles the total charge for the two companies is the same. If you compare the charge for 1 mile, Aardvark charges $5 and Baboon charges $6.75. Clearly Aardvark is cheaper for distances up to 8 miles, but Baboon Taxi is cheaper for distances greater than 8 miles.

This problem can also be solved by graphing the 2 equations.
$$y = 1x + 4 \qquad y = .75x + 6$$

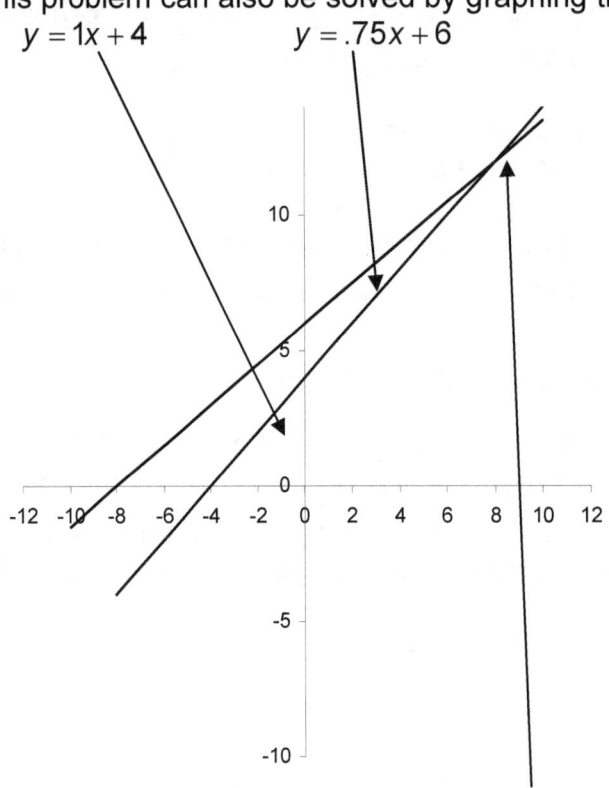

The lines intersect at (8, 12), therefore at 8 miles, both companies charge $12. At values less than 8 miles, Aardvark Taxi charges less (the graph is below Baboon). Greater than 8 miles, Aardvark charges more (the graph is above Baboon).

Example: Sharon's Bike Shoppe can assemble a 3 speed bike in 30 minutes or a 10 speed bike in 60 minutes. The profit on each bike sold is $60 for a 3 speed or $75 for a 10 speed bike. How many of each type of bike should they assemble during an 8 hour day (480 minutes) to make the maximum profit? Total daily profit must be at least $300.

Let x = number of 3 speed bikes.
y = number of 10 speed bikes.

Since there are only 480 minutes to use each day,

$30x + 60y \leq 480$ is the first inequality.

Since the total daily profit must be at least $300,

$60x + 75y \geq 300$ is the second inequality.

$30x + 60y \leq 480$ solves to $y \leq 8 - 1/2\,x$
$60y \leq -30x + 480$

$$y \leq -\frac{1}{2}x + 8$$

$60x + 75y \geq 300$ solves to $y \geq 4 - 4/5\,x$
$75y + 60x \geq 300$
$75y \geq -60x + 300$

$$y \geq -\frac{4}{5}x + 4$$

Graph these 2 inequalities:

$$y \le 8 - 1/2\,x$$
$$y \ge 4 - 4/5\,x$$

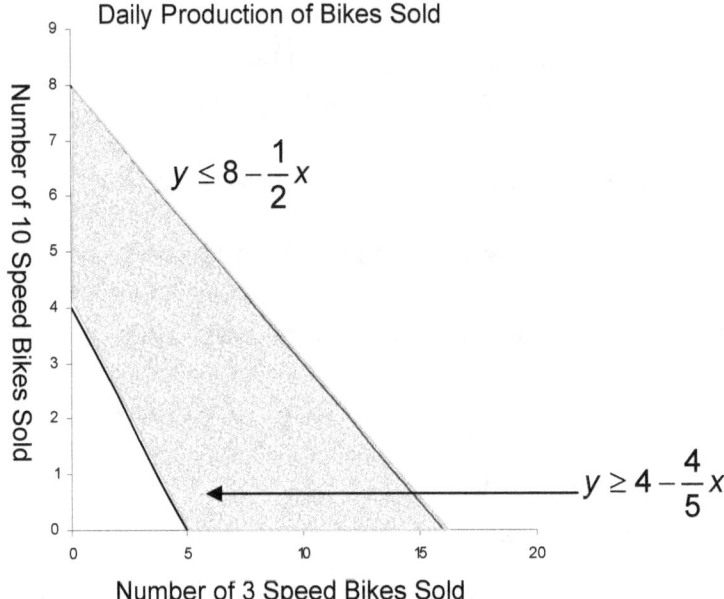

Realize that $x \ge 0$ and $y \ge 0$, since the number of bikes assembled can not be a negative number. Graph these as additional constraints on the problem. The number of bikes assembled must always be an integer value, so points within the shaded area of the graph must have integer values. The maximum profit will occur at or near a corner of the shaded portion of this graph. Those points occur at (0,4), (0,8), (16,0), or (5,0).

Since profits are $60/3-speed or $75/10-speed, the profit would be:

(0,4) $60(0) + 75(4) = 300$
(0,8) $60(0) + 75(8) = 600$
(16,0) $60(16) + 75(0) = 960$ ← Maximum profit
(5,0) $60(5) + 75(0) = 300$

The maximum profit would occur if 16 3-speed bikes are made daily.

Word problems can sometimes be solved by using a system of two equations in 2 unknowns. This system can then be solved using **substitution**, or the **addition-subtraction method**.

Example: Farmer Greenjeans bought 4 cows and 6 sheep for $1700. Mr. Ziffel bought 3 cows and 12 sheep for $2400. If all the cows were the same price and all the sheep were another price, find the price charged for a cow or for a sheep.

Let x = price of a cow
Let y = price of a sheep

Then Farmer Greenjeans' equation would be: $4x + 6y = 1700$
Mr. Ziffel's equation would be: $3x + 12y = 2400$

To solve by **addition-subtraction**:

Multiply the first equation by -2: $-2(4x + 6y = 1700)$
Keep the other equation the same: $(3x + 12y = 2400)$
By doing this, the equations can be added to each other to eliminate one variable and solve for the other variable.

$$-8x - 12y = -3400$$
$$3x + 12y = 2400 \quad \text{Add these equations.}$$
$$-5x = -1000$$

$x = 200 \leftarrow$ the price of a cow was $200.
Solving for y, $y = 150 \leftarrow$ the price of a sheep was $150.

To solve by **substitution**:

Solve one of the equations for a variable. (Try to make an equation without fractions if possible.) Substitute this expression into the equation that you have not yet used. Solve the resulting equation for the value of the remaining variable.

$$4x + 6y = 1700$$
$$3x + 12y = 2400 \leftarrow \text{Solve this equation for } x.$$

It becomes $x = 800 - 4y$. Now substitute $800 - 4y$ in place of x in the OTHER equation. $4x + 6y = 1700$ now becomes:
$$4(800 - 4y) + 6y = 1700$$
$$3200 - 16y + 6y = 1700$$
$$3200 - 10y = 1700$$
$$-10y = -1500$$
$$y = 150, \text{ or } \$150 \text{ for a sheep.}$$

Substituting 150 back into an equation for y, find x.
$$4x + 6(150) = 1700$$
$$4x + 900 = 1700$$
$$4x = 800 \text{ so } x = 200 \text{ for a cow.}$$

To solve by **determinants**:

Let x = price of a cow
Let y = price of a sheep

Then Farmer Greenjeans' equation would be: $4x + 6y = 1700$
Mr. Ziffel's equation would be: $3x + 12y + 2400$

To solve this system using determinants, make one 2 by 2 determinant divided by another 2 by 2 determinant. The bottom determinant is filled with the x and y term coefficients. The top determinant is almost the same as this bottom determinant. The only difference is that when you are solving for x, the x-coefficients are replaced with the constants. Likewise, when you are solving for y, the y-coefficients are replaced with the constants. The value of a 2 by 2 determinant, $\begin{pmatrix} a & b \\ c & d \end{pmatrix}$, is found by $ad - bc$.

$$x = \frac{\begin{pmatrix} 1700 & 6 \\ 2400 & 12 \end{pmatrix}}{\begin{pmatrix} 4 & 6 \\ 3 & 12 \end{pmatrix}} = \frac{1700(12) - 6(2400)}{4(12) - 6(3)} = \frac{20400 - 14400}{48 - 18} = \frac{6000}{30} = 200$$

$$y = \frac{\begin{pmatrix} 4 & 1700 \\ 3 & 2400 \end{pmatrix}}{\begin{pmatrix} 4 & 6 \\ 3 & 12 \end{pmatrix}} = \frac{2400(4) - 3(1700)}{4(12) - 6(3)} = \frac{9600 - 5100}{48 - 18} = \frac{4500}{30} = 150$$

NOTE: The bottom determinant is always the same value for each letter.

To graph an inequality, solve the inequality for y. This gets the inequality in **slope intercept form**, (for example $y < mx + b$). The point (0, b) is the y-intercept and m is the line's slope.

If the inequality solves to $x \geq$ **any number**, then the graph includes a **vertical line**.

If the inequality solves to $y \leq$ **any number**, then the graph includes a **horizontal line**.

When graphing a linear inequality, the line will be dotted if the inequality sign is $<$ or $>$. If the inequality signs are either $\geq$ or $\leq$, the line on the graph will be a solid line. Shade above the line when the inequality sign is $\geq$ or $>$. Shade below the line when the inequality sign is $<$ or $\leq$. For inequalities of the forms $x >$ number, $x \leq$ number, $x <$ number, or $x \geq$ number, draw a vertical line (solid or dotted). Shade to the right for $>$ or $\geq$. Shade to the left for $<$ or $\leq$.

Use these rules to graph and shade each inequality. The solution to a system of linear inequalities consists of the part of the graph that is shaded for each inequality. For instance, if the graph of one inequality is shaded with red, and the graph of another inequality is shaded with blue, then the overlapping area would be shaded purple. The purple area would be the points in the solution set of this system.

Example: Solve by graphing:

$$x + y \leq 6$$
$$x - 2y \leq 6$$

Solving the inequalities for y, they become:

$$y \leq -x + 6 \quad (y\text{-intercept of 6 and slope} = -1)$$
$$y \geq 1/2\, x - 3 \quad (y \text{ intercept of } -3 \text{ and slope} = 1/2)$$

A graph with shading is shown below:

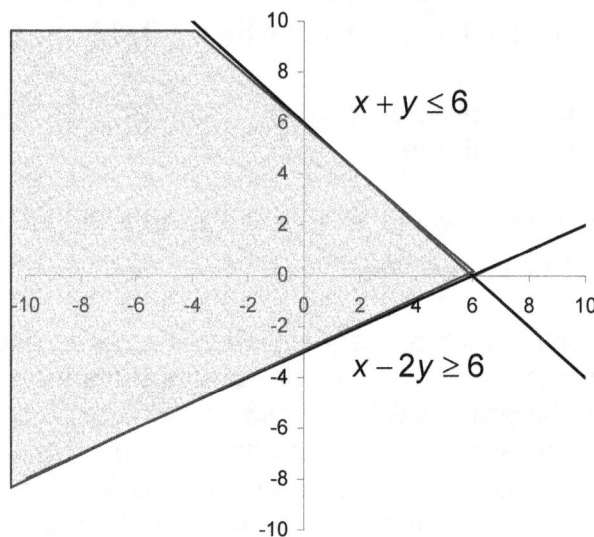

Competency 006   **The teacher understands and uses nonlinear functions and relations to model and solve problems.**

Definition: A function f is even if $f(^-x) = f(x)$ and odd if $f(^-x) = {}^-f(x)$ for all x in the domain of f.

Sample problems:

Determine if the given function is even, odd, or neither even nor odd.

1. $f(x) = x^4 - 2x^2 + 7$
   $f(^-x) = (^-x)^4 - 2(^-x)^2 + 7$
   $f(^-x) = x^4 - 2x^2 + 7$

   f(x) is an even function.

   1. Find $f(^-x)$.
   2. Replace x with ^-x.
   3. Since $f(^-x) = f(x)$, $f(x)$ is an even function.

2. $f(x) = 3x^3 + 2x$
   $f(^-x) = 3(^-x)^3 + 2(^-x)$
   $f(^-x) = {}^-3x^3 - 2x$

   $^-f(x) = {}^-(3x^3 + 2x)$
   $^-f(x) = {}^-3x^3 - 2x$

   f(x) is an odd function.

   1. Find $f(^-x)$.
   2. Replace x with ^-x.
   3. Since $f(x)$ is not equal to $f(^-x)$, $f(x)$ is not an even function.
   4. Try $^-f(x)$.
   5. Since $f(^-x) = {}^-f(x)$, $f(x)$ is an odd function.

3. $g(x) = 2x^2 - x + 4$
   $g(^-x) = 2(^-x)^2 - (^-x) + 4$
   $g(^-x) = 2x^2 + x + 4$

   $^-g(x) = {}^-(2x^2 - x + 4)$
   $^-g(x) = {}^-2x^2 + x - 4$

   $g(x)$ is neither even nor odd.

   1. First find $g(^-x)$.
   2. Replace x with ^-x.
   3. Since $g(x)$ does not equal $g(^-x)$, $g(x)$ is not an even function.
   4. Try $^-g(x)$.
   5. Since $^-g(x)$ does not equal $g(^-x)$, $g(x)$ is not an odd function.

The discriminant of a quadratic equation is the part of the quadratic formula that is usually inside the radical sign, $b^2 - 4ac$.

$$x = \frac{-b \pm \sqrt{b^2 - 4ac}}{2a}$$

The radical sign is NOT part of the discriminant!! Determine the value of the discriminant by substituting the values of a, b, and c from $ax^2 + bx + c = 0$.

-If the value of the discriminant is **any negative number**, then there are **two complex roots** including "i."
-If the value of the discriminant is **zero**, then there is only **1 real rational root**. This would be a double root.
-If the value of the discriminant is **any positive number that is also a perfect square**, then there are **two real rational roots**. (There are no longer any radical signs.)
-If the value of the discriminant is **any positive number that is NOT a perfect square**, then there are **two real irrational roots**. (There are still unsimplified radical signs.)

Example:

Find the value of the discriminant for the following equations. Then determine the number and nature of the solutions of that quadratic equation.

A) $\quad 2x^2 - 5x + 6 = 0$

$a = 2$, $b = {}^-5$, $c = 6$ so $b^2 - 4ac = ({}^-5)^2 - 4(2)(6) = 25 - 48 = {}^-23$.

Since ${}^-23$ is a negative number, there are **two complex roots** including "i".

$$x = \frac{5}{4} + \frac{i\sqrt{23}}{4}, \quad x = \frac{5}{4} - \frac{i\sqrt{23}}{4}$$

B) $\quad 3x^2 - 12x + 12 = 0$

$a = 3$, $b = {}^-12$, $c = 12$ so $b^2 - 4ac = ({}^-12)^2 - 4(3)(12) = 144 - 144 = 0$

Since 0 is the value of the discriminant, there is only **1 real rational root**.

$$x = 2$$

C) $\qquad 6x^2 - x - 2 = 0$

$a = 6$, $b = {}^-1$, $c = {}^-2$ so $b^2 - 4ac = ({}^-1)^2 - 4(6)({}^-2) = 1 + 48 = 49$.

Since 49 is positive and is also a perfect square $(\sqrt{49}) = 7$, then there are **two real rational roots.**

$$x = \frac{2}{3}, \quad x = -\frac{1}{2}$$

Try these:

1. $6x^2 - 7x - 8 = 0$
2. $10x^2 - x - 2 = 0$
3. $25x^2 - 80x + 64 = 0$

A **quadratic equation** is written in the form $ax^2 + bx + c = 0$. To solve a quadratic equation by factoring, at least one of the factors must equal zero.

Example:

Solve the equation.

$x^2 + 10x - 24 = 0$
$(x + 12)(x - 2) = 0$ $\qquad$ Factor.
$x + 12 = 0$ or $x - 2 = 0$ $\qquad$ Set each factor equal to 0.
$x = -12 \qquad x = 2$ $\qquad$ Solve.

Check:
$x^2 + 10x - 24 = 0$
$(-12)^2 + 10(-12) - 24 = 0 \qquad (2)^2 + 10(2) - 24 = 0$
$144 - 120 - 24 = 0 \qquad\qquad 4 + 20 - 24 = 0$
$0 = 0 \qquad\qquad\qquad\qquad\quad 0 = 0$

A quadratic equation that cannot be solved by factoring can be solved by **completing the square**.

Example:

Solve the equation.

$x^2 - 6x + 8 = 0$	
$x^2 - 6x = {}^-8$	Move the constant to the right side.
$x^2 - 6x + 9 = {}^-8 + 9$	Add the square of half the coefficient of x to both sides.
$(x - 3)^2 = 1$	Write the left side as a perfect square.
$x - 3 = \pm\sqrt{1}$	Take the square root of both sides.
$x - 3 = 1 \quad x - 3 = {}^-1$	Solve.
$x = 4 \qquad\;\; x = 2$	

Check:

$x^2 - 6x + 8 = 0$

$4^2 - 6(4) + 8 = 0$	$2^2 - 6(2) + 8 = 0$
$16 - 24 + 8 = 0$	$4 - 12 + 8 = 0$
$0 = 0$	$0 = 0$

To solve a quadratic equation using the quadratic formula, be sure that your equation is in the form $ax^2 + bx + c = 0$. Substitute these values into the formula:

$$x = \frac{-b \pm \sqrt{b^2 - 4ac}}{2a}$$

Example:

Solve the equation.

$3x^2 = 7 + 2x \rightarrow 3x^2 - 2x - 7 = 0$

$a = 3 \quad b = -2 \quad c = -7$

$x = \dfrac{-(-2) \pm \sqrt{(-2)^2 - 4(3)(-7)}}{2(3)}$

$x = \dfrac{2 \pm \sqrt{4 + 84}}{6}$

$x = \dfrac{2 \pm \sqrt{88}}{6}$

$x = \dfrac{2 \pm 2\sqrt{22}}{6}$

$x = \dfrac{1 \pm \sqrt{22}}{3}$

An **exponential function** is a function defined by the equation $y = ab^x$, where a is the starting value, b is the growth factor, and x tells how many times to multiply by the growth factor.

Example: $y = 100(1.5)^x$

x	y
0	100
1	150
2	225
3	337.5
4	506.25

This is an **exponential** or multiplicative pattern of growth.

Two or more quantities can vary directly or inversely to each other. To convert the statements to equations, we introduce another quantity which is constant, called the constant of proportionality.

1) If x varies directly as y, then $x = ky$, where k is the constant of proportionality, or, just a constant.

2) If x varies inversely as y, then $x = \dfrac{k}{y}$, where k is the constant.

Procedure to solve proportionality problems:

1) Translate the problem into a mathematical equation.

   Note: k is always in the numerator.
   -the variable that follows the word directly is also in the numerator.
   -the variable that follows the word inversely is in the denominator.

2) Substitute the given complete set of values in the equation to find k.

   Note: Since k is constant, it will stay the same for any values associated with the other variables.

3) Substitute the known values of the variables in the equation and solve for the missing variable.

Example: If x varies directly as y, and $x = 6$ when $y = 8$, find y when $x = 48$.

x varies directly as $y \rightarrow x = ky$ (equation 1)

Substitute $x = 6$ and $y = 8$ into equation 1 to get

$$6 = 8k \rightarrow k = \dfrac{6}{8} \rightarrow k = \dfrac{3}{4}$$

Substitute $k = \dfrac{3}{4}$ into equation 1 for $x = 48$, we get:

$48 = \dfrac{3}{4}y$ Multiply both sides by $\dfrac{4}{3}$.

$\dfrac{4}{3}(48) = y$

$y = 64$

Example: A varies inversely as the square of R. When A = 2, R = 4. Find A if R = 10.

A varies inversely as the square of R.

$$A = \frac{k}{R^2} \quad \text{(equation 1)}, k \text{ is a constant.}$$

Use equation 1 to find k when A = 2 and R = 4.

$$2 = \frac{k}{4^2} \rightarrow 2 = \frac{k}{16} \rightarrow k = 32.$$

Substituting k = 32 into equation 1 with R = 10, we get:

$$A = \frac{32}{10^2} \rightarrow A = \frac{32}{100} \rightarrow A = 0.32$$

Example: x varies directly as the cube root of y and inversely as the square of z. When x = 2, y = 27 and z = 1. Find y when z = 2 and x = 1.

x varies directly as the cube root of y and inversely as the square of z.

$$x = k \cdot \frac{\sqrt[3]{y}}{z^2} \quad \text{(equation 1)}, k \text{ is constant.}$$

Substituting in equation 1 to solve for k when x = 2, y = 27, and z = 1 we get:

$$2 = k \cdot \frac{\sqrt[3]{27}}{1^2} \rightarrow 2 = k \cdot \frac{3}{1} \rightarrow 2 = 3k$$

$$k = \frac{2}{3}$$

To solve for y when z = 2 and x = 1, we substitute in equation 1 using the value we found for k to get:

$$1 = \frac{2}{3} \cdot \frac{\sqrt[3]{y}}{2^2} \rightarrow 1 = \frac{2 \cdot \sqrt[3]{y}}{3(4)} \rightarrow 1 = \frac{2 \cdot \sqrt[3]{y}}{12}$$

$$1 = \frac{\sqrt[3]{y}}{6} \rightarrow 6 = \sqrt[3]{y} \quad \text{Cube both sides.}$$

$$6^3 = \left(\sqrt[3]{y}\right)^3$$

$$y = 216$$

Different types of function transformations affect the graph and characteristics of a function in predictable ways. The basic types of transformation are horizontal and vertical shift (translation), horizontal and vertical scaling (dilation), and reflection. As an example of the types of transformations, we will consider transformations of the functions $f(x) = x^2$.

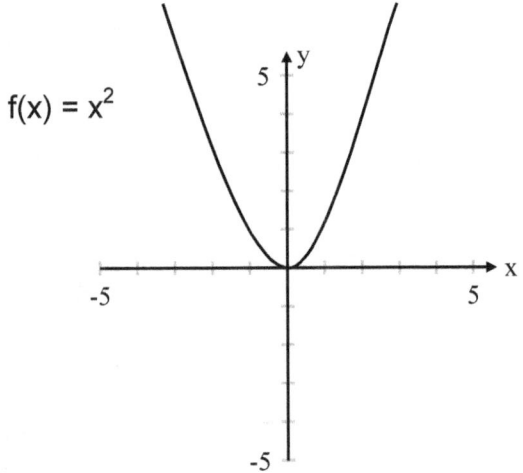

Horizontal shifts take the form $g(x) = f(x \pm c)$. For example, we obtain the graph of the function $g(x) = (x + 2)^2$ by shifting the graph of $f(x) = x^2$ two units to the left. The graph of the function $h(x) = (x - 2)^2$ is the graph of $f(x) = x^2$ shifted two units to the right.

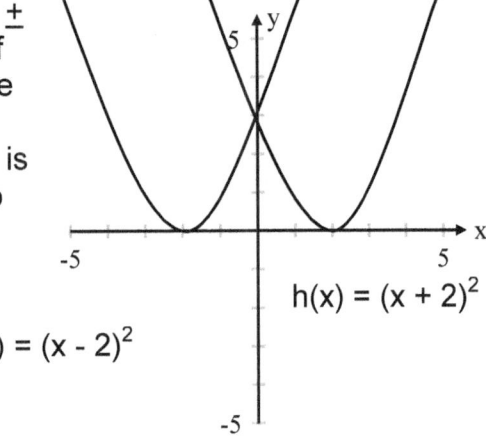

Vertical shifts take the form $g(x) = f(x) \pm c$. For example, we obtain the graph of the function $g(x) = (x^2) - 2$ by shifting the graph of $f(x) = x^2$ two units down. The graph of the function $h(x) = (x^2) + 2$ is the graph of $f(x) = x^2$ shifted two units up.

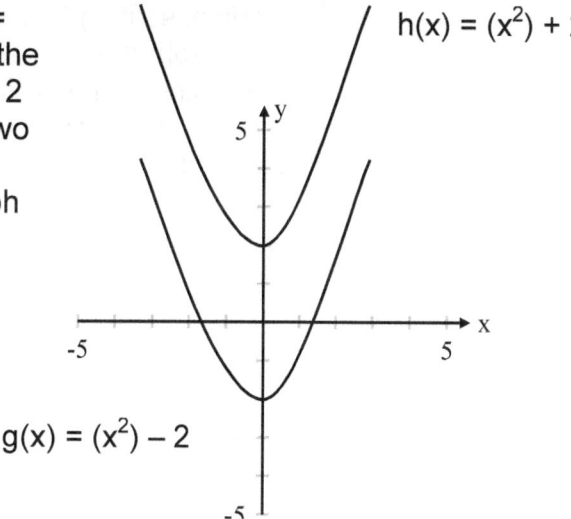

Horizontal scaling takes the form $g(x) = f(cx)$. For example, we obtain the graph of the function $g(x) = (2x)^2$ by compressing the graph of $f(x) = x^2$ in the x-direction by a factor of two. If $c > 1$ the graph is compressed in the x-direction, while if $1 > c > 0$ the graph is stretched in the x-direction.

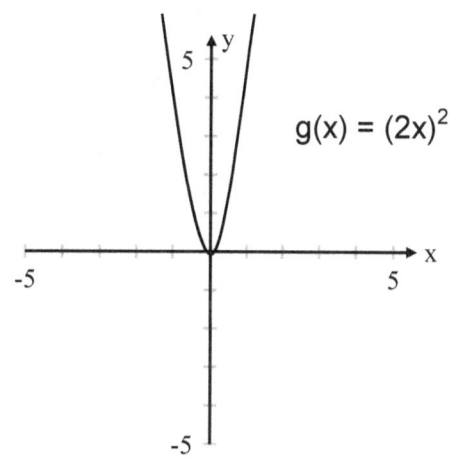

Vertical scaling takes the form $g(x) = cf(x)$. For example, we obtain the graph of $g(x) = 1/2(x^2)$ by compressing the graph of $f(x) = x^2$ in the y-direction by a factor of 1/2. If $c > 1$ the graph is stretched in the y-direction while if $1 > c > 0$ the graph is compressed in the y-direction.

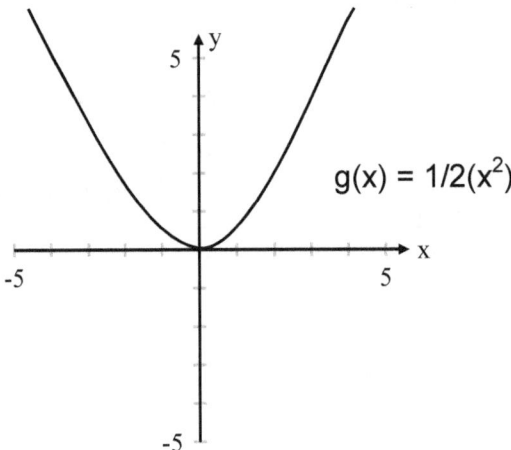

Related to scaling is reflection, in which the graph of a function flips across either the x or y-axis. Reflections take the form of $g(x) = f(-x)$, horizontal reflection, and $g(x) = -f(x)$, vertical reflection. For example, we obtain the graph of $g(x) = -(x^2)$ by reflecting the graph of $f(x) = x^2$ across the x-axis. Note that in the case of $f(x) = x^2$, horizontal reflection produces the same graph because the function is horizontally symmetrical.

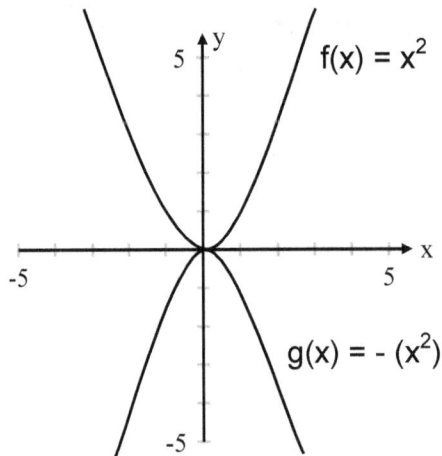

A final type of transformation is absolute value. Taking the absolute value of the outputs of a function produces a v-shaped graph where all y-values are positive. Conversely, taking the absolute value of the inputs of a function produces a v-shaped graph where all x-values are positive. For example, compare the graphs of $f(x) = x$ and $g(x) = |x|$.

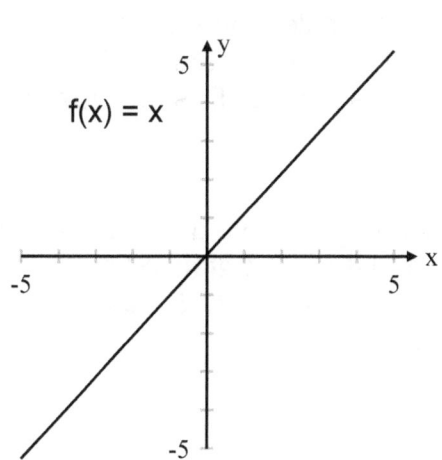
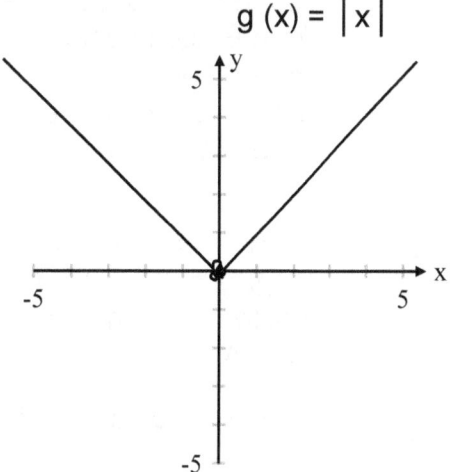

Some word problems will give a quadratic equation to be solved. When the quadratic equation is found, set it equal to zero and solve the equation by factoring or the quadratic formula. Examples of this type of problem follow.

Example:
Alberta (A) is a certain distance north of Boston (B). The distance from Boston east to Carlisle (C) is 5 miles more than the distance from Alberta to Boston. The distance from Alberta to Carlisle is 10 miles more than the distance from Alberta to Boston. How far is Alberta from Carlisle?

Solution:
Since north and east form a right angle, these distances are the lengths of the legs of a right triangle. If the distance from Alberta to Boston is x, then from Boston to Carlisle is $x+5$, and the distance from Alberta to Carlisle is $x+10$.

The equation is: $AB^2 + BC^2 = AC^2$

$$x^2 + (x+5)^2 = (x+10)^2$$
$$x^2 + x^2 + 10x + 25 = x^2 + 20x + 100$$
$$2x^2 + 10x + 25 = x^2 + 20x + 100$$
$$x^2 - 10x - 75 = 0$$
$$(x-15)(x+5) = 0 \quad \text{Distance cannot be negative.}$$
$$x = 15 \quad \text{Distance from Alberta to Boston.}$$
$$x + 5 = 20 \quad \text{Distance from Boston to Carlisle.}$$
$$x + 10 = 25 \quad \text{Distance from Alberta to Carlisle.}$$

Example:
The square of a number is equal to 6 more than the original number. Find the original number.

Solution: If x = original number, then the equation is:

$x^2 = 6 + x$ Set this equal to zero.

$x^2 - x - 6 = 0$ Now factor.

$(x-3)(x+2) = 0$

$x = 3$ or $x = -2$ There are two solutions, 3 or -2.

Try these:

1. One side of a right triangle is 1 less than twice the shortest side, while the third side of the triangle is 1 more than twice the shortest side. Find all 3 sides.

2. Twice the square of a number equals 2 less than 5 times the number. Find the number(s).

Some word problems can be solved by setting up a quadratic equation or inequality. Examples of this type could be problems that deal with finding a maximum area.

Example:

A family wants to enclose 3 sides of a rectangular garden with 200 feet of fence. In order to have a garden with an area of **at least** 4800 square feet, find the dimensions of the garden. Assume that a wall or a fence already borders the fourth side of the garden

Solution:
Let $x =$ distance from the wall

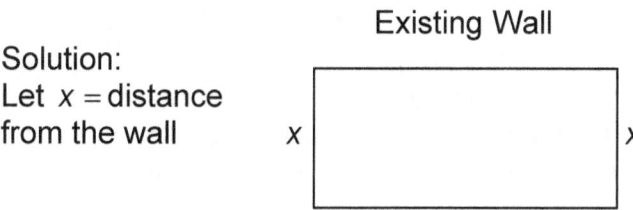

Then $2x$ feet of fence is used for these 2 sides. The remaining side of the garden would use the rest of the 200 feet of fence, that is, $200-2x$ feet of fence. Therefore the width of the garden is x feet and the length is $200-2x$ ft.

The area, $200x - 2x^2$, needs to be greater than or equal to 4800 sq. ft. So, this problem uses the inequality $4800 \leq 200x - 2x^2$. This becomes $2x^2 - 200x + 4800 \leq 0$. Solving this, we get:

$$200x - 2x^2 \geq 4800$$
$$-2x^2 + 200x - 4800 \geq 0$$
$$2\left(-x^2 + 100x - 2400\right) \geq 0$$
$$-x^2 + 100x - 2400 \geq 0$$
$$(-x + 60)(x - 40) \geq 0$$
$$-x + 60 \geq 0$$
$$-x \geq -60$$
$$x \leq 60$$
$$x - 40 \geq 0$$
$$x \geq 40$$

So the area will be at least 4800 square feet if the width of the garden is from 40 up to 60 feet. (The length of the rectangle would vary from 120 feet to 80 feet depending on the width of the garden.)

Example:

he height of a projectile fired upward at a velocity of v meters per second from an original height of h meters is $y = h + vx - 4.9x^2$. If a rocket is fired from an original height of 250 meters with an original velocity of 4800 meters per second, find the approximate time the rocket would drop to sea level (a height of 0).

Solution:

The equation for this problem is: $y = 250 + 4800x - 4.9x^2$. If the height at sea level is zero, then $y = 0$ so $0 = 250 + 4800x - 4.9x^2$. Solving this for x could be done by using the quadratic formula. In addition, the approximate time in x seconds until the rocket would be at sea level could be estimated by looking at the graph. When the y value of the graph goes from positive to negative then there is a root (also called solution or x–intercept) in that interval.

$$x = \frac{-4800 \pm \sqrt{4800^2 - 4(-4.9)(250)}}{2(-4.9)} \approx 980 \text{ or } -0.05 \text{ seconds}$$

Since the time has to be positive, it will be about 980 seconds until the rocket is at sea level.

To graph an inequality, graph the quadratic as if it was an equation; however, if the inequality has just a > or < sign, then make the curve itself dotted. Shade above the curve for > or ≥. Shade below the curve for < or ≤.

Examples:

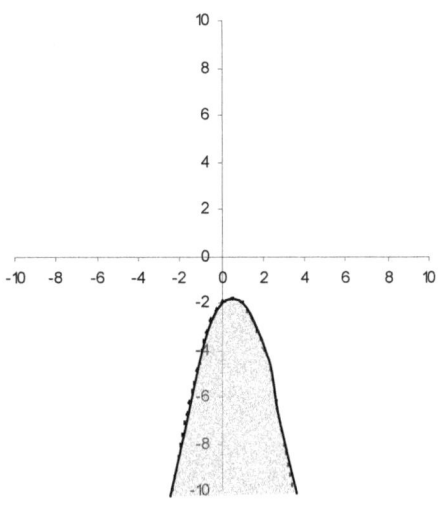

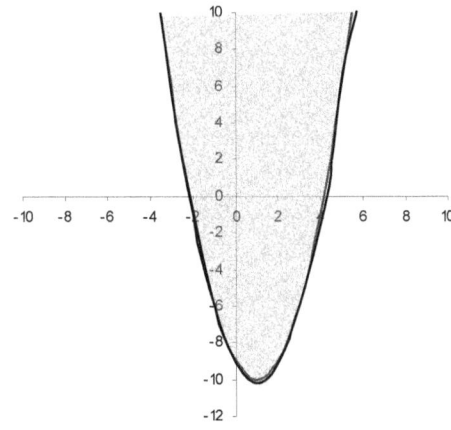

The **absolute value function** for a 1st degree equation is of the form: $y = m(x - h) + k$. Its graph is in the shape of a V. The point (h,k) is the location of the maximum/minimum point on the graph. "$\pm m$" are the slopes of the 2 sides of the V. The graph opens up if m is positive and down if m is negative.

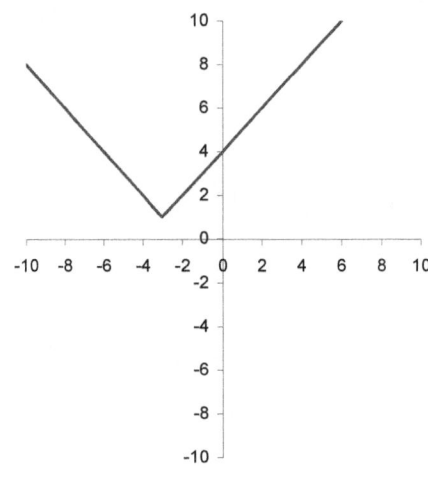

$$y = |x + 3| + 1$$

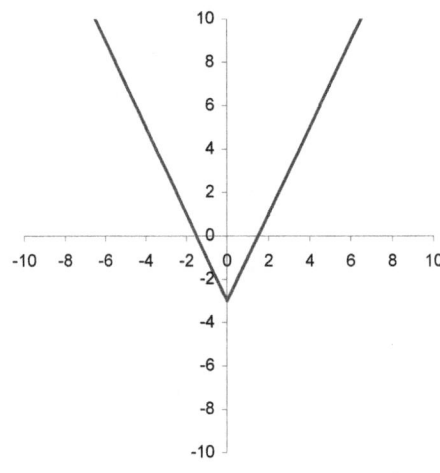

$$y = 2|x| - 3$$

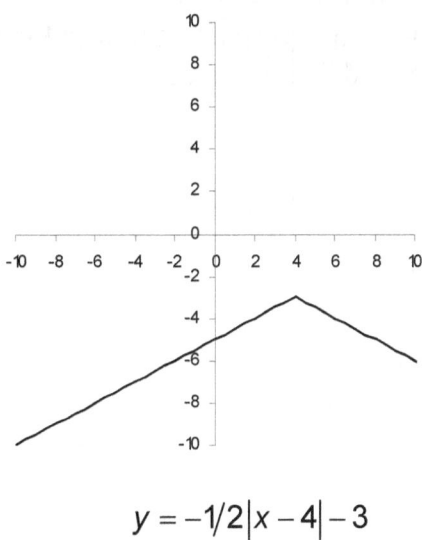

$$y = -1/2|x-4|-3$$

Note that on the first graph, the graph opens up since m is positive 1. It has (−3,1) as its minimum point. The slopes of the two upward rays are ± 1.

The second graph also opens up since m is positive. (0,−3) is its minimum point. The slopes of the 2 upward rays are ± 2.

The third graph is a downward ∧ because m is −1/2. The maximum point on the graph is at (4,−3). The slopes of the two downward rays are ±1/2.

The **identity function** is the linear equation $y = x$. Its graph is a line going through the origin (0,0) and through the first and third quadrants at a 45° degree angle.

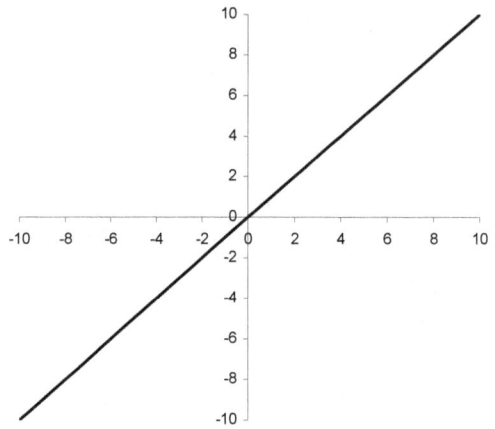

The **greatest integer function** or **step function** has the equation: $f(x) = j[rx - h] + k$ or $y = j[rx - h] + k$. (h,k) is the location of the left endpoint of one step. j is the vertical jump from step to step. r is the reciprocal of the length of each step. If (x,y) is a point of the function, then when x is an integer, its y value is the same integer. If (x,y) is a point of the function, then when x is not an integer, its y value is the first integer less than x. Points on $y = [x]$ would include:

(3,3), (−2,−2), (0,0), (1.5,1), (2.83,2), (−3.2,−4), (−.4,−1).

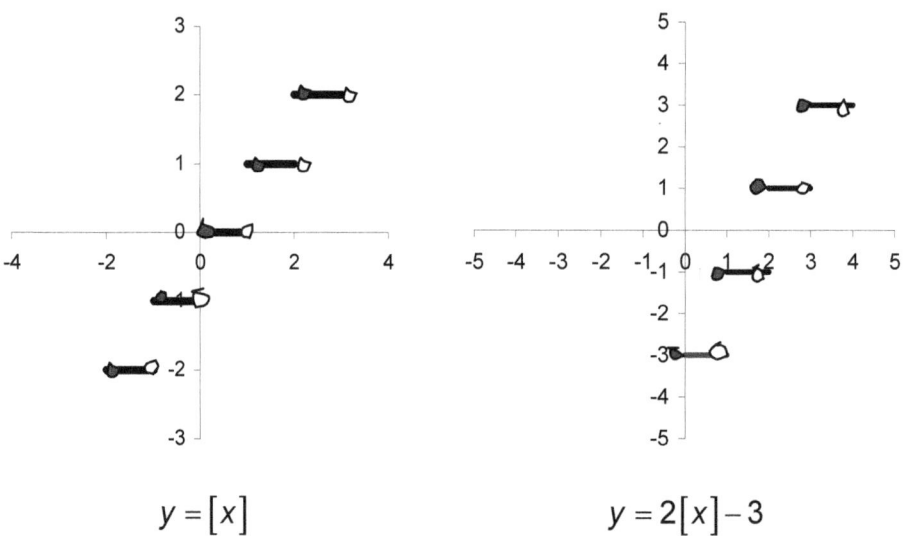

$y = [x]$ $y = 2[x] - 3$

Note that in the graph of the first equation, the steps are going up as they move to the right. Each step is one space wide (inverse of r) with a solid dot on the left and a hollow dot on the right where the jump to the next step occurs. Each step is one square higher (j = 1) than the previous step. One step of the graph starts at (0,0) ← values of (h,k).

In the second graph, the graph goes up to the right. One step starts at the point (0,−3) ← values of (h,k). Each step is one square wide (r = 1) and each step is two squares higher than the previous step (j = 2).

Practice: Graph the following equations:

1. $f(x) = x$
2. $y = -|x-3| + 5$
3. $y = 3[x]$
4. $y = 2/5|x-5| - 2$

A **rational function** is given in the form $f(x) = p(x)/q(x)$. In the equation, $p(x)$ and $q(x)$ both represent polynomial functions where $q(x)$ does not equal zero. The branches of rational functions approach asymptotes. Setting the denominator equal to zero and solving will give the value(s) of the vertical asymptotes(s) since the function will be undefined at this point. If the value of $f(x)$ approaches b as the $|x|$ increases, the equation $y = b$ is a horizontal asymptote. To find the horizontal asymptote it is necessary to make a table of value for x that are to the right and left of the vertical asymptotes. The pattern for the horizontal asymptotes will become apparent as the $|x|$ increases.

If there are more than one vertical asymptotes, remember to choose numbers to the right and left of each one in order to find the horizontal asymptotes and have sufficient points to graph the function.

Sample problem:

1. Graph $f(x) = \dfrac{3x+1}{x-2}$.

$x - 2 = 0$
$x = 2$

1. Set denominator $= 0$ to find the vertical asymptote.

x	f(x)
3	10
10	3.875
100	3.07
1000	3.007
1	⁻4
⁻10	2.417
⁻100	2.93
⁻1000	2.99

2. Make table choosing numbers to the right and left of the vertical asymptote.

3. The pattern shows that as the $|x|$ increases $f(x)$ approaches the value 3, therefore a horizontal asymptote exists at $y = 3$

Sketch the graph.

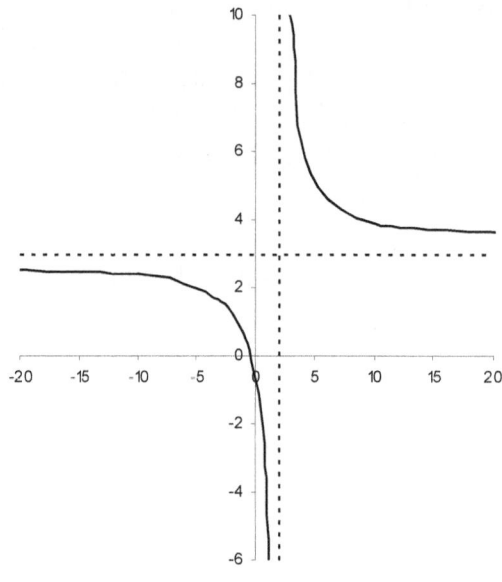

Functions defined by two or more formulas are **piecewise functions**. The formula used to evaluate piecewise functions varies depending on the value of x. The graphs of piecewise functions consist of two or more pieces, or intervals, and are often discontinuous.

Example 1:

$f(x) = x + 1$ if $x > 2$
$ x - 2$ if $x \leq 2$

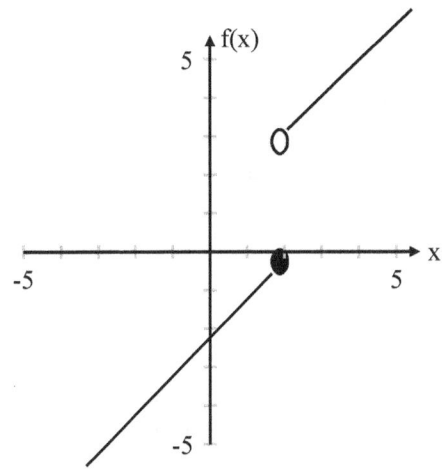

Example 2:

$f(x) = x$ if $x \geq 1$
$ x^2$ if $x < 1$

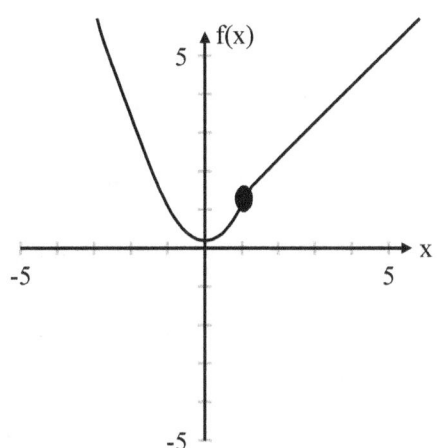

When graphing or interpreting the graph of piecewise functions it is important to note the points at the beginning and end of each interval because the graph must clearly indicate what happens at the end of each interval. Note that in the graph of Example 1, point (2, 3) is not part of the graph and is represented by an empty circle. On the other hand, point (2, 0) is part of the graph and is represented as a solid circle. Note also that the graph of Example 2 is continuous despite representing a piecewise function.

Practice: Graph the following piecewise equations.

1. $f(x) = x^2$ if $x > 0$
   $ = x + 4$ if $x \leq 0$

2. $f(x) = x^2 - 1$ if $x > 2$
   $ = x^2 + 2$ if $x \leq 2$

Logarithmic Functions

Logarithmic functions of base a are of the basic form

$f(x) = \log_a x$, where $a > 0$ and not equal to 1.

The domain of the function, f, is $(0, +\inf.)$ and the range is $(-\inf., +\inf.)$. The x-intercept of the logarithmic function is $(1,0)$ because any number raised to the power of 0 is equal to one. The graph of the function, f, has a vertical asymptote at $x = 0$. As the value of $f(x)$ approaches negative infinity, x becomes closer and closer to 0.

Example:

Graph the function $f(x) = \log_2(x + 1)$.

The domain of the function is all values of x such that $x + 1 > 0$.

Thus, the domain of $f(x)$ is $x > -1$.

The range of $f(x)$ is $(-\inf., +\inf.)$.

The vertical asymptote of $f(x)$ is the value of x that satisfies the equation $x + 1 = 0$. Thus, the vertical asymptote is $x = -1$. Note that we can find the vertical asymptote of a logarithmic function by setting the product of the logarithm (containing the variable) equal to 0.

The x-intercept of $f(x)$ is the value of x that satisfies the equation $x + 1 = 1$ because $2^0 = 1$. Thus, the x-intercept of $f(x)$ is $(0,0)$. Note that we can find the x-intercept of a logarithmic function by setting the product of the logarithm equal to 1.

Finally, we find two additional values of $f(x)$, one between the vertical asymptote and the x-intercept and the other to the right of the x-intercept. For example, $f(-0.5) = -1$ and $f(3) = 2$.

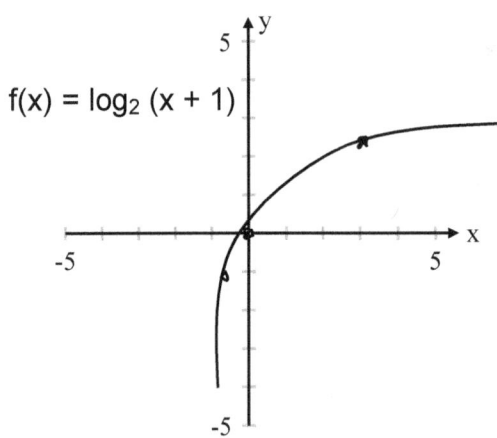

Exponential Functions

The inverse of a logarithmic function is an exponential function. Exponential functions of base a take the basic form

$f(x) = a^x$, where $a > 0$ and not equal to 1.

The domain of the function, f, is (-inf., +inf.). The range is the set of all positive real numbers. If $a < 1$, f is a decreasing function and if $a > 1$ f is an increasing function. The y-intercept of $f(x)$ is (0,1) because any base raised to the power of 0 equals 1. Finally, $f(x)$ has a horizontal asymptote at $y = 0$.

Example:

Graph the function $f(x) = 2^x - 4$.

The domain of the function is the set of all real numbers and the range is $y > -4$. Because the base is greater than 1, the function is increasing. The y-intercept of $f(x)$ is (0,-3). The x-intercept of $f(x)$ is (2,0). The horizontal asymptote of $f(x)$ is $y = -4$.

Finally, to construct the graph of $f(x)$ we find two additional values for the function. For example, $f(-2) = -3.75$ and $f(3) = 4$.

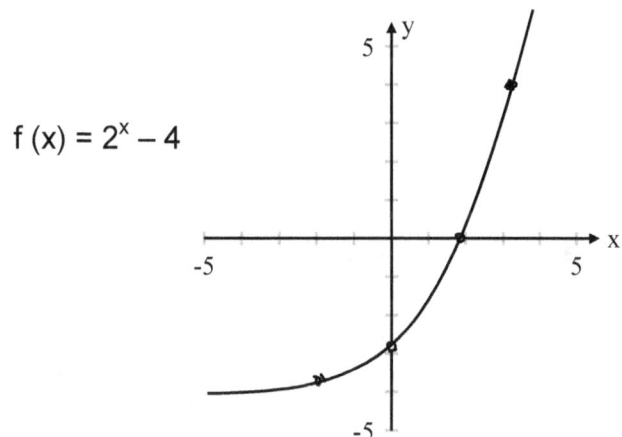

Note that the horizontal asymptote of any exponential function of the form $g(x) = a^x + b$ is $y = b$. Note also that the graph of such exponential functions is the graph of $h(x) = a^x$ shifted b units up or down. Finally, the graph of exponential functions of the form $g(x) = a^{(x+b)}$ is the graph of $h(x) = a^x$ shifted b units left or right.

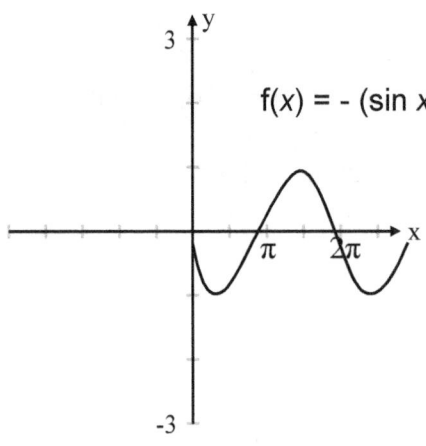

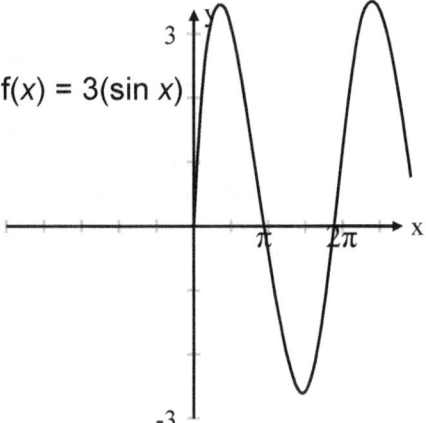

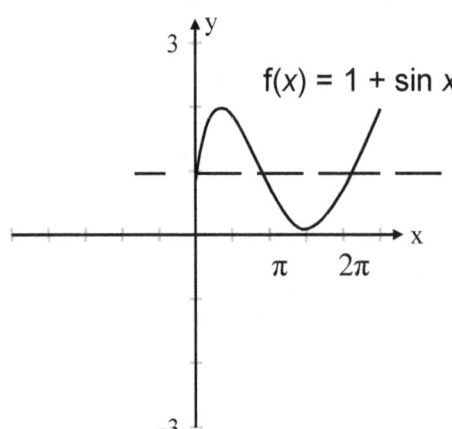

Note the graph of $f(x) = -(\sin x)$ is a reflection of $f(x) = \sin x$ about the x-axis. The graph of $f(x) = 3(\sin x)$ is a vertical stretch of $f(x) = \sin x$ by a factor of 3 that produces an increase in amplitude or range. The graph of $f(x) = 1 + \sin x$ is a 1 unit vertical shift of $f(x) = \sin x$. In all cases, the period of the graph is 2π, the same as the period of $f(x) = \sin x$.

Competency 007   **The teacher uses and understands the conceptual foundations of calculus related to topics in middle school mathematics.**

When teaching sequences and series, middle school teachers should introduce the concept of limit because understanding of limits is essential to the understanding of topics in pre-calculus and calculus.

A sequence is a list of numbers that follow a specific pattern. For example, the following list of numbers represents a sequence defined by the formula $\frac{n}{n+1}$.

$$\frac{1}{2}, \frac{2}{3}, \frac{3}{4}, \frac{4}{5}, \frac{5}{6}, \ldots \frac{99}{100}, \ldots$$

We say the limit of the sequence is 1, because as n approaches infinity, the sequence approaches 1. Thus, a limit is the upper or lower boundary of a sequence; the value that the sequence will never pass.

A series is a sum of numbers. Convergent series possess a numerical limit. For example, the sum of the series $\frac{1}{2^n}$ starting with n = 1, has a limit of 1. We represent the series as follows:

$$\sum_{n=1}^{\infty} \frac{1}{2^n} = \frac{1}{2} + \frac{1}{4} + \frac{1}{8} + \frac{1}{16} + \frac{1}{32} + \ldots$$

The symbol $\sum$ (sigma) represents summation and the numbers to the right of the sigma symbol are the beginning and end values of the series. Note, that no matter how many values we add in the series, the total sum approaches, but never reaches 1.

The **difference quotient** is the average rate of change over an interval. For a function f, the **difference quotient** is represented by the formula:

$$\frac{f(x+h)-f(x)}{h}.$$

This formula computes the slope of the secant line through two points on the graph of f. These are the points with x-coordinates x and $x+h$.

Example: Find the difference quotient for the function $f(x)=2x^2+3x-5$.

$$\frac{f(x+h)-f(x)}{h} = \frac{2(x+h)^2+3(x+h)-5-(2x^2+3x-5)}{h}$$

$$= \frac{2(x^2+2hx+h^2)+3x+3h-5-2x^2-3x+5}{h}$$

$$= \frac{2x^2+4hx+2h^2+3x+3h-5-2x^2-3x+5}{h}$$

$$= \frac{4hx+2h^2+3h}{h}$$

$$= 4x+2h+3$$

The **derivative** is the slope of a tangent line to a graph $f(x)$, and is usually denoted $f'(x)$. This is also referred to as the instantaneous rate of change.

The derivative of $f(x)$ at $x=a$ is given by taking the limit of the average rates of change (computed by the difference quotient) as h approaches 0.

$$f'(a) = \lim_{h \to 0} \frac{f(a+h)-f(a)}{h}$$

Example: Suppose a company's annual profit (in millions of dollars) is represented by the above function $f(x) = 2x^2 + 3x - 5$ and x represents the number of years in the interval. Compute the rate at which the annual profit was changing over a period of 2 years.

$$f'(a) = \lim_{h \to 0} \frac{f(a+h) - f(a)}{h}$$

$$= f'(2) = \lim_{h \to 0} \frac{f(2+h) - f(2)}{h}$$

Using the difference quotient we computed above, $4x + 2h + 3$, we get

$$f'(2) = \lim_{h \to 0}(4(2) + 2h + 3)$$
$$= 8 + 3$$
$$= 11.$$

We have, therefore, determined that the annual profit for the company has increased at the average rate of $11 million per year over the two-year period.

Taking the integral of a function and evaluating it from one x value to another provides the total **area under the curve** (i.e. between the curve and the x axis). Remember, though, that regions above the x axis have "positive" area and regions below the x axis have "negative" area. You must account for these positive and negative values when finding the area under curves. Follow these steps.

1. Determine the x values that will serve as the left and right boundaries of the region.
2. Find all x values between the boundaries that are either solutions to the function or are values which are not in the domain of the function. These numbers are the interval numbers.
3. Integrate the function.
4. Evaluate the integral once for each of the intervals using the boundary numbers.
5. If any of the intervals evaluates to a negative number, make it positive (the negative simply tells you that the region is below the x axis).
6. Add the value of each integral to arrive at the area under the curve.

Example:

Find the area under the following function on the given intervals.
$f(x) = \sin x$; $(0, 2\pi)$

$\sin x = 0$ Find any roots to f(x) on $(0, 2\pi)$.

$x = \pi$

$(0, \pi)$ $(\pi, 2\pi)$ Determine the intervals using the boundary numbers and the roots.

$\int \sin x \, dx = {}^- \cos x$ Integrate f(x). We can ignore the constant c because we have numbers to use to evaluate the integral.

${}^- \cos x \big]_{x=0}^{x=\pi} = {}^- \cos \pi - ({}^- \cos 0)$

${}^- \cos x \big]_{x=0}^{x=\pi} = {}^- (-1) + (1) = 2$

${}^- \cos x \big]_{x=\pi}^{x=2\pi} = {}^- \cos 2\pi - ({}^- \cos \pi)$

${}^- \cos x \big]_{x=\pi}^{x=2\pi} = {}^- 1 + ({}^- 1) = {}^- 2$ The ${}^- 2$ means that for $(\pi, 2\pi)$, the region is below the x axis, but the area is still 2.

Area = 2 + 2 = 4   Add the 2 integrals together to get the area.

The derivative of a function has two basic interpretations.

  I. Instantaneous rate of change
  II. Slope of a tangent line at a given point

If a question asks for the rate of change of a function, take the derivative to find the equation for the rate of change. Then plug in for the variable to find the instantaneous rate of change.

The following is a list summarizing some of the more common quantities referred to in rate of change problems.

area	height	profit
decay	population growth	sales
distance	position	temperature
frequency	pressure	volume

Pick a point, say $x = {}^-3$, on the graph of a function. Draw a tangent line at that point. Find the derivative of the function and plug in $x = {}^-3$. The result will be the slope of the tangent line.

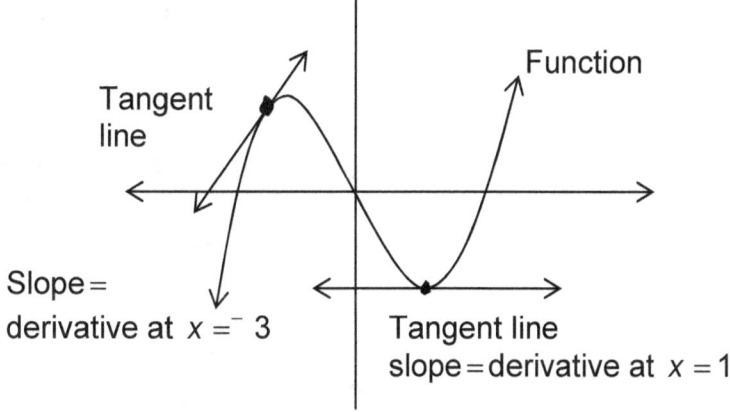

Tangent line

Function

Slope = derivative at $x = {}^-3$

Tangent line slope = derivative at $x = 1$

A function is said to be increasing if it is rising from left to right and decreasing if it is falling from left to right. Lines with positive slopes are increasing, and lines with negative slopes are decreasing. If the function in question is something other than a line, simply refer to the slopes of the tangent lines as the test for increasing or decreasing. Take the derivative of the function and plug in an x value to get the slope of the tangent line; a positive slope means the function is increasing and a negative slope means it is decreasing. If an interval for x values is given, just pick any point between the two values to substitute.

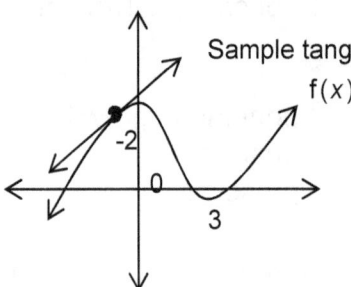

Sample tangent line on $(^-2, 0)$

On the interval $(^-2, 0)$, $f(x)$ is increasing. The tangent lines on this part of the graph have positive slopes.

Example:

The growth of a certain bacteria is given by $f(x) = x + \dfrac{1}{x}$. Determine if the rate of growth is increasing or decreasing on the time interval $(^-1, 0)$.

$$f'(x) = 1 + \dfrac{^-1}{x^2}$$

To test for increasing or decreasing, find the slope of the tangent line by taking the derivative.

$$f'\left(\dfrac{^-1}{2}\right) = 1 + \dfrac{^-1}{(^-1/2)^2}$$

Pick any point on $(^-1, 0)$ and substitute into the derivative.

$$f'\left(\dfrac{^-1}{2}\right) = 1 + \dfrac{^-1}{1/4}$$

$$= 1 - 4$$

$$= ^-3$$

The slope of the tangent line at $x = \dfrac{^-1}{2}$ is $^-3$. The exact value of the slope is not important. The important fact is that the slope is negative.

Substituting an x value into a function produces a corresponding y value. The coordinates of the point (x,y), where y is the largest of all the y values, is said to be a maximum point. The coordinates of the point (x,y), where y is the smallest of all the y values, is said to be a minimum point. To find these points, only a few x values must be tested. First, find all of the x values that make the derivative either zero or undefined. Substitute these values into the original function to obtain the corresponding y values. Compare the y values. The largest y value is a maximum; the smallest y value is a minimum. If the question asks for the maxima or minima on an interval, be certain to also find the y values that correspond to the numbers at either end of the interval.

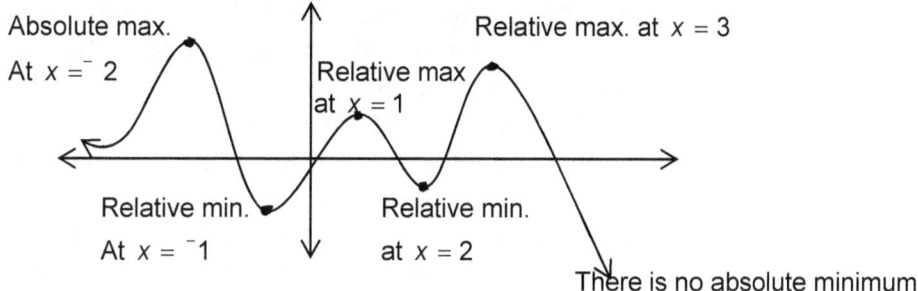

Example:

Find the maxima and minima of $f(x) = 2x^4 - 4x^2$ at the interval $(^-2, 1)$.

$f'(x) = 8x^3 - 8x$
$8x^3 - 8x = 0$
$8x(x^2 - 1) = 0$
$8x(x - 1)(x + 1) = 0$
$x = 0, x = 1, \text{ or } x = {}^-1$
$f(0) = 2(0)^4 - 4(0)^2 = 0$
$f(1) = 2(1)^4 - 4(1)^2 = {}^-2$
$f(^-1) = 2(^-1)^4 - 4(^-1)^2 = {}^-2$
$f(^-2) = 2(^-2)^4 - 4(^-2)^2 = 16$

Take the derivative first. Find all the x values (critical values) that make the derivative zero or undefined. In this case, there are no x values that make the derivative undefined. Substitute the critical values into the original function. Also, plug in the endpoint of the interval. Note that 1 is a critical point and an endpoint.

The maximum is at (–2, 16) and there are minima at (1, –2) and (–1, –2). (0,0) is neither the maximum or minimum on (–2, 1) but it is still considered a relative extra point.

The first derivative reveals whether a curve is rising or falling (increasing or decreasing) from the left to the right. In much the same way, the second derivative relates whether the curve is concave up or concave down. Curves which are concave up are said to "collect water;" curves which are concave down are said to "dump water." To find the intervals where a curve is concave up or concave down, follow the following steps.

1. Take the second derivative (i.e. the derivative of the first derivative).
2. Find the critical x values.
   -Set the second derivative equal to zero and solve for critical x values.
   -Find the x values that make the second derivative undefined (i.e. make the denominator of the second derivative equal to zero).
   Such values may not always exist.
3. Pick sample values which are both less than and greater than each of the critical values.
4. Substitute each of these sample values into the second derivative and determine whether the result is positive or negative.
   -If the sample value yields a positive number for the second derivative, the curve is concave up on the interval where the sample value originated.
   -If the sample value yields a negative number for the second derivative, the curve is concave down on the interval where the sample value originated.

Example:

Find the intervals where the curve is concave up and concave down for $f(x) = x^4 - 4x^3 + 16x - 16$.

$f'(x) = 4x^3 - 12x^2 + 16$ Take the second derivative.
$f''(x) = 12x^2 - 24x$

Find the critical values by setting the second derivative equal to zero.

$12x^2 - 24x = 0$

$12x(x - 2) = 0$

$x = 0$ or $x = 2$

There are no values that make the second derivative undefined. Set up a number line with the critical values.

Sample values: −1, 1, 3

$f''(-1) = 12(-1)^2 - 24(-1) = 36$
$f''(1) = 12(1)^2 - 24(1) = -12$
$f''(3) = 12(3)^2 - 24(3) = 36$

Pick sample values in each of the 3 intervals. If the sample value produces a negative number, the function is concave down.

If the value produces a positive number, the curve is concave up. If the value produces a zero, the function is linear.

Therefore when $x < 0$ the function is concave up,
when $0 < x < 2$ the function is concave down,
when $x > 2$ the function is concave up.

A **point of inflection** is a point where a curve changes from being concave up to concave down or vice versa. To find these points, follow the steps for finding the intervals where a curve is concave up or concave down. A critical value is part of an inflection point if the curve is concave up on one side of the value and concave down on the other. The critical value is the x coordinate of the inflection point. To get the y coordinate, plug the critical value into the original function.

Example: Find the inflection points of $f(x) = 2x - \tan x$ where $\dfrac{-\pi}{2} < x < \dfrac{\pi}{2}$.

$f(x) = 2x - \tan x \quad \dfrac{-\pi}{2} < x < \dfrac{\pi}{2}$ Note the restriction on x.

$f'(x) = 2 - \sec^2 x$ Take the second derivative.

$f''(x) = 0 - 2 \bullet \sec x \bullet (\sec x \tan x)$ Use the Power rule.

$= -2 \bullet \dfrac{1}{\cos x} \bullet \dfrac{1}{\cos x} \bullet \dfrac{\sin x}{\cos x}$

The derivative of $\sec x$ is $(\sec x \tan x)$.

$f''(x) = \dfrac{-2 \sin x}{\cos^3 x}$

Find critical values by solving for the second derivative equal to zero.

$$0 = \frac{-2\sin x}{\cos^3 x}$$

No x values on $\left(\frac{-\pi}{2}, \frac{\pi}{2}\right)$ make the denominator zero.

$-2\sin x = 0$
$\sin x = 0$
$x = 0$

$\longleftarrow\!\!\!\!\!\!\!+\!\!\!\!\!\!\!\longrightarrow$
$\quad\quad\; 0$

Pick sample values on each side of the critical value $x = 0$.

Sample values: $x = \frac{-\pi}{4}$ and $x = \frac{\pi}{4}$

$$f''\left(\frac{-\pi}{4}\right) = \frac{-2\sin(-\pi/4)}{\cos^3(\pi/4)} = \frac{-2(-\sqrt{2}/2)}{(\sqrt{2}/2)^3} = \frac{\sqrt{2}}{(\sqrt{8}/8)} = \frac{8\sqrt{2}}{\sqrt{8}} = \frac{8\sqrt{2}}{\sqrt{8}} \cdot \frac{\sqrt{8}}{\sqrt{8}}$$

$$= \frac{8\sqrt{16}}{8} = 4$$

$$f''\left(\frac{\pi}{4}\right) = \frac{-2\sin(\pi/4)}{\cos^3(\pi/4)} = \frac{-2(\sqrt{2}/2)}{(\sqrt{2}/2)^3} = \frac{-\sqrt{2}}{(\sqrt{8}/8)} = \frac{-8\sqrt{2}}{\sqrt{8}} = -4$$

The second derivative is positive on $(0, \infty)$ and negative on $(-\infty, 0)$. So the curve changes concavity at $x = 0$. Use the original equation to find the y value that inflection occurs at.

$f(0) = 2(0) - \tan 0 = 0 - 0 = 0$ The inflection point is $(0,0)$.

Extreme value problems are also known as max-min problems. Extreme value problems require using the first derivative to find values which either maximize or minimize some quantity such as area, profit, or volume. Follow these steps to solve an extreme value problem.

1. Write an equation for the quantity to be maximized or minimized.
2. Use the other information in the problem to write secondary equations.
3. Use the secondary equations for substitutions, and rewrite the original equation in terms of only one variable.
4. Find the derivative of the primary equation (step 1) and the critical values of this derivative.
5. Substitute these critical values into the primary equation.

The value which produces either the largest or smallest value is used to find the solution.

Example:

A manufacturer wishes to construct an open box from the piece of metal shown below by cutting squares from each corner and folding up the sides. The square piece of metal is 12 feet on a side. What are the dimensions of the squares to be cut out which will maximize the volume?

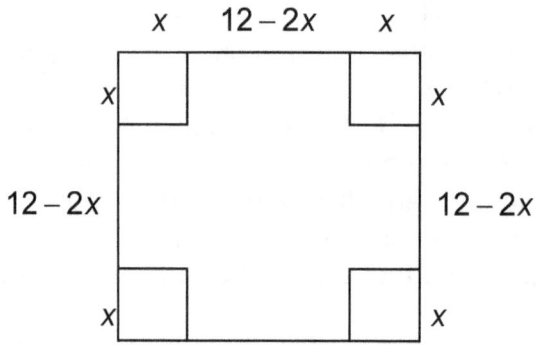

Volume = lwh Primary equation.
$l = 12 - 2x$
$w = 12 - 2x$ Secondary equations.
$h = x$
$V = (12 - 2x)(12 - 2x)(x)$ Make substitutions.
$V = (144x - 48x^2 + 4x^3)$ Take the derivative.
$\dfrac{dV}{dx} = 144 - 96x + 12x^2$

Find critical values by setting the derivative equal to zero.
$0 = 12(x^2 - 8x + 12)$
$0 = 12(x - 6)(x - 2)$
$x = 6$ and $x = 2$ Substitute critical values into volume equation.

$V = 144(6) - 48(6)^2 + 4(6)^3$ $V = 144(2) - 48(2)^2 + 4(2)^3$
$V = 0$ ft^3 when $x = 6$ $V = 128$ ft^3 when $x = 2$

Therefore, the manufacturer can maximize the volume if the squares to be cut out are 2 feet by 2 feet ($x = 2$).

If a particle (or a car, a bullet, etc.) is moving along a line, then the distance that the particle travels can be expressed by a function in terms of time.

1. The first derivative of the distance function will provide the velocity function for the particle. Substituting a value for time into this expression will provide the instantaneous velocity of the particle at the time. Velocity is the rate of change of the distance traveled by the particle. Taking the absolute value of the derivative provides the speed of the particle. A positive value for the velocity indicates that the particle is moving forward, and a negative value indicates the particle is moving backwards.

2. The second derivative of the distance function (which would also be the first derivative of the velocity function) provides the acceleration function. The acceleration of the particle is the rate of change of the velocity. If a value for time produces a positive acceleration, the particle is speeding up; if it produces a negative value, the particle is slowing down. If the acceleration is zero, the particle is moving at a constant speed.

To find the time when a particle stops, set the first derivative (i.e. the velocity function) equal to zero and solve for time. This time value is also the instant when the particle changes direction.

Example:

The motion of a particle moving along a line is according to the equation:

$s(t) = 20 + 3t - 5t^2$ where s is in meters and t is in seconds. Find the position, velocity, and acceleration of a particle at $t = 2$ seconds.

$s(2) = 20 + 3(2) - 5(2)^2$ $= 6$ meters	Plug $t = 2$ into the original equation to find the position.
$s'(t) = v(t) = 3 - 10t$	The derivative of the first function gives the velocity.
$v(2) = 3 - 10(2) = {}^-17$ m/s	Plug $t = 2$ into the velocity function to find the velocity. ${}^-17$ m/s indicates the particle is moving backwards.
$s''(t) = a(t) = {}^-10$ $a(2) = {}^-10$ m/s^2	The second derivation of position gives the acceleration. Substitute $t = 2$, yields an acceleration of ${}^-10$ m/s^2 which indicates the particle is slowing down.

Finding the rate of change of one quantity (for example distance, volume, etc.) with respect to time it is often referred to as a rate of change problem. To find an instantaneous rate of change of a particular quantity, write a function in terms of time for that quantity; then take the derivative of the function. Substitute in the values at which the instantaneous rate of change is sought.

Functions which are in terms of more than one variable may be used to find related rates of change. These functions are often not written in terms of time. To find a related rate of change, follow these steps.

1. Write an equation which relates all the quantities referred to in the problem.
2. Take the derivative of both sides of the equation with respect to time.

   Follow the same steps as used in implicit differentiation. This means take the derivative of each part of the equation remembering to multiply each term by the derivative of the variable involved with respect to time. For example, if a term includes the variable v for volume, take the derivative of the term remembering to multiply by dv/dt for the derivative of volume with respect to time. dv/dt is the rate of change of the volume.

3. Substitute the known rates of change and quantities, and solve for the desired rate of change.

Example:

1. What is the instantaneous rate of change of the area of a circle where the radius is 3 cm?

$A(r) = \pi r^2$	Write an equation for area.
$A'(r) = 2\pi r$	Take the derivative to find the rate of change.
$A'(3) = 2\pi(3) = 6\pi$	Substitute in $r = 3$ to arrive at the instantaneous rate of change.

DOMAIN III.   GEOMETRY AND MEASUREMENT

Competency 008   The teacher understands measurement as a process.

Measurements of length (Metric system)

kilometer (km)	=	1000 meters (m)
hectometer (hm)	=	100 meters (m)
decameter (dam)	=	10 meters (m)
meter (m)	=	1 meter (m)
decimeter (dm)	=	1/10 meter (m)
centimeter (cm)	=	1/100 meter (m)
millimeter (mm)	=	1/1000 meter (m)

Conversion of length from English to Metric

1 inch	=	2.54 centimeters
1 foot	≈	30.48 centimeters
1 yard	≈	0.91 meters
1 mile	≈	1.61 kilometers

Measurements of weight (English system)

28.35 grams (g)	=	1 ounce (oz)
16 ounces (oz)	=	1 pound (lb)
2000 pounds (lb)	=	1 ton (t) (short ton)
1.1 ton (t)	=	1 metric ton (t)

Measurements of weight (Metric system)

kilogram (kg)	=	1000 grams (g)
gram (g)	=	1 gram (g)
milligram (mg)	=	1/1000 gram (g)

Conversion of weight from English to metric

1 ounce	≈	28.35 grams
1 pound	≈	0.454 kilogram
1.1 ton	=	1 metric ton

Measurement of volume (English system)

8 fluid ounces (oz)	=	1 cup (c)
2 cups (c)	=	1 pint (pt)
2 pints (pt)	=	1 quart (qt)
4 quarts (qt)	=	1 gallon (gal)

Measurement of volume (Metric system)

kiloliter (kl)	=	1000 liters (l)
liter (l)	=	1 liter (l)
milliliter (ml)	=	1/1000 liter (ml)

Conversion of volume from English to metric

1 teaspoon (tsp)	≈	5 milliliters
1 fluid ounce	≈	29.56 milliliters
1 cup	≈	0.24 liters
1 pint	≈	0.47 liters
1 quart	≈	0.95 liters
1 gallon	≈	3.8 liters

Note: (') represents feet and (") represents inches.

Square units can be derived with knowledge of basic units of length by squaring the equivalent measurements.

> 1 square foot (sq. ft.) = 144 sq. in.
> 1 sq. yd. = 9 sq. ft.
> 1 sq. yd. = 1296 sq. in.

Example: 14 sq. yd. = _____ sq. ft.
 14 × 9 = 126 sq. ft.

There are many methods for converting measurements within a system. One method is to multiply the given measurement by a conversion factor. This conversion factor is the ratio of:

$$\frac{\text{new units}}{\text{old units}} \quad \text{OR} \quad \frac{\text{what you want}}{\text{what you have}}$$

Sample problems:

1. Convert 3 miles to yards.

$$\frac{3 \text{ miles}}{1} \times \frac{1{,}760 \text{ yards}}{1 \text{ mile}} = \underline{} \text{ yards}$$

$$= 5{,}280 \text{ yards}$$

1. multiply by the conversion factor
2. cancel the miles units
3. solve

2. Convert 8,750 meters to kilometers.

$$\frac{8{,}750 \text{ meters}}{1} \times \frac{1 \text{ kilometer}}{1000 \text{ meters}} = \underline{} \text{ km}$$

$$= 8.75 \text{ kilometers}$$

1. multiply by the conversion factor
2. cancel the meters units
3. solve

The **precision** of a measurement is related to the unit that is used. The smaller the unit, the more precise the measurement will be. Thus, 42 mm is more accurate than 4 cm.

The amount of precision or **greatest possible error (GPE)** is equal to one-half of the smallest unit of measurement that is used. For example, if the smallest unit was mm, then the greatest possible error would be $\pm \frac{1}{2}$ mm.

Similarly, if the smallest unit was $\frac{1}{2}$ inch, then the greatest possible error would be $\frac{1}{2} \times \frac{1}{2} = \pm \frac{1}{4}$ inch.

The greatest possible error can be used to find the range of a measurement by adding and subtracting the greatest possible error from that measurement.

Example 1: If the measurement is 93 miles, the GPE is $\pm \frac{1}{2}$ mile and the range of measurement is between $92\frac{1}{2}$ miles and $93\frac{1}{2}$ miles.

Example 2: If the measurement is 18.3 cm, the GPE is ±0.05 cm and the range is 18.25 - 18.35 cm.

The Pythagorean Theorem

Given any right-angles triangle, $\triangle ABC$, the square of the hypotenuse is equal to the sum of the squares of the other two sides.

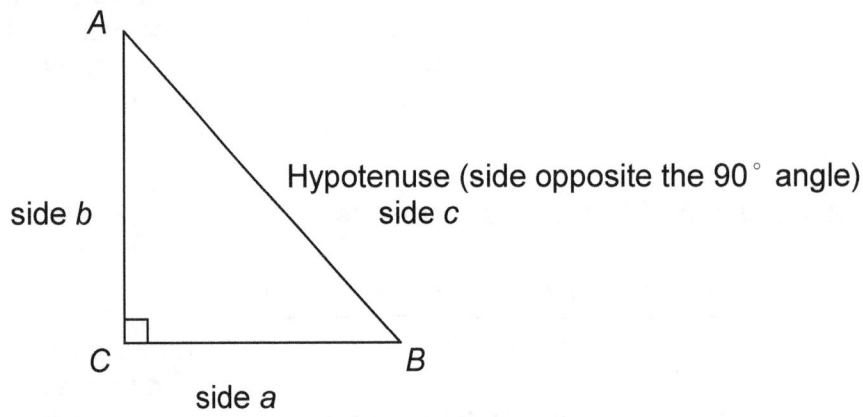

This theorem says that $(AB)^2 = (BC)^2 + (AC)^2$
or
$c^2 = a^2 + b^2$

Example: Two old cars leave a road intersection at the same time. One car traveled due north at 55 mph while the other car traveled due east. After 3 hours, the cars were 180 miles apart. Find the speed of the second car.

Using a right triangle to represent the problem we get the figure:

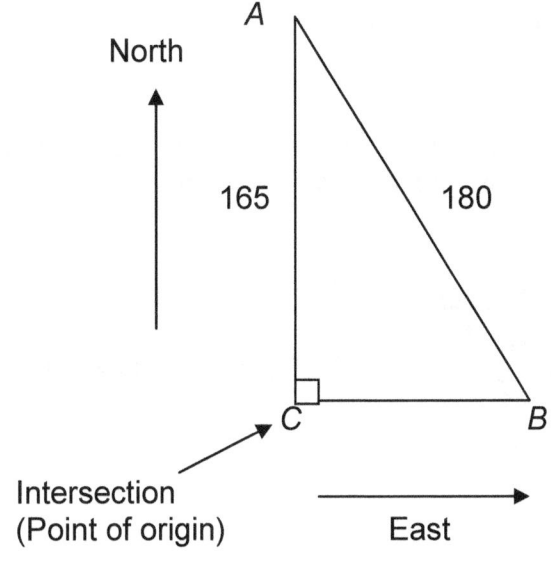

Traveling at 55 mph for 3 hours, the northbound car has driven (55)(3)=165 miles. This is the side AC.

We are given that the cars are 180 miles apart. This is side AB.

Since $\triangle ABC$ is a right triangle, then, by Pythagorean Theorem, we get:

$$(AB)^2 = (BC)^2 + (AC)^2 \text{ or}$$
$$(BC)^2 = (AB)^2 - (AC)^2$$

$$(BC)^2 = 180^2 - 165^2$$
$$(BC)^2 = 32400 - 27225$$
$$(BC)^2 = 5175$$

Take the square root of both sides to get:

$$\sqrt{(BC)^2} = \sqrt{5175} \approx 71.935 \text{ miles}$$

Since the east bound car has traveled 71.935 miles in 3 hours, then the average speed is:

$$\frac{71.935}{3} \approx 23.97 \text{ mph}$$

Use the basic trigonometric ratios of sine, cosine, and tangent to solve for the missing sides of right triangles when given at least one of the acute angles.

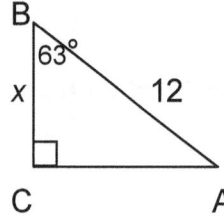

In the triangle ABC, an acute angle is 63 degrees and the length of the hypotenuse, 12. The missing side is the one adjacent to the given angle.

The appropriate trigonometric ratio to use would be cosine since we are looking for the adjacent side and we have the length of the hypotenuse.

$\cos x = \dfrac{\text{adjacent}}{\text{hypotenuse}}$ 1. Write formula.

$\cos 63 = \dfrac{x}{12}$ 2. Substitute known values.

$0.454 = \dfrac{x}{12}$

$x = 5.448$ 3. Solve.

Example: Find the missing angle.

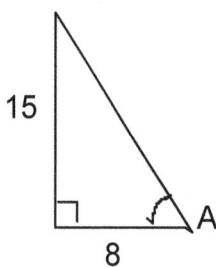

$\tan A = \dfrac{opp}{adj}$ $\tan A = \dfrac{15}{8} = 1.875$

Looking on the trigonometric chart, the angle whose tangent is closest to 1.875 is 62°. Thus $\angle A \approx 62°$

Example: Find the missing side.

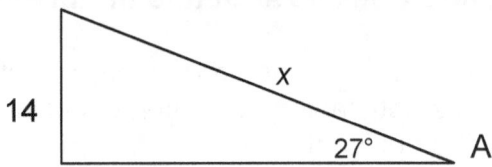

$\sin A = \dfrac{opp}{hyp}$

$\sin 27° = \dfrac{14}{x}$

$0.4540 \approx \dfrac{14}{x}$

$x \approx \dfrac{14}{.454}$

$x \approx 30.8$

Example: Find the missing side.

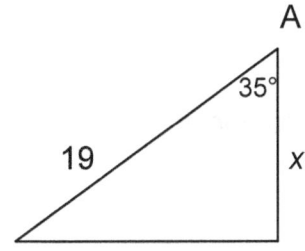

$\cos A = \dfrac{adj}{opp}$

$\cos 35° = \dfrac{x}{19}$

$x \approx 19 \times .5636$

$x \approx 10.7$

Competency 009 The teacher understands the geometric relationships and axiomatic structure of Euclidian geometry.

A point, a line and a plane are actually undefined terms since we cannot give a satisfactory definition using simple defined terms. However, their properties and characteristics give a clear understanding of what they are.

A **point** indicates place or position. It has no length, width or thickness.

•   point A
A

A **line** is considered a set of points. Lines may be straight or curved, but the term line commonly denotes a straight line. Lines extend indefinitely.

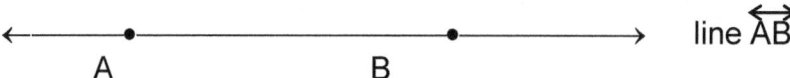

A **plane** is a set of points composing a flat surface. A plane also has no boundaries.

plane A

A **line segment** has two endpoints.

•————————•   segment $\overline{AB}$
A   B

A **ray** has exactly one endpoint. It extend indefinitely in one direction.

An **angle** is formed by the intersection of two rays.

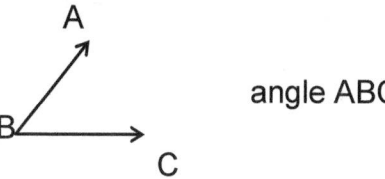

angle ABC

Angles are measured in degrees. $1° = \frac{1}{360}$ of a circle.

A **right angle** measures 90°.

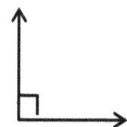

An **acute angle** measures more than 0° and less than 90°.

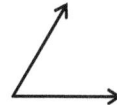

An **obtuse angle** measures more than 90° and less than 180°.

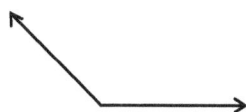

A **straight angle** measures 180°.

A **reflexive angle** measures more than 180° and less than 360°.

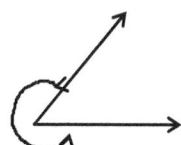

An infinite number of lines can be drawn through any point.

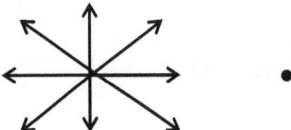

Exactly one line can be drawn through two points.

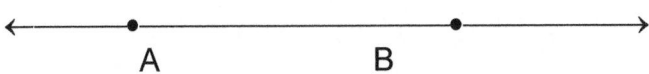

Two lines intersect at exactly one point.

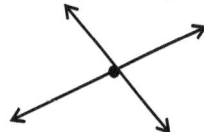

Two planes intersect to form a line.

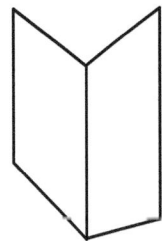

The <u>intersection</u> of two sets of points is those points that belong to <u>both</u> sets.

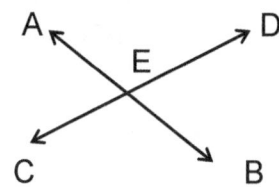

$\overleftrightarrow{AB} \cap \overleftrightarrow{CD} = \{E\}$ Line $\overleftrightarrow{AB}$ intersects line $\overleftrightarrow{CD}$ at point E.

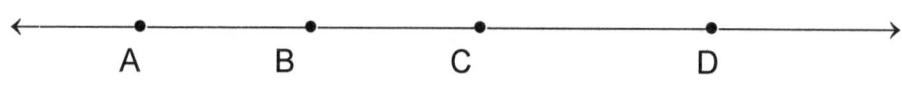

$\overline{AC} \cap \overline{BD} = \overline{BC}$

Segment AC intersects segment BD at segment BC.

The <u>union</u> of two sets of points is those points that belong to <u>either</u> or <u>both</u> sets.

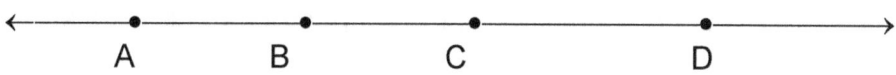

$\overline{AC} \cup \overline{BD} = \overline{AD}$ The union of segment AC and segment BD is segment AD.

Two lines intersect at exactly one point. Two lines are **perpendicular** if their intersection forms right angles.

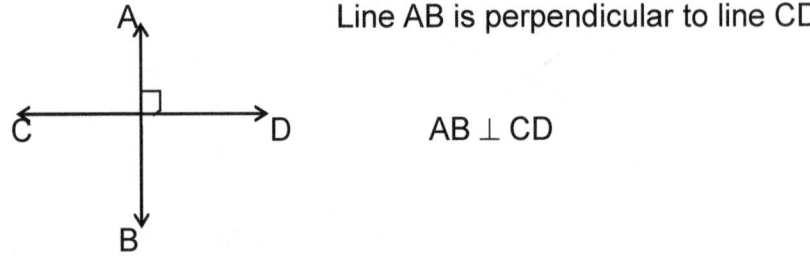

Line AB is perpendicular to line CD.

AB ⊥ CD

Two lines in the same plane that do not intersect are **parallel**. Parallel lines are everywhere equidistant.

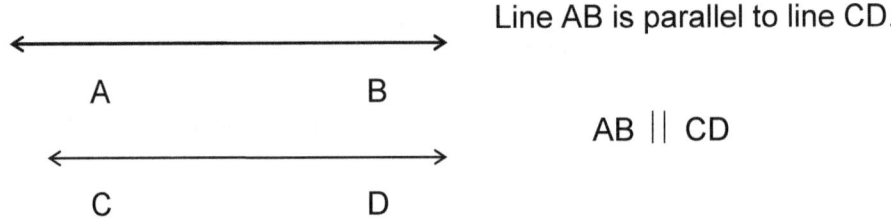

Line AB is parallel to line CD.

AB ∥ CD

Two triangles are congruent if each of the three angles and three sides of one triangle match up in a one-to-one fashion with congruent angles and sides of the second triangle. In order to see how the sides and angles match up, it is sometimes necessary to imagine rotating or reflecting one of the triangles so the two figures are oriented in the same position.

There are shortcuts to the above procedure for proving two triangles congruent.

Side-Side-Side (SSS) Congruence—If the three sides of one triangle match up in a one-to-one congruent fashion with the three sides of the other triangle, then the two triangles are congruent. With SSS, it is not necessary to compare the angles; they will automatically be congruent.

Angle-Side-Angle (ASA) Congruence—If two angles of one triangle match up in a one-to-one congruent fashion with two angles in the other triangle and if the sides between the two angles are also congruent, then the two triangles are congruent. With ASA, the sides that are used for congruence must be located between the two angles used in the first part of the proof.

Side-Angle-Side (SAS) Congruence—If two sides of one triangle match up in a one-to-one congruent fashion with two sides in the other triangle and if the angles between the two sides are also congruent, then the two triangles are congruent. With SAS, the angles that are used for congruence must be located between the two sides used in the first part of the proof.

Angle-Angle-Side (AAS)—If two angles of one triangle match up in a one-to-one congruent fashion with two angles in the other triangle and if two sides that are not between the aforementioned sets of angles are also congruent, then the triangles are congruent. ASA and AAS are very similar; the only difference is where the congruent sides are located. If the sides are between the congruent sets of angles, use ASA. If the sides are not located between the congruent sets of angles, use AAS.

Hypotenuse-Leg (HL) is a congruence shortcut which can only be used with right triangles. If the hypotenuse and leg of one right triangle are congruent to the hypotenuse and leg of the other right triangle, then the two triangles are congruent.

Two triangles are overlapping if a portion of the interior region of one triangle is shared in common with all or a part of the interior region of the second triangle.

The most effective method for proving two overlapping triangles congruent is to draw the two triangles separated. Separate the two triangles and label all of the vertices using the labels from the original overlapping figures. Once the separation is complete, apply one of the congruence shortcuts: SSS, ASA, SAS, AAS, or HL.

A geometric construction is a drawing made using only a compass and straightedge. A construction consists of only segments, arcs, and points. The easiest construction to make is to duplicate a given line segment. Given segment AB, construct a segment equal in length to segment AB by following these steps.

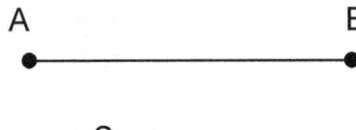

1. Place a point anywhere in the plane to anchor the duplicate segment. Call this point S.

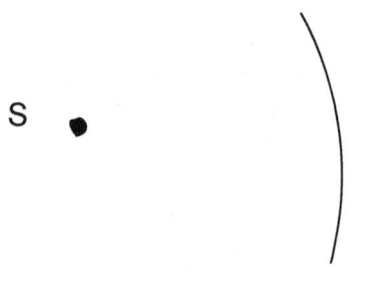

2. Open the compass to match the length of segment AB. Keeping the compass rigid, swing an arc from S.

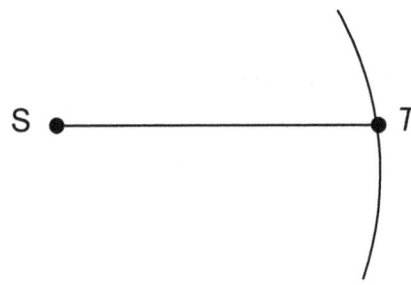

3. Draw a segment from S to any point on the arc. This segment will be the same length as AB.

Samples:

Construct segments congruent to the given segments.

1. 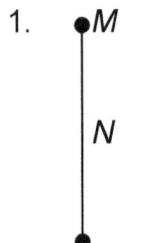   2.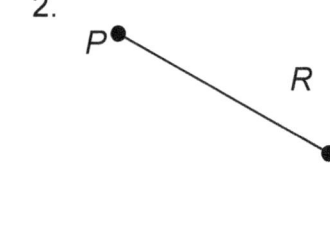

Given a line such as line $\overline{AB}$ and a point K on the line, follow these steps to construct a perpendicular line to line l through K.

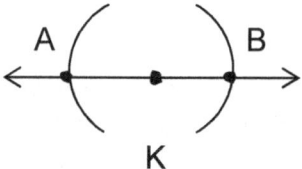

1. Swing an arc of any radius from point K so that it intersects line $\overline{AB}$ in two points, A and B.

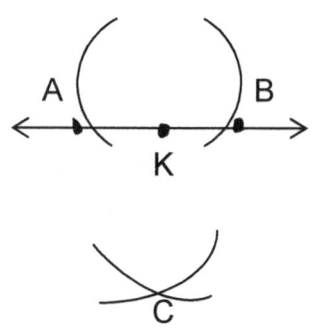

2. Open the compass to any length and swing one arc from B and another from A so that the two arcs intersect at point C.

3. Connect K and C to form line KC which is perpendicular to line $\overline{AB}$.

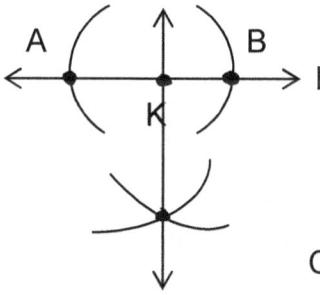

Given a line segment with two endpoints such as A and B, follow these steps to construct the line which both bisects and is perpendicular to the line given segment.

1. Swing an arc of any radius from point A. Swing another arc of the same radius from B. The arcs will intersect at two points. Label these points C and D.

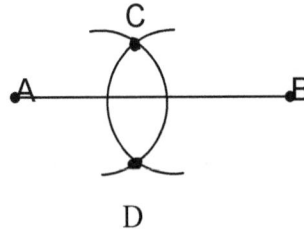

2. Connect C and D to form the perpendicular bisector of segment AB

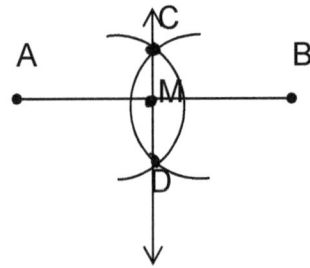

3. The point M where line $\overline{CD}$ and segment $\overline{AB}$ intersect is the midpoint of segment $\overline{AB}$.

Euclid wrote a set of 13 books around 330 B.C. called the Elements. He outlined ten axioms and then deduced 465 theorems. Euclidean geometry is based on the undefined concept of the point, line and plane.

The fifth of Euclid's axioms (referred to as the parallel postulate) was not as readily accepted as the other nine axioms. Many mathematicians throughout the years have attempted to prove that this axiom is not necessary because it could be proved by the other nine. Among the many who attempted to prove this was Carl Friedrich Gauss. His works led to the development of hyperbolic geometry. Elliptical or Reimannian geometry was suggested by G.F. Berhard Riemann. He based his work on the theory of surfaces and used models as physical interpretations of the undefined terms that satisfy the axioms.

The chart below lists the fifth axiom (parallel postulate) as it is given in each of the three geometries.

EUCLIDEAN	ELLIPTICAL	HYPERBOLIC
Given a line and a point not on that line, one and only one line can be drawn through the given point parallel to the given line.	Given a line and a point not on that line, no line can be drawn through the given point parallel to the given line.	Given a line and a point not on that line, two or more lines can be drawn through the point parallel to the given line.

Competency 010 The teacher analyzes the properties of two- and three-dimensional figures.

The **perimeter** of any polygon is the sum of the lengths of the sides.

$$P = \text{sum of sides}$$

Since the opposite sides of a rectangle are congruent, the perimeter of a rectangle equals twice the sum of the length and width or

$$P_{rect} = 2l + 2w \text{ or } 2(l + w)$$

Similarly, since all the sides of a square have the same measure, the perimeter of a square equals four times the length of one side or

$$P_{square} = 4s$$

The **area** of a polygon is the number of square units covered by the figure.

$$A_{rect} = l \times w$$
$$A_{square} = s^2$$

Example: Find the perimeter and the area of this rectangle.

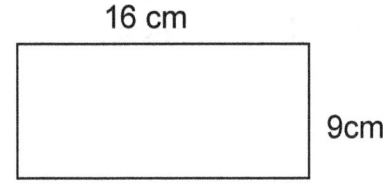

$$\begin{aligned} P_{rect} &= 2l + 2w \\ &= 2(16) + 2(9) \\ &= 32 + 18 = 50 \text{ cm} \end{aligned} \qquad \begin{aligned} A_{rect} &= l \times w \\ &= 16(9) \\ &= 144 \text{ cm}^2 \end{aligned}$$

Example: Find the perimeter and area of this square.

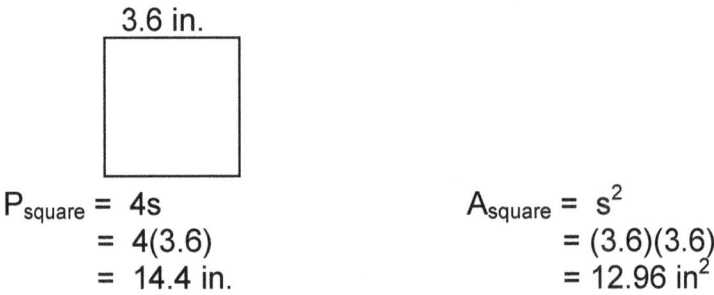

The distance around a circle is the **circumference**. The ratio of the circumference to the diameter is represented by the Greek letter pi. $\Pi \sim 3.14 \sim \frac{22}{7}$.

The circumference of a circle is found by the formula $C = 2\Pi r$ or $C = \Pi d$ where r is the radius of the circle and d is the diameter.

The **area** of a circle is found by the formula $A = \Pi r^2$.

Example: Find the circumference and area of a circle whose radius is 7 meters.

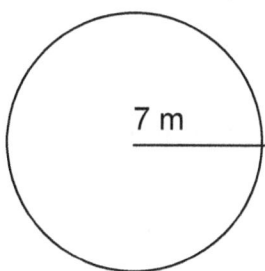

$C = 2\Pi r$
$= 2(3.14)(7)$
$= 43.96$ m

$A = \Pi r^2$
$= 3.14(7)(7)$
$= 153.86$ m^2

The **lateral** area is the area of the faces excluding the bases.

The **surface area** is the total area of all the faces, including the bases.

The **volume** is the number of cubic units in a solid. This is the amount of space a figure holds.

Right prism

$V = Bh$ (where B = area of the base of the prism and h = the height of the prism)

Rectangular right prism

$S = 2(lw + hw + lh)$ (where l = length, w = width and h = height)
$V = lwh$

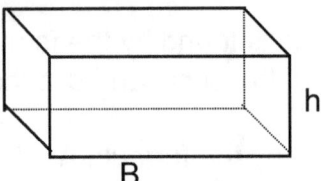

Example: Find the height of a box whose volume is 120 cubic meters and the area of the base is 30 square meters.

$$V = Bh$$
$$120 = 30h$$
$$h = 4 \text{ meters}$$

Regular pyramid

$V = 1/3 Bh$

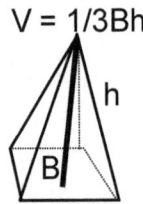

Right circular cylinder

$S = 2\Pi r(r + h)$ (where r is the radius of the base)
$V = \Pi r^2 h$

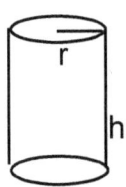

Right circular cone

$V = \frac{1}{3} Bh$

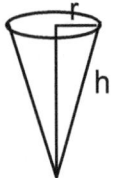

Sphere

$S = 4\Pi r^2$
$V = \frac{4}{3}\Pi r^3$

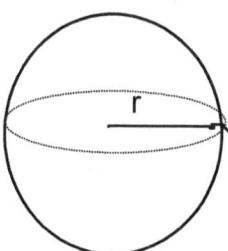

Similar solids share the same shape but are not necessarily the same size. The ratio of any two corresponding measurements of similar solids is the scale factor. For example, the scale factor for two square pyramids, one with a side measuring 2 inches and the other with a side measuring 4 inches, is 2:4.

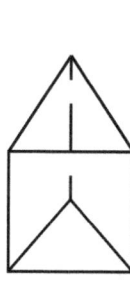
2

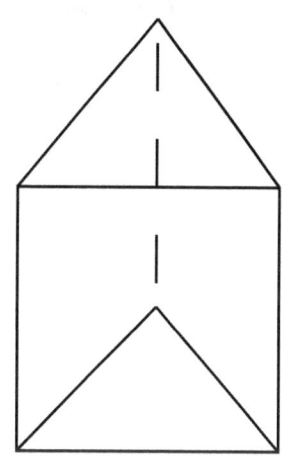
4

The base perimeter, the surface area, and the volume of similar solids are directly related to the scale factor. If the scale factor of two similar solids is a:b, then the...

    ratio of base perimeters = a:b
    ratio of areas = $a^2:b^2$
    ratio of volumes = $a^3:b^3$

Thus, for the above example the...

    ratio of base perimeters = 2:4
    ratio of areas = $2^2:4^2$ = 4:16
    ratio of volumes = $2^3:4^3$ = 8:64

Sample problems:

1. What happens to the volume of a square pyramid when the length of the sides of the base are doubled?

   scale factor = a:b = 1:2

ratio of volume = $1^3:2^3$ = 1:8 (The volume is increased 8 times.)

2. Given the following measurements for two similar cylinders with a scale factor of 2:5 (Cylinders A to Cylinder B), determine the height, radius, and volume of each cylinder.

   Cylinder A: r = 2
   Cylinder B: h = 10

   Solution:

   Cylinder A –

   $$\frac{h_a}{10} = \frac{2}{5}$$
   $5h_a = 20$ Solve for h_a
   $h_a = 4$

   Volume of Cylinder a = $\pi r^2 h = \pi(2)^2 4 = 16\pi$

   Cylinder B –

   $$\frac{2}{r_b} = \frac{2}{5}$$
   $2r_b = 10$ Solve for r_b
   $r_b = 5$

   Volume of Cylinder b = $\pi r^2 h = \pi(5)^2 10 = 250\pi$

The union of all points on a simple closed surface and all points in its interior form a space figure called a **solid**. The five regular solids, or **polyhedra**, are the cube, tetrahedron, octahedron, icosahedron, and dodecahedron. A **net** is a two-dimensional figure that can be cut out and folded up to make a three-dimensional solid. Below are models of the five regular solids with their corresponding face polygons and nets.

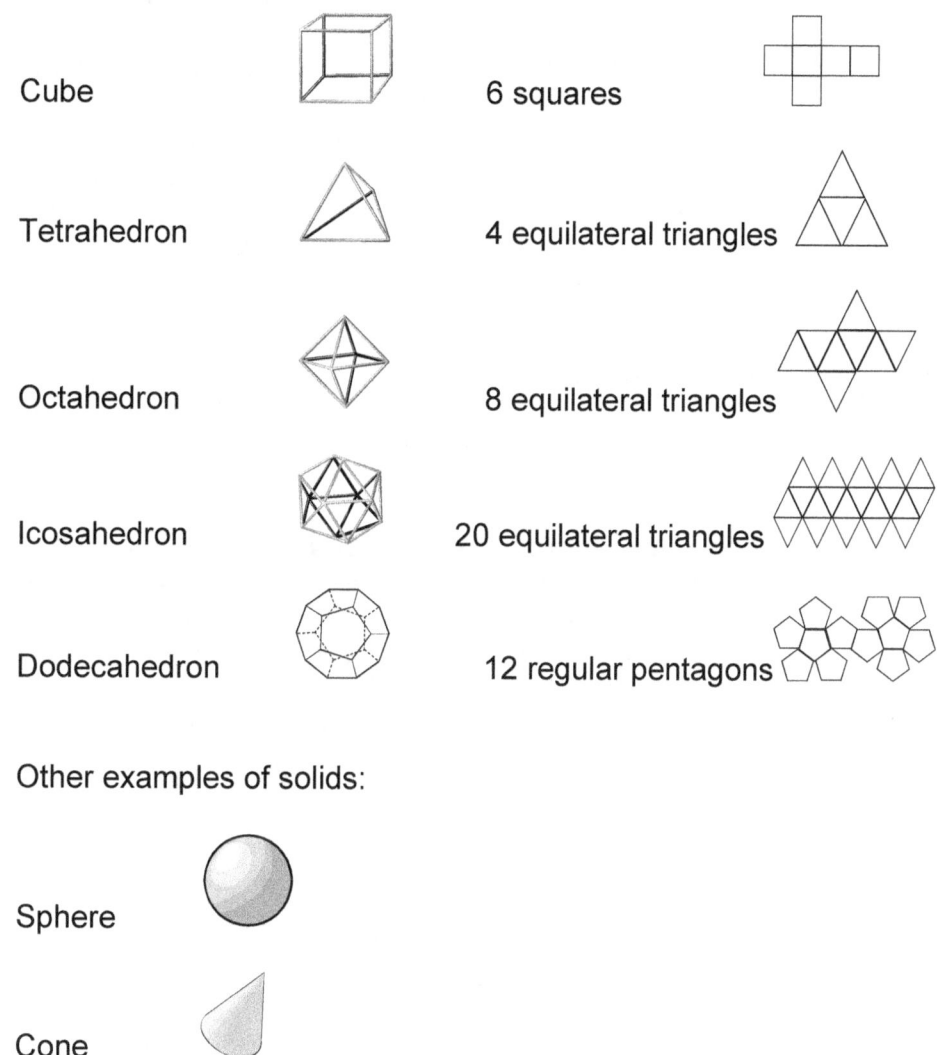

Cube — 6 squares

Tetrahedron — 4 equilateral triangles

Octahedron — 8 equilateral triangles

Icosahedron — 20 equilateral triangles

Dodecahedron — 12 regular pentagons

Other examples of solids:

Sphere

Cone

Competency 011 The teacher understands transformational geometry and related algebra to geometry and trigonometry using the Cartesian coordinate system.

Reflection devices and other technologies, like overhead projectors, transform geometric constructions in predictable ways. Students should have the ability to recognize the patterns and properties of geometric constructions made with these technologies.

The most common reflection device is a mirror. Mirrors reflect geometric constructions across a given axis. For example, if we place a mirror on the side AC of the triangle (below) the composite image created by the original figure and the reflection is a square.

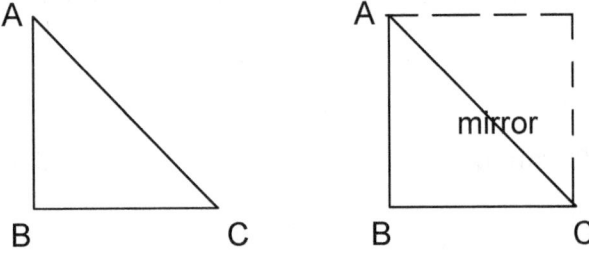

Projection devices, like overhead projectors, often expand geometric figures. These expansions are proportional, meaning the ratio of the measures of the figure remains the same. Consider the following projection of the rectangle ABCD.

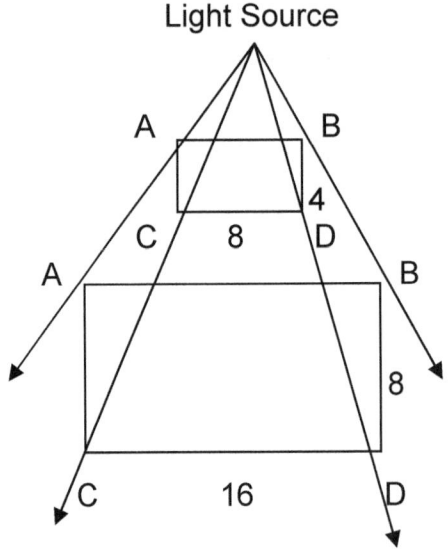

A **transformation** is a change in the position, shape, or size of a geometric figure. **Transformational geometry** is the study of manipulating objects by flipping, twisting, turning, and scaling. **Symmetry** is exact similarity between two parts or halves, as if one were a mirror image of the other.

A **translation** is a transformation that "slides" an object a fixed distance in a given direction. The original object and its translation have the same shape and size, and they face in the same direction.

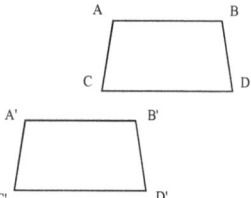

An example of a translation in architecture would be stadium seating. The seats are the same size and the same shape and face in the same direction.

A **rotation** is a transformation that turns a figure about a fixed point called the center of rotation. An object and its rotation are the same shape and size, but the figures may be turned in different directions. Rotations can occur in either a clockwise or a counterclockwise direction.

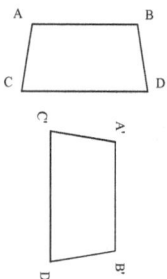

Rotations can be seen in wallpaper and art, and a Ferris wheel is an example of rotation.

An object and its **reflection** have the same shape and size, but the figures face in opposite directions.

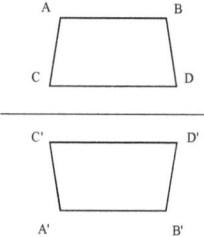

The line (where a mirror may be placed) is called the **line of reflection**. The distance from a point to the line of reflection is the same as the distance from the point's image to the line of reflection.

A **glide reflection** is a combination of a reflection and a translation.

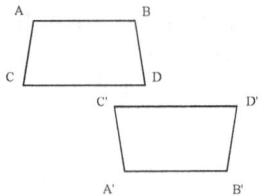

Another type of transformation is **dilation**. Dilation is a transformation that "shrinks" or "makes it bigger."

Example:

Using dilation to transform a diagram.

Starting with a triangle whose center of dilation is point P,

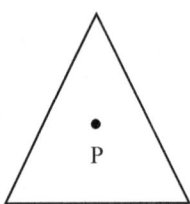

we dilate the lengths of the sides by the same factor to create a new triangle.

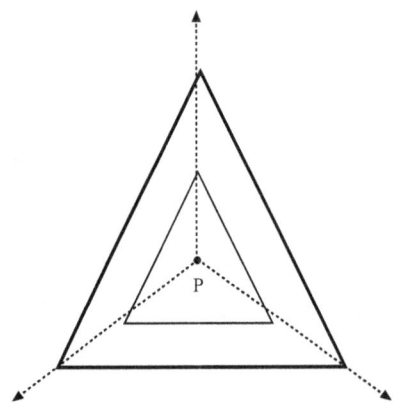

A **Tessellation** is an arrangement of closed shapes that completely covers the plane without overlapping or leaving gaps. Unlike **tilings**, tessellations do not require the use of regular polygons. In art the term is used to refer to pictures or tiles mostly in the form of animals and other life forms, which cover the surface of a plane in a symmetrical way without overlapping or leaving gaps. M. C. Escher is known as the "Father" of modern tessellations. Tessellations are used for tiling, mosaics, quilts, and art. If you look at a completed tessellation, you will see the original motif repeats in a pattern. There are 17 possible ways that a pattern can be used to tile a flat surface or "wallpaper."

There are four basic transformational symmetries that can be used in tessellations: **translation, rotation, reflection,** and **glide reflection**. The transformation of an object is called its image. If the original object was labeled with letters, such as $ABCD$, the image may be labeled with the same letters followed by a prime symbol, $A'B'C'D'$.

The tessellation below is a combination of the four types of transformational symmetry we have discussed:

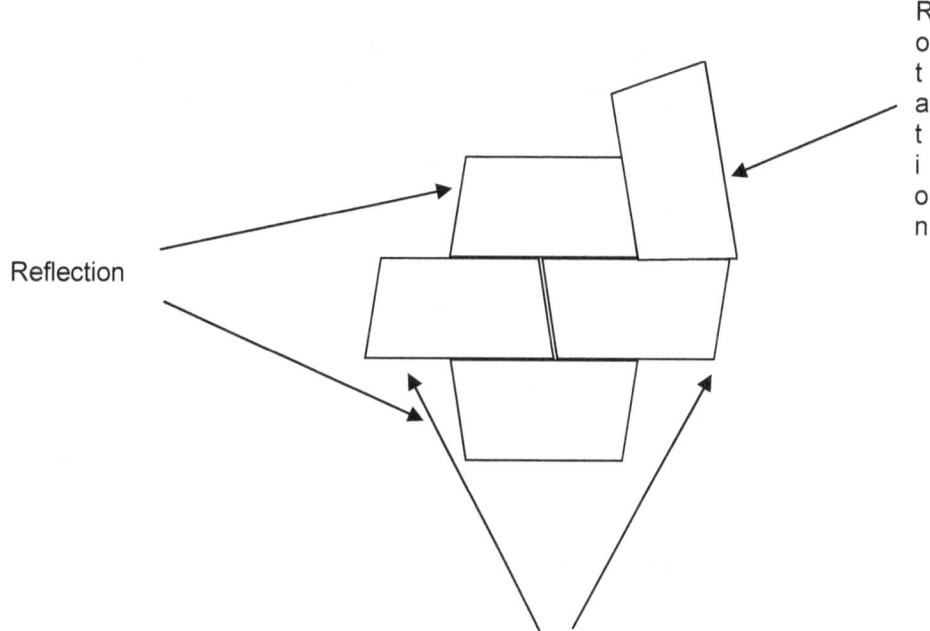

Applies concepts and properties of slope, midpoint, parallelism, and distance in the coordinate plane to explore properties of geometric figures and solve problems.

The equation of a graph can be found by finding its slope and its y-intercept. To find the slope, find two points on the graph where coordinates are integer values. Using points: (x_1, y_1) and (x_2, y_2).

$$\text{slope} = \frac{y_2 - y_1}{x_2 - x_1}$$

The y-intercept is the y-coordinate of the point where a line crosses the y-axis. The equation can be written in slope-intercept form, which is $y = mx + b$, where m is the slope and b is the y-intercept. To rewrite the equation into some other form, multiply each term by the common denominator of all the fractions. Then rearrange terms as necessary.

If the graph is a **vertical line**, then the equation solves to **x = the x-coordinate of any point on the line**.

If the graph is a **horizontal line**, then the equation solves to **y = the y-coordinate of any point on the line**.

The key to applying the **distance formula** is to understand the problem before beginning.

$$D = \sqrt{(x_2 - x_1)^2 + (y_2 - y_1)^2}$$

Sample Problem:

1. Find the perimeter of a figure with vertices at (4, 5), $(-4, 6)$ and $(-5, -8)$.

The figure being described is a triangle. Therefore, the distance for all three sides must be found. Carefully identify all three sides before beginning.

Side 1 = (4,5) to (−4,6)
Side 2 = (−4,6) to (−5,−8)
Side 3 = (−5,−8) to (4,5)

$$D_1 = \sqrt{(-4-4)^2 + (6-5)^2} = \sqrt{65}$$

$$D_2 = \sqrt{(-5-(-4))^2 + (-8-6)^2} = \sqrt{197}$$

$$D_3 = \sqrt{(4-(-5))^2 + (5-(-8))^2} = \sqrt{250} \text{ or } 5\sqrt{10}$$

$$\text{Perimeter} = \sqrt{65} + \sqrt{197} + 5\sqrt{10}$$

Midpoint Definition:

If a line segment has endpoints of (x_1, y_1) and (x_2, y_2), then the midpoint can be found using:

$$\left(\frac{x_1 + x_2}{2}, \frac{y_1 + y_2}{2}\right)$$

Sample problems:

1. Find the center of a circle with a diameter whose endpoints are (3, 7) and (−4,−5).

$$\text{Midpoint} = \left(\frac{3+(-4)}{2}, \frac{7+(-5)}{2}\right)$$

$$\text{Midpoint} = \left(\frac{-1}{2}, 1\right)$$

2. Find the midpoint given the two points $(5, 8\sqrt{6})$ and $(9, -4\sqrt{6})$.

$$\text{Midpoint} = \left(\frac{5+9}{2}, \frac{8\sqrt{6}+(-4\sqrt{6})}{2}\right)$$

$$\text{Midpoint} = (7, 2\sqrt{6})$$

A figure on a coordinate plane can be translated by changing the ordered pairs.

Example:

Plot the given ordered pairs on a coordinate plane and join them in the given order, then join the first and last points.

(-3, -2), (3, -2), (5, -4), (5, -6), (2, -4), (-2, -4), (-5, -6), (-5, -4)

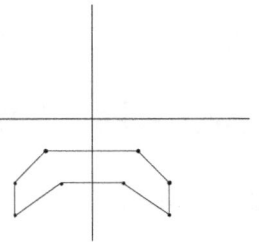

Increase all y-coordinates by 6.
(-3, 4), (3, 4), (5, 2), (5, 0), (2, 2), (-2, 2), (-5, 0), (-5, 2)

Plot the points and join them to form a second figure.

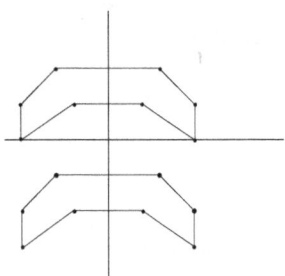

The unit circle is a circle with a radius of one centered at (0,0) on the coordinate plane. Thus, any ray from the origin to a point on the circle forms an angle, t, with the positive x-axis. In addition, the ray from the origin to a point on the circle in the first quadrant forms a right triangle with a hypotenuse measuring one unit. Applying the Pythagorean theorem, $x^2 + y^2 = 1$. Because the reflections of the triangle about both the x- and y-axis are on the unit circle and $x^2 = (-x)^2$ for all x values, the formula holds for all points on the unit circle.

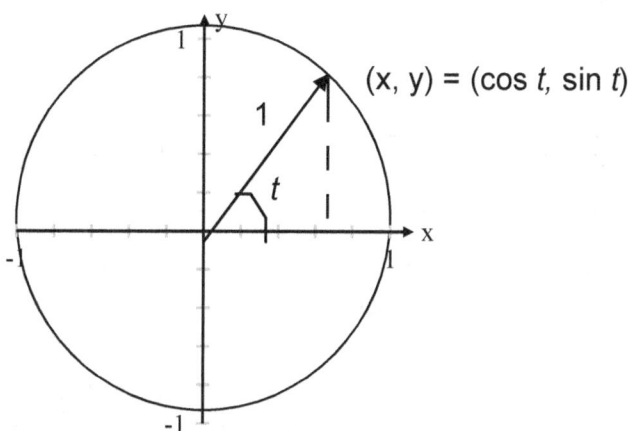

Note that because the length of the hypotenuse of the right triangles formed in the unit circle is one, the point on the unit circle that forms the triangle is (cos t, sin t). In other words, for any point on the unit circle, the x value represents the cosine of t, the y value represents the sine of t, and the ratio of the y-coordinate to the x-coordinate is the tangent of t.

The unit circle illustrates several properties of trigonometric functions. For example, applying the Pythagorean theorem to the unit circle yields the equation $\cos^2(t) + \sin^2(t) = 1$. In addition, the unit circle reveals the periodic nature of trigonometric functions. When we increase the angle t beyond 2π radians or 360 degrees, the values of x and y coordinates on the unit circle remain the same. Thus, the sine and cosine values repeat with each revolution. The unit circle also reveals the range of the sine and cosine functions. The values of sine and cosine are always between one and negative one. Finally, the unit circle shows that when the x coordinate of a point on the circle is zero, the tangent function is undefined. Thus, tangent is undefined at the angles $\frac{\pi}{2}, \frac{3\pi}{2}$ and the corresponding angles in all subsequent revolutions.

TEACHER CERTIFICATION STUDY GUIDE

DOMAIN IV. **PROBABILITY AND STATISTICS**

Competency 012 The teacher understands how to use graphical and numerical techniques to explore data, characterize patterns, and describe departures from patterns.

	Test 1	Test 2	Test 3	Test 4	Test 5
Evans, Tim	75	66	80	85	97
Miller, Julie	94	93	88	97	98
Thomas, Randy	81	86	88	87	90

Bar graphs are used to compare various quantities (Graph corresponds to table above).

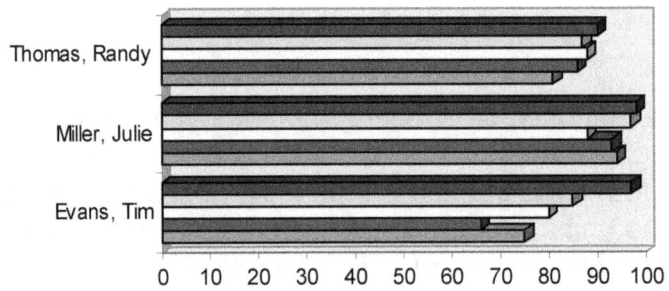

Line graphs are used to show trends, often over a period of time.

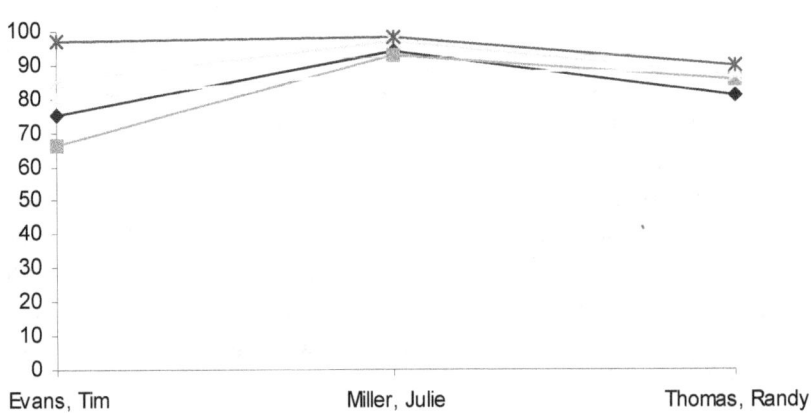

MATHEMATICS-SCIENCE 4-8   116

A **pictograph** shows comparison of quantities using symbols. Each symbol represents a number of items.

¶ ¶ ¶ ¶
¶ ¶ ¶ ¶ ¶ ¶ ¶
¶ ¶ ¶
¶ ¶
¶ ¶ ¶ ¶ ¶

Circle graphs show the relationship of various parts to each other and the whole. Percents are used to create circle graphs.

Julie spends 8 hours each day in school, 2 hours doing homework, 1 hour eating dinner, 2 hours watching television, 10 hours sleeping and the rest of the time doing other things.

Stem and leaf plots are visually similar to line plots. The **stems** are the digits in the greatest place value of the data values, and the **leaves** are the digits in the next greatest place values. Stem and leaf plots are best suited for small sets of data and are especially useful for comparing two sets of data. The following is an example using test scores:

4	9
5	4 9
6	1 2 3 4 6 7 8 8
7	0 3 4 6 6 6 7 7 7 8 8 8 8
8	3 5 5 7 8
9	0 0 3 4 5
10	0 0

Histograms are used to summarize information from large sets of data that can be naturally grouped into intervals. The vertical axis indicates **frequency** (the number of times any particular data value occurs), and the horizontal axis indicates data values or ranges of data values. The number of data values in any interval is the **frequency of the interval**.

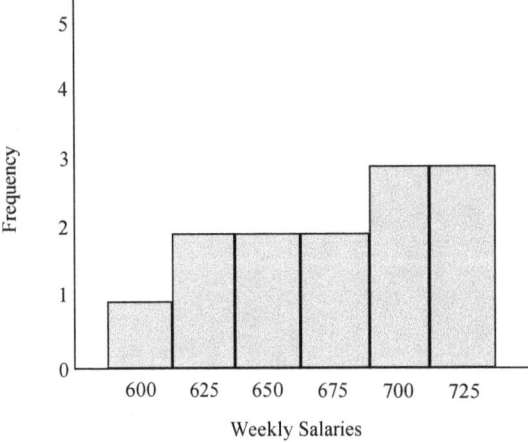

Weekly Salaries

Basic statistical concepts can be applied without computations. For example, inferences can be drawn from a graph or statistical data. A bar graph could display which grade level collected the most money. Student test scores would enable the teacher to determine which units need to be remediated.

Mean, median, and mode are three measures of central tendency. The **mean** is the average of the data items. The **median** is found by putting the data items in order from smallest to largest and selecting the item in the middle (or the average of the two items in the middle). The **mode** is the most frequently occurring item.

Range is a measure of variability. It is found by subtracting the smallest value from the largest value.

Sample problem:

Find the mean, median, mode, and range of the test scores listed below:

85	77	65
92	90	54
88	85	70
75	80	69
85	88	60
72	74	95

Mean (X) = sum of all scores ÷ number of scores
= 1404 ÷ 18
= 78

Median = put numbers in order from smallest to largest. Pick middle number.

54, 60, 65, 69, 70, 72, 74, 75, 77, 80, 85, 85, 85, 88, 88, 90, 92, 95

The numbers 77 and 80 are both in the middle. Therefore, the median is the average of the 2 numbers in the middle or 78.5.

Mode = most frequent number
= 85

Range = largest number minus the smallest number
= 95 − 54
= 41

The shape of a data distribution is described as symmetrical or asymmetrical. In a symmetrical distribution, Mean = Mode = Median. The opposite of a symmetrical distribution is one that demonstrates **skewness**, meaning the distribution is asymmetrical.

An understanding of the definitions is important in determining the validity and uses of statistical data. All definitions and applications in this section apply to ungrouped data.

Data item: each piece of data is represented by the letter X.

Mean: the average of all data represented by the symbol $\overline{X}$.

Range: difference between the highest and lowest value of data items.

Sum of the Squares: sum of the squares of the differences between each item and the mean.

$$Sx^2 = (X - \overline{X})^2$$

Variance: the sum of the squares quantity divided by the number of items.

(the lower case Greek letter sigma (σ) squared represents variance).

$$\frac{Sx^2}{N} = \sigma^2$$

The larger the value of the variance the larger the spread

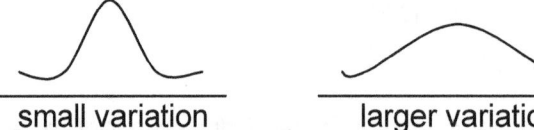

small variation   larger variation

Standard Deviation: the square root of the variance. The lower case Greek letter sigma (σ) is used to represent standard deviation.

$$\sigma = \sqrt{\sigma^2}$$

Most statistical calculators have standard deviation keys on them and should be used when asked to calculate statistical functions. It is important to become familiar with the calculator and the location of the keys needed.
Sample Problem:

Given the ungrouped data below, calculate the mean, range, standard deviation and the variance.

$$\begin{array}{cccccc} 15 & 22 & 28 & 25 & 34 & 38 \\ 18 & 25 & 30 & 33 & 19 & 23 \end{array}$$

Mean ($\overline{X}$) = 25.8333333
Range: $38 - 15 = 23$
standard deviation (σ) = 6.699137
Variance (σ^2) = 48.87879

Center and spread are concepts that describe the characteristics of a data set. Measures of central tendency define the center of a data set and measures of dispersion define the amount of spread. Wide dispersion in a data set indicates the presence of data gaps and narrow dispersion indicates data clusters. Graphical representations of center and spread, like box-and-whisker plots, allow identification of data outliers.

The most common measures of central tendency that define the center of a data set are mean, median and mode. Mean is the average value of a data set; Median is the middle value of a data set; Mode is the value that appears the most in a data set. The mean is the most descriptive value for tightly clustered data with few outliers. Outlier data, values in a data set that are unusually high or low, can greatly distort the mean of a data set. Median, on the other hand, may better describe widely dispersed data and data sets with outliers because outliers and dispersion have little effect on the median value.

The most common measures of spread that define the dispersion of a data set are range, variance, standard deviation and quantiles. Range is the difference between the highest and lowest values in a data set. The variance is the average squared distance from each value of a data set to the mean. The standard deviation is the square root of the variance. A data set clustered around the center has a small variance and standard deviation, while a disperse data set with many gaps has a large variance and standard deviation. Quantiles or percentiles, divide a data set into equal sections. For example, the 50th quantile is the median value of a data set.

Finally, graphical representations of data sets, like box-and-whisker plots, help relate the measures of central tendency to data outliers, clusters and gaps. Consider the hypothetical box-and-whisker plot with one outlier value on each end of the distribution.

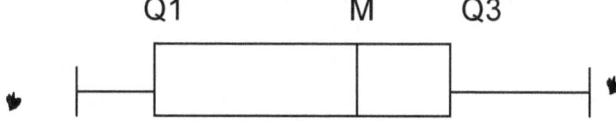

Note the beginning of the box is the value of the first quartile of the data set and the end is the value of the third quartile. We represent the median as a vertical line in the box. The "whiskers" extend to the last point that is not an outlier (i.e. within 3/2 times the range between Q1 and Q3). The points beyond the figure represent outlier values.

Percentiles divide data into 100 equal parts. A person whose score falls in the 65th percentile has outperformed 65 percent of all those who took the test. This does not mean that the score was 65 percent out of 100 nor does it mean that 65 percent of the questions answered were correct. It means that the grade was higher than 65 percent of all those who took the test.

Stanine "standard nine" scores combine the understandability of percentages with the properties of the normal curve of probability. Stanines divide the bell curve into nine sections, the largest of which stretches from the 40th to the 60th percentile and is the "Fifth Stanine" (the average of taking into account error possibilities).

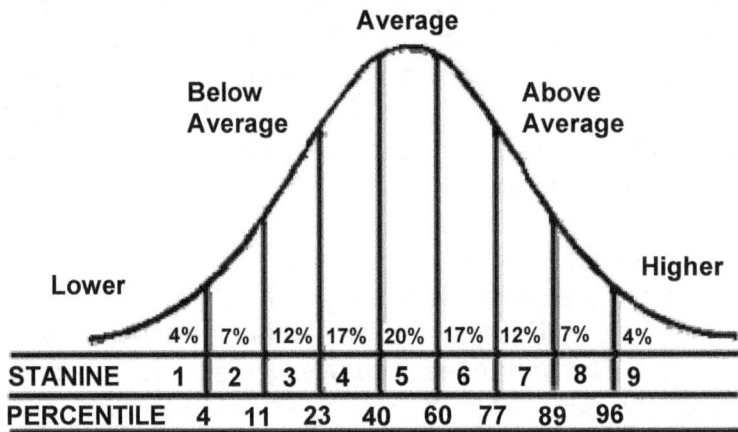

Quartiles divide the data into 4 parts. First find the median of the data set (Q2), then find the median of the upper (Q3) and lower (Q1) halves of the data set. If there are an odd number of values in the data set, include the median value in both halves when finding quartile values. For example, given the data set: {1, 4, 9, 16, 25, 36, 49, 64, 81} first find the median value, which is 25 this is the second quartile. Since there are an odd number of values in the data set (9), we include the median in both halves. To find the quartile values, we much find the medians of: {1, 4, 9, 16, 25} and {25, 36, 49, 64, 81}. Since each of these subsets had an odd number of elements (5), we use the middle value. Thus the first quartile value is 9 and the third quartile value is 49. If the data set had an even number of elements, average the middle two values. The quartile values are always either one of the data points, or exactly half way between two data points.

Sample problem:

1. Given the following set of data, find the percentile of the score 104.

   70, 72, 82, 83, 84, 87, 100, 104, 108, 109, 110, 115

Solution: Find the percentage of scores below 104.

7/12 of the scores are less than 104. This is 58.333%; therefore, the score of 104 is in the 58th percentile.

2. Find the first, second and third quartile for the data listed.

   6, 7, 8, 9, 10, 12, 13, 14, 15, 16, 18, 23, 24, 25, 27, 29, 30, 33, 34, 37

Quartile 1: The 1st Quartile is the median of the lower half of the data set, which is 11.

Quartile 2: The median of the data set is the 2nd Quartile, which is 17.

Quartile 3: The 3rd Quartile is the median of the upper half of the data set, which is 28.

TEACHER CERTIFICATION STUDY GUIDE

Competency 013   The teacher understands the theory of probability.

Probability measures the chances of an event occurring. The probability of an event that *must* occur, a certain event, is **one**. When no outcome is favorable, the probability of an impossible event is **zero**.

$$P(event) = \frac{\text{number of favorable outcomes}}{\text{number of favorable outcomes}}$$

Example: Given one die with faces numbered 1 - 6, the probability of tossing an even number on one throw of the die is $\frac{3}{6}$ or $\frac{1}{2}$ since there are 3 favorable outcomes (even faces) and a total of 6 possible outcomes (faces).

If A and B are **independent** events then the probability both A and B will occur is the product of their individual probabilities.

Example 1: Given two dice, the probability of tossing a 3 on each of them simultaneously is the probability of a 3 on the first die, or $\frac{1}{6}$, times the probability of tossing a 3 on the second die, also $\frac{1}{6}$.

$$\frac{1}{6} \times \frac{1}{6} = \frac{1}{36}$$

Example 2: Given a jar containing 10 marbles, 3 red, 5 black, and 2 white. What is the probability of drawing a red marble and then a white marble if the marble is returned to the jar after choosing?

$$\frac{3}{10} \times \frac{2}{10} = \frac{6}{100} = \frac{3}{50}$$

When the outcome of the first event affects the outcome of the second event, the events are **dependent.** Any two events that are not independent are dependent. This is also known as conditional probability.

$$\text{Probability of (A and B)} = P(A) \times P(B \text{ given } A)$$

Example: Two cards are drawn from a deck of 52 cards, without replacement; that is, the first card is not returned to the deck before the second card is drawn. What is the probability of drawing a diamond?

       A = drawing a diamond first
       B = drawing a diamond second

$$P(A) = \tfrac{13}{52} = \tfrac{1}{4} \qquad\qquad P(B) = \tfrac{12}{51} = \tfrac{4}{17}$$
$$P(A + B) = \tfrac{1}{4} \times \tfrac{4}{17} = \tfrac{1}{17}$$

MATHEMATICS-SCIENCE 4-8

The Addition Principle of Counting states:

If A and B are events, $n(AorB) = n(A) + n(B) - n(A \cap B)$.

Example:

In how many ways can you select a black card or a Jack from an ordinary deck of playing cards?

Let B denote the set of black cards and let J denote the set of Jacks. Then,

$n(B) = 26, n(J) = 4, n(B \cap J) = 2$
$= 26 + 4 - 2$
$= 28$

The Addition Principle of Counting for Mutually Exclusive Events states:

If A and B are mutually exclusive events, $n(AorB) = n(A) + n(B)$.

Example:

A travel agency offers 40 possible trips: 14 to Asia, 16 to Europe and 10 to South America. In how many ways can you select a trip to Asia or Europe through this agency?

Let A denote trips to Asia and let E denote trips to Europe. Then, $A \cap E = \varnothing$ and

$$n(AorE) = 14 + 16 = 30.$$

Therefore, the number of ways you can select a trip to Asia or Europe is 30.

The Multiplication Principle of Counting for Dependent Events states:

Let A be a set of outcomes of Stage 1 and B a set of outcomes of Stage 2. Then the number of ways $n(AandB)$, that A and B can occur in a two-stage experiment is given by:

$$n(AandB) = n(A)n(B|A),$$

where $n(B|A)$ denotes the number of ways B can occur given that A has already occurred.

Example:

How many ways from an ordinary deck of 52 cards can 2 Jacks be drawn in succession if the first card is drawn but not replaced in the deck and then the second card is drawn?

This is a two-stage experiment for which we wish to compute $n(A \text{ and } B)$, where A is the set of outcomes for which a Jack is obtained on the first draw and B is the set of outcomes for which a Jack is obtained on the second draw.

If the first card drawn is a Jack, then there are only three remaining Jacks left to choose from on the second draw. Thus, drawing two cards without replacement means the events A and B are dependent.

$$n(A \text{ and } B) = n(A)n(B|A) = 4 \cdot 3 = 12$$

The Multiplication Principle of Counting for Independent Events states:

Let A be a set of outcomes of Stage 1 and B a set of outcomes of Stage 2. If A and B are independent events, then the number of ways $n(A \text{ and } B)$, that A and B can occur in a two-stage experiment is given by:

$$n(A \text{ and } B) = n(A)n(B).$$

Example:

How many six-letter code "words" can be formed if repetition of letters is not allowed?

Since these are code words, a word does not have to look like a word; for example, abcdef could be a code word. Since we must choose a first letter *and* a second letter *and* a third letter *and* a fourth letter *and* a fifth letter *and* a sixth letter, this experiment has six stages.

Since repetition is not allowed there are 26 choices for the first letter; 25 for the second; 24 for the third; 23 for the fourth; 22 for the fifth; and 21 for the sixth. Therefore, we have:

n(six-letter code words without repetition of letters)

A **Bernoulli trial** is an experiment whose outcome is random and can be either of two possible outcomes, called "success" or "failure." Tossing a coin would be an example of a Bernoulli trial. We make the outcomes into a random variable by assigning the number 0 to one outcome and the number 1 to the other outcome. Traditionally, the "1" outcome is considered the "success" and the "0" outcome is considered the "failure." The probability of success is represented by p, with the probability of failure being $1-p$, or q.

Bernoulli trials can be applied to any real-life situation in which there are just two possible outcomes. For example, concerning the birth of a child, the only two possible outcomes for the sex of the child are male or female.

In probability, the **sample space** is a list of all possible outcomes of an experiment. For example, the sample space of tossing two coins is the set {HH, HT, TT, TH}, the sample space of rolling a six-sided die is the set {1, 2, 3, 4, 5, 6}, and the sample space of measuring the height of students in a class is the set of all real numbers {R}.

When conducting experiments with a large number of possible outcomes, it is important to determine the size of the sample space. The size of the sample space can be determined by using the fundamental counting principle and the rules of combinations and permutations.

The **fundamental counting principle** states that if there are *m* possible outcomes for one task and *n* possible outcomes of another, there are (*m* x *n*) possible outcomes of the two tasks together.

A **permutation** is the number of possible arrangements of items, without repetition, where order of selection is important.

A **combination** is the number of possible arrangements, without repetition, where order of selection is not important.

Examples:

1. Find the size of the sample space of rolling two six-sided dice and flipping two coins.

   Solution:
   List the possible outcomes of each event:
   each dice: {1, 2, 3, 4, 5, 6}
   each coin: {Heads, Tails}

   Apply the fundamental counting principle:
   size of sample space = 6 x 6 x 2 x 2 = 144

2. Find the size of the sample space of selecting three playing cards at random from a standard fifty-two card deck.

Solution:
Use the rule of combination –
$$_{52}C_3 = \frac{52!}{(52-3)!3!} = 22100$$

Probability is the study and quantification of the likelihood of an event. We use combinations and permutations to determine the number of possible outcomes of a given selection of objects. The key difference between combinations and permutations is that we use combinations when the order of the objects chosen does not matter and permutations when the order does matter. In addition, geometric probability describes the likelihood of events involving shapes and measures.

Permutations are the number of possible arrangements of objects when the order of the objects matters. For example, we can find the number of 3 letter arrangements we can make from the first ten letters of the alphabet by determining the number of permutations of size 3 taken from a set of 10. We represent this situation symbolically as $_{10}P_3$. The general formula for determining the value of a permutation is

$$_nP_r = \frac{n!}{(n-r)!},$$

where n is the number of objects in the set and r is the number of objects chosen. Thus, the number of possible three-letter arrangements chosen from the first 10 letters of the alphabet is

$$_{10}P_3 = \frac{10!}{(10-3)!} = \frac{10!}{7!} = 10 \times 9 \times 8 = 720.$$

Combinations are the number of possible groups of objects when the order of the objects does not matter. For example, we can determine the number of different 4 card hands we can deal from a deck of 52 cards by determining a combination, because the order of the cards is not important. We present this situation symbolically as $_{52}C_4$. The general formula for determining the value of a combination is

$$_nC_r = \frac{n!}{r!(n-r)!},$$

where n is the total number of objects and r is the number of objects chosen. Thus, the number of possible four-card hands in a 52-card deck is

$$_{52}C_4 = \frac{52!}{4!(52-4)!} = \frac{52!}{4!(48!)} = \frac{52 \times 51 \times 50 \times 49}{4 \times 3 \times 2 \times 1} = 270725.$$

Geometric probability describes situations that involve shapes and measures. For example, given a 10-inch string, we can determine the probability of cutting the string so that one piece is at least 8 inches long. If the cut occurs in the first or last two inches of the string, one of the pieces will be at least 8 inches long.

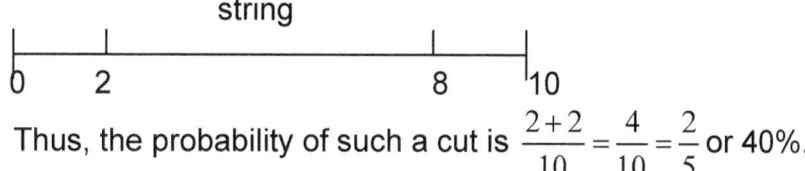

Thus, the probability of such a cut is $\frac{2+2}{10} = \frac{4}{10} = \frac{2}{5}$ or 40%.

Other geometric probability problems involve the ratio of areas. For example, to determine the likelihood of randomly hitting a defined area of a dartboard (pictured below) we determine the ratio of the target area to the total area of the board.

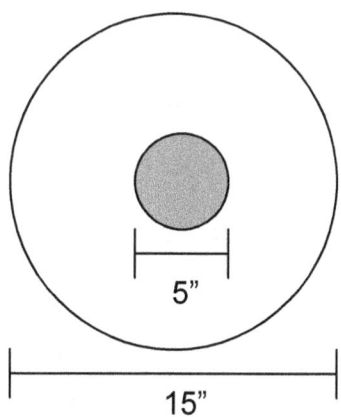

Given that a randomly thrown dart lands somewhere on the board, the probability that it hits the target area is the ratio of the areas of the two circles. Thus, the probability, P, of hitting the target is

$$P = \frac{(2.5)^2 \pi}{(7.5)^2 \pi} \times 100 = \frac{6.25}{56.25} \times 100 = 11.1\%.$$

The **binomial distribution** is a sequence of probabilities with each probability corresponding to the likelihood of a particular event occurring. It is called a binomial distribution because each trial has precisely two possible outcomes. An **event** is defined as a sequence of Bernoulli trials that has within it a specific number of successes. The order of success is not important.

Note: There are two parameters to consider in a binomial distribution:

1. p = the probability of a success
2. n = the number of Bernoulli trials (i.e., the length of the sequence).

Example:

Toss a coin two times. Each toss is a Bernoulli trial as discussed above. Consider heads to be success. One event is one sequence of two coin tosses. Order does not matter.

There are two possibilities for each coin toss. Therefore, there are four (2·2) possible subevents: 00, 01, 10, 11 (where 0 = tail and 1 = head).

According to the multiplication rule, each subevent has a probability of $\frac{1}{4}\left(\frac{1}{2} \cdot \frac{1}{2}\right)$.

One subevent has zero heads, so the event of zero heads in two tosses is $p(h=0) = \frac{1}{4}.$

Two subevents have one head, so the event of one head in two tosses is $p(h=1) = \frac{2}{4}.$

One subevent has two heads, so the event of two heads in two tosses is $p(h=2) = \frac{1}{4}.$

So the binomial distribution for two tosses of a fair coin is:

$$p(h=0) = \frac{1}{4}, \ p(h=1) = \frac{2}{4}, \ p(h=2) = \frac{1}{4}.$$

A **normal distribution** is the distribution associated with most sets of real-world data. It is frequently called a **bell curve**. A normal distribution has a **random variable** X with mean μ and variance σ^2.

Example:

Albert's Bagel Shop's morning customer load follows a normal distribution, with **mean** (average) 50 and **standard deviation** 10. The standard deviation is the measure of the variation in the distribution. Determine the probability that the number of customers tomorrow will be less than 42.

First convert the raw score to a **z-score**. A z-score is a measure of the distance in standard deviations of a sample from the mean.

The z-score = $\dfrac{X_i - \bar{X}}{s} = \dfrac{42 - 50}{10} = \dfrac{-8}{10} = -.8$

Next, use a table to find the probability corresponding to the z-score. The table gives us .2881. Since our raw score is negative, we subtract the table value from .5.

$$.5 - .2881 = .2119$$

We can conclude that $P(x < 42) = .2119$. This means that there is about a 21% chance that there will be fewer than 42 customers tomorrow morning.

Example:

The scores on Mr. Rogers' statistics exam follow a normal distribution with mean 85 and standard deviation 5. A student is wondering what the probability is that she will score between a 90 and a 95 on her exam.

We wish to compute $P(90 < x < 95)$.

Compute the z-scores for each raw score.

$$\frac{90-85}{5} = \frac{5}{5} = 1 \text{ and } \frac{95-85}{5} = \frac{10}{5} = 2.$$

Now we want $P(1 < z < 2)$.

Since we are looking for an occurrence between two values, we subtract:

$$P(1 < z < 2) = P(z < 2) - P(z < 1).$$

We use a table to get :

$P(1 < z < 2) = .9772 - .8413 = .1359$. (Remember that since the z-scores are positive, we add .5 to each probability.)

We can then conclude that there is a 13.6% chance that the student will score between a 90 and a 95 on her exam.

Competency 014 The teacher understands the relationship among probability theory, sampling, and statistical inference, and how statistical inference is used in making and evaluating predictions.

The four main types of measurement scales used in statistical analysis are nominal, ordinal, interval, and ratio. The type of variable measured and the research questions asked determine the appropriate measurement scale. The different measurement scales have distinctive qualities and attributes.

The nominal measurement scale is the most basic measurement scale. When measuring using the nominal scale, we simply label or classify responses into categories. Examples of variables measured on the nominal scale are gender, religion, ethnicity, and marital status. The essential attribute of the nominal scale is that the classifications have no numerical or comparative value. For example, when classifying people by marital status, there is no sense in which "single" is greater or less than "married". The only measure of central tendency applicable to the nominal scale is mode and the only applicable arithmetic operation is counting.

The ordinal measurement scale is more descriptive than the nominal scale in that the ordinal scale allows comparison between categories. Examples of variables measured on the ordinal scale are movie ratings, consumer satisfaction surveys, and the rank or order of anything. While we can compare categories of responses on the ordinal scale (e.g. "highly satisfied" indicates a higher level of satisfaction than "somewhat satisfied"), we cannot determine anything about the difference between the categories. In other words, we cannot presume that the difference between two categories is the same as the distance between two other categories. Even if the responses are in numeric form (e.g. 1 = good, 2 = fair, 3 = poor), we can presume nothing about the intervals separating the groups. Ordinal scales allow greater or less-than comparisons and the applicable measures of central tendency are range and median.

The next measurement scale is the interval scale. Interval scales are numeric scales where intervals have a fixed, uniform value throughout the scale. An example of an interval scale is the Fahrenheit temperature scale. On the Fahrenheit scale, the difference between 40 degrees and 50 degrees is the same as the difference between 70 degrees and 80 degrees. The major limitation of interval scales is that there is no fixed zero point.

For example, while the Fahrenheit scale has a value of zero degrees, this assignment is arbitrary because the measurement does not represent the absence of temperature. Because interval scales lack a true zero point, ratio comparison of values has no meaning. Thus, the arithmetic operations addition and subtraction are applicable to interval scales while multiplication and division are not. The measures of central tendency applicable to interval scales are mode, median, and arithmetic mean.

The final, and most informative, measurement scale is the ratio scale. The ratio scale is essentially an interval scale with a true zero point. Examples of ratio scales are measurement of length (meters, inches, etc.), monetary systems, and degrees Kelvin. The zero value of each of these scales represents the absence of length, money, and temperature, respectively. The presence of a true zero point allows proportional comparisons. For example, we can say that someone with one dollar has twice as much money as someone with fifty cents. Because ratios have meaning on ratio scales, we can apply the arithmetic operations of multiplication and division to the data sets.

Random sampling is the process of studying an aspect of a population by selecting and gathering data from a segment of the population and making inferences and generalizations based on the results. Two main types of random sampling are simple and stratified. With simple random sampling, each member of the population has an equal chance of selection to the sample group. With stratified random sampling, each member of the population has a known but unequal chance of selection to the sample group, as the study selects a random sample from each population demographic. In general, stratified random sampling is more accurate because it provides a more representative sample group. Sample statistics are important generalizations about the entire sample such as mean, median, mode, range, and sampling error (standard deviation). Various factors affect the accuracy of sample statistics and the generalizations made from them about the larger population.

Sample size is one important factor in the accuracy and reliability of sample statistics. As sample size increases, sampling error (standard deviation) decreases. Sampling error is the main determinant of the size of the confidence interval. Confidence intervals decrease in size as sample size increases.

A confidence interval gives an estimated range of values, which is likely to include a particular population parameter. The confidence level associated with a confidence interval is the probability that the interval contains the population parameter. For example, a poll reports 60% of a sample group prefers candidate A with a margin of error of $\pm$ 3% and a confidence level of 95%. In this poll, there is a 95% chance that the preference for candidate A in the whole population is between 57% and 63%.

The ultimate goal of sampling is to make generalizations about a population based on the characteristics of a random sample. Estimators are sample statistics used to make such generalizations. For example, the mean value of a sample is the estimator of the population mean. Unbiased estimators, on average, accurately predict the corresponding population characteristic. Biased estimators, on the other hand, do not exactly mirror the corresponding population characteristic. While most estimators contain some level of bias, limiting bias to achieve accurate projections is the goal of statisticians.

The **law of large numbers and the central limit theorem** are two fundamental concepts in statistics. The law of large numbers states that the larger the sample size, or the more times we measure a variable in a population, the closer the sample mean will be to the population mean. For example, the average weight of 40 apples out of a population of 100 will more closely approximate the population average weight than will a sample of 5 apples. The central limit theorem expands on the law of large numbers. The central limit theorem states that as the number of samples increases, the distribution of sample means (averages) approaches a normal distribution. This holds true regardless of the distribution of the population. Thus, as the number of samples taken increases the sample mean becomes closer to the population mean. This property of statistics allows us to analyze the properties of populations of unknown distribution.

In conclusion, the law of large numbers and central limit theorem show the importance of large sample size and large number of samples to the process of statistical inference. As sample size and the number of samples taken increase, the accuracy of conclusions about the population drawn from the sample data increases.

Correlation is a measure of association between two variables. It varies from −1 to 1, with 0 being a random relationship, 1 being a perfect positive linear relationship, and −1 being a perfect negative linear relationship.

The **correlation coefficient** (r) is used to describe the strength of the association between the variables and the direction of the association.

Example:

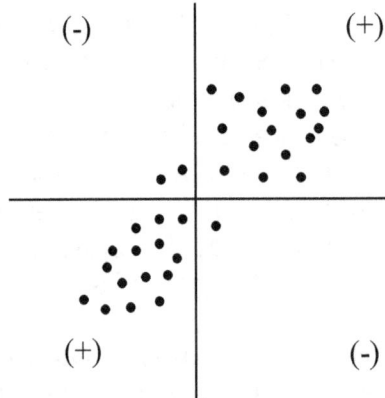

Horizontal and vertical lines are drawn through the point of averages which is the point on the averages of the x and y values. This divides the scatter plot into four quadrants. If a point is in the lower left quadrant, the product of two negatives is positive; in the upper right, the product of two positives is positive. The positive quadrants are depicted with the positive sign (+). In the two remaining quadrants (upper left and lower right), the product of a negative and a positive is negative. The negative quadrants are depicted with the negative sign (−). If r is positive, then there are more points in the positive quadrants and if r is negative, then there are more points in the two negative quadrants.

Regression is a form of statistical analysis used to predict a dependent variable (y) from values of an independent variable (x). A regression equation is derived from a known set of data.

The simplest regression analysis models the relationship between two variables using the following equation: $y = a + bx$, where y is the dependent variable and x is the independent variable. This simple equation denotes a linear relationship between x and y. This form would be appropriate if, when you plotted a graph of x and y, you tended to see the points roughly form along a straight line.

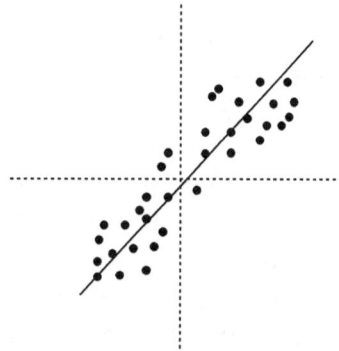

The line can then be used to make predictions.

If all of the data points fell on the line, there would be a perfect correlation ($r = 1.0$) between the x and y data points. These cases represent the best scenarios for prediction. A positive or negative r value represents how y varies with x. When r is positive, y increases as x increases. When r is negative y decreases as x increases.

DOMAIN V. MATHEMATICAL PROCESSES AND PERSPECTIVES

Competency 015 The teacher understands mathematical reasoning and problem solving.

In a **2 column proof**, the left side of the proof should be the given information, or statements that could be proved by deductive reasoning. The right column of the proof consists of the reasons used to determine that each statement to the left was verifiably true. The right side can identify given information, or state theorems, postulates, definitions or algebraic properties used to prove that particular line of the proof is true.

Assume the opposite of the conclusion. Keep your hypothesis and given information the same. Proceed to develop the steps of the proof, looking for a statement that contradicts your original assumption or some other known fact. This contradiction indicates that the assumption you made at the beginning of the proof was incorrect; therefore, the original conclusion has to be true.

Conditional statements can be diagrammed using a **Venn diagram**. A diagram can be drawn with one circle inside another circle. The inner circle represents the hypothesis. The outer circle represents the conclusion. If the hypothesis is taken to be true, then you are located inside the inner circle. If you are located in the inner circle then you are also inside the outer circle, so that proves the conclusion is true. Sometimes that conclusion can then be used as the hypothesis for another conditional, which can result in a second conclusion.

Suppose that these statements were given to you, and you are asked to try to reach a conclusion. The statements are:

All swimmers are athletes.
All athletes are scholars.

In "if-then" form, these would be:
If you are a swimmer, then you are an athlete.
If you are an athlete, then you are a scholar.

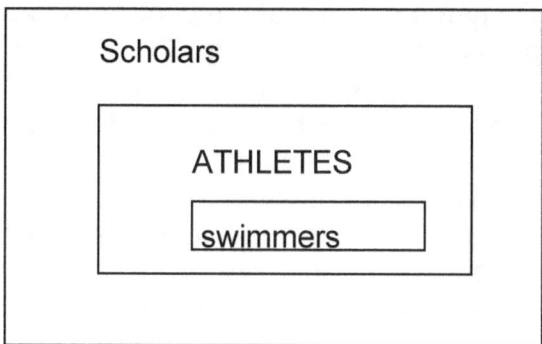

Clearly, if you are a swimmer, then you are also an athlete. This includes you in the group of scholars.

Suppose that these statements were given to you, and you are asked to try to reach a conclusion. The statements are:

All swimmers are athletes.
All wrestlers are athletes.

In "if-then" form, these would be:
If you are a swimmer, then you are an athlete.
If you are a wrestler, then you are an athlete.

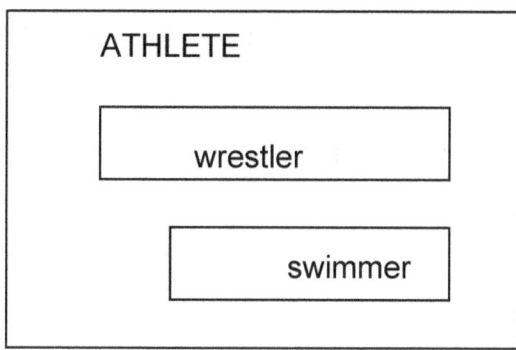

Clearly, if you are a swimmer or a wrestler, then you are also an athlete. This does NOT allow you to come to any other conclusions.

A swimmer may or may NOT also be a wrestler. Therefore, NO CONCLUSION IS POSSIBLE.

Suppose that these statements were given to you, and you are asked to try to reach a conclusion. The statements are:

All rectangles are parallelograms.
Quadrilateral ABCD is not a parallelogram.

In "if-then" form, the first statement would be:
If a figure is a rectangle, then it is also a parallelogram.

Note that the second statement is the negation of the conclusion of statement one. Remember also that the contrapositive is logically equivalent to a given conditional. That is, **"If ~ q, then ~ p"**. Since "ABCD is NOT a parallelogram " is like saying **"If ~ q,"** then you can come to the conclusion **"then ~ p"**. Therefore, the conclusion is ABCD is not a rectangle. Looking at the Venn diagram below, if all rectangles are parallelograms, then rectangles are included as part of the parallelograms. Since quadrilateral ABCD is not a parallelogram, it is excluded from anywhere inside the parallelogram box. This allows you to conclude that ABCD cannot be a rectangle either.

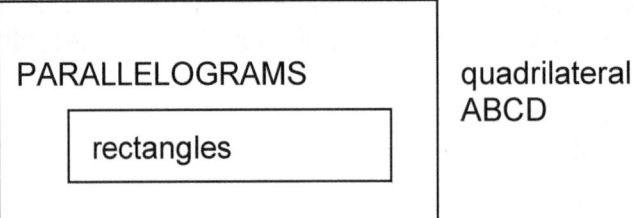

Try These. (The answers are in the Answer Key to Practice Problems):

What conclusion, if any, can be reached? Assume each statement is true, regardless of any personal beliefs.

1. If the Red Sox win the World Series, I will die.
   I died.

2. If an angle's measure is between 0° and 90°, then the angle is acute. Angle B is not acute.

3. Students who do well in geometry will succeed in college.
   Annie is doing extremely well in geometry.

4. Left-handed people are witty and charming.
   You are left-handed.

Inductive thinking is the process of finding a pattern from a group of examples. That pattern is the conclusion that this set of examples seems to indicate. It may be a correct conclusion or it may be an incorrect conclusion, because other examples may not follow the predicted pattern.

Deductive thinking is the process of arriving at a conclusion based on other statements that are all known to be true, such as theorems, axioms, or postulates. Conclusions found by deductive thinking based on true statements will **always** be true.

Examples :

Suppose:
On Monday, Mr. Peterson eats breakfast at McDonalds.
On Tuesday, Mr. Peterson eats breakfast at McDonalds.
On Wednesday, Mr. Peterson eats breakfast at McDonalds.
On Thursday, Mr. Peterson eats breakfast at McDonalds again.

Conclusion: On Friday, Mr. Peterson will eat breakfast at McDonalds again.

This is a conclusion based on inductive reasoning. Based on several days' observations, you conclude that Mr. Peterson will eat at McDonalds. This may or may not be true, but it is a conclusion arrived at by inductive thinking.

The **questioning technique** is a mathematic process skill in which students devise questions to clarify the problem, eliminate possible solutions, and simplify the problem solving process. By developing and attempting to answer simple questions, students can tackle difficult and complex problems.

Observation-inference is another mathematic process skill that used regularly in statistics. We can use the data gathered or observed from a sample of the population to make inferences about traits and qualities of the population as a whole. For example, if we observe that 40% of voters in our sample favors Candidate A, then we can infer that 40% of the entire voting population favors Candidate A. Successful use of observation-inference depends on accurate observation and representative sampling.

Estimation and testing for **reasonableness** are related skills students should employ prior to and after solving a problem. These skills are particularly important when students use calculators to find answers.

Example:

Find the sum of 4387 + 7226 + 5893.
4300 + 7200 + 5800 = 17300   Estimation.
4387 + 7226 + 5893 = 17506   Actual sum.

By comparing the estimate to the actual sum, students can determine that their answer is reasonable.

Successful math teachers introduce their students to multiple problem solving strategies and create a classroom environment where free thought and experimentation are encouraged. Teachers can promote problem solving by allowing multiple attempts at problems, giving credit for reworking test or homework problems, and encouraging the sharing of ideas through class discussion. There are several specific problem solving skills with which teachers should be familiar.

The **guess-and-check** strategy calls for students to make an initial guess at the solution, check the answer, and use the outcome of to guide the next guess. With each successive guess, the student should get closer to the correct answer. Constructing a table from the guesses can help organize the data.

Example:

There are 100 coins in a jar. 10 are dimes. The rest are pennies and nickels. There are twice as many pennies as nickels. How many pennies and nickels are in the jar?

There are 90 total nickels and pennies in the jar (100 coins – 10 dimes).

There are twice as many pennies as nickels. Make guesses that fulfill the criteria and adjust based on the answer found. Continue until we find the correct answer, 60 pennies and 30 nickels.

Number of Pennies	Number of Nickels	Total Number of Pennies and Nickels
40	20	60
80	40	120
70	35	105
60	30	90

When solving a problem where the final result and the steps to reach the result are given, students must **work backwards** to determine what the starting point must have been.
Example:

John subtracted seven from his age, and divided the result by 3. The final result was 4. What is John's age?

    Work backward by reversing the operations.
    4 x 3 = 12;
    12 + 7 = 19
    John is 19 years old.

Competency 016   **The teacher understands mathematical connections within and outside of mathematics and how to communicate mathematical ideas and concepts.**

The coordinate plane consists of a plane divided into 4 quadrants by the intersection of two axis, the *x*-axis (horizontal axis), and the *y*-axis (vertical axis).

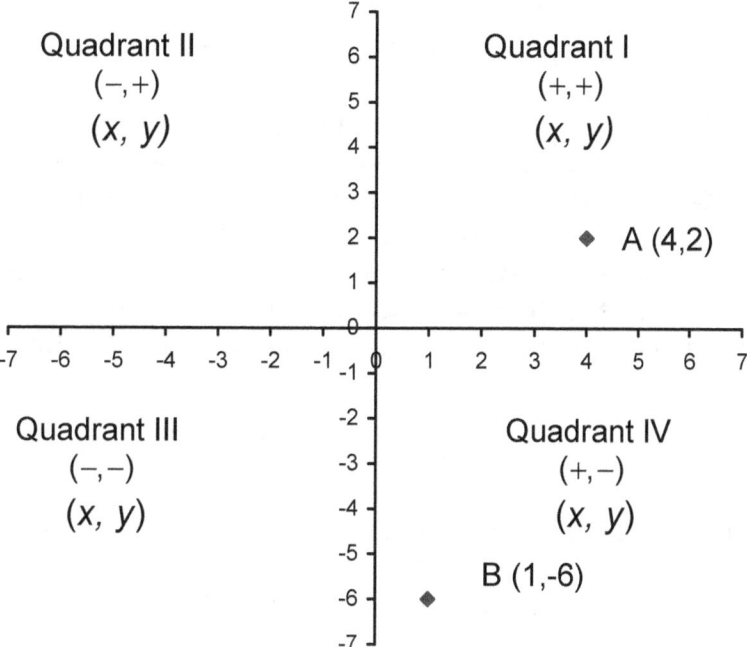

A point is represented by the order pair (*x* ,*y*) with the proper sign of the *x* and *y* components.

Example: (Refer to points in above graph). Point A has the ordered pair (4,2); Point B has the ordered pair (1,-6).

Artists, musicians, scientists, social scientists, and business people use mathematical modeling to solve problems in their disciplines. These disciplines rely on the tools and symbology of mathematics to model natural events and manipulate data. Mathematics is a key aspect of visual art.

Artists use the geometric properties of shapes, ratios, and proportions in creating paintings and sculptures. For example, mathematics is essential to the concept of perspective. Artists must determine the appropriate lengths and heights of objects to portray three-dimensional distance in two dimensions.

Mathematics is also an important part of music. Many musical terms have mathematical connections. For example, the musical octave contains twelve notes and spans a factor of two in frequency. In other words, the frequency, the speed of vibration that determines tone and sound quality, doubles from the first note in an octave to the last. Thus, starting from any note we can determine the frequency of any other note with the following formula.

$$\text{Freq} = \text{note} \times 2^{N/12}$$

Where N is the number of notes from the starting point and note is the frequency of the starting note. Mathematical understanding of frequency plays an important role in the tuning of musical instruments.

In addition to the visual and auditory arts, mathematics is an integral part of most scientific disciplines. The uses of mathematics in science are almost endless. The following are but a few examples of how scientists use mathematics. Physical scientists use vectors, functions, derivatives, and integrals to describe and model the movement of objects. Biologists and ecologists use mathematics to model ecosystems and study DNA. Finally, chemists use mathematics to study the interaction of molecules and to determine proper amounts and proportions of reactants.

Many social science disciplines use mathematics to model and solve problems. Economists, for example, use functions, graphs, and matrices to model the activities of producers, consumers, and firms. Political scientists use mathematics to model the behavior and opinions of the electorate. Finally, sociologists use mathematical functions to model the behavior of humans and human populations.

Finally, mathematical problem solving and modeling is essential to business planning and execution. For example, businesses rely on mathematical projections to plan business strategy. Additionally, stock market analysis and accounting rely on mathematical concepts.

Mathematical concepts and procedures can take many different forms. Students of mathematics must be able to recognize different forms of equivalent concepts.

For example, we can represent the slope of a line graphically, algebraically, verbally, and numerically. A line drawn on a coordinate plane will show the slope. In the equation of a line, $y = mx + b$, the term m represents the slope. We can define the slope of a line several different ways. The slope of a line is the change in the value of the y divided by the change in the value of x over a given interval. Alternatively, the slope of a line is the ratio of "rise" to "run" between two points. Finally, we can calculate the numeric value of the slope by using the verbal definitions and the algebraic representation of the line.

Examples, illustrations, and symbolic representations are useful tools in explaining and understanding mathematical concepts. The ability to create examples and alternative methods of expression allows students to solve real world problems and better communicate their thoughts.

Concrete examples are real world applications of mathematical concepts. For example, measuring the shadow produced by a tree or building is a real world application of trigonometric functions, acceleration or velocity of a car is an application of derivatives, and finding the volume or area of a swimming pool is a real world application of geometric principles.

Pictorial illustrations of mathematic concepts help clarify difficult ideas and simplify problem solving.

Examples:

1. Rectangle R represents the 300 students in School A. Circle P represents the 150 students that participated in band. Circle Q represents the 170 students that participated in a sport. 70 students participated in both band and a sport.

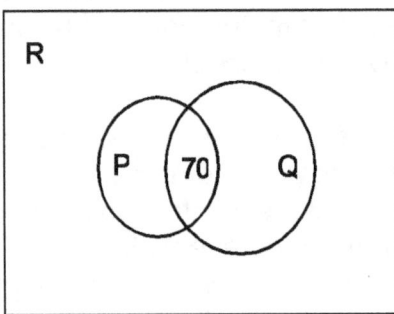

Pictorial representation of above situation.

2. A ball rolls up an incline and rolls back to its original position. Create a graph of the velocity of the ball.

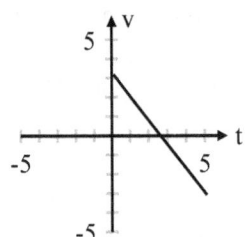

Velocity starts out at its maximum as the ball begins to roll, decreases to zero at the top of the incline, and returns to the maximum in the opposite direction at the bottom of the incline.

Symbolic representation is the basic language of mathematics. Converting data to symbols allows for easy manipulation and problem solving. Students should have the ability to recognize what the symbolic notation represents and convert information into symbolic form. For example, from the graph of a line, students should have the ability to determine the slope and intercepts and derive the line's equation from the observed data. Another possible application of symbolic representation is the formulation of algebraic expressions and relations from data presented in word problem form.

To read a bar graph or a pictograph, read the explanation of the scale that was used in the legend. Compare the length of each bar with the dimensions on the axes and calculate the value each bar represents. On a pictograph, count the number of pictures used in the chart and calculate the value of all the pictures.

To read a circle graph, find the total of the amounts represented on the entire circle graph. To determine the actual amount that each sector of the graph represents, multiply the percent in a sector times the total amount number.

To read a chart, read the row and column headings on the table. Use this information to evaluate the given information in the chart.

In order to understand mathematics and solve problems, one must know the definitions of **basic mathematic terms and concepts**. For a list of definitions and explanations of basic math terms, visit the following website: http://home.blarg.net/~math/deflist.html

Additionally, one must use the language of mathematics correctly and precisely to accurately communicate concepts and ideas.

For example, the statement "minus ten times minus five equals plus fifty" is incorrect because minus and plus are arithmetic operations not numerical modifiers. The statement should read "negative ten times negative five equals positive 50".

Students need to use the proper mathematical terms and expressions. When reading decimals, they need to read 0.4 as "four tenths" to promote better understanding of the concepts. They should do their work in a neat and organized manner. Students need to be encouraged to verbalize their strategies, both in computation and word problems. Additionally, writing original word problems fosters understanding of math language. Another idea is requiring students to develop their own glossary of mathematical glossary. Knowing the answers and being able to communicate them are equally important.

DOMAIN VI.	MATHEMATICAL LEARNING, INSTRUCTION, AND ASSESSMENT

Competency 017   **The teacher understands how children learn and develop mathematical skills, procedures, and concepts.**

Teachers can use theories of learning to plan curriculum and instructional activities. Research indicates that students learn math more easily in an applied, project-based setting. In addition, prior knowledge, learning, and self-taught understanding are important factors that dictate a student's ability to learn and preferred method of learning.

Many educators believe that the best method of teaching math is situated learning. Proponents of situated learning argue that learning is largely a function of the activity, context, and environment in which learning occurs. According to situated learning theory, students learn more easily from instruction involving relevant, real-world situations and applications rather than abstract thoughts and ideas. Research or project-based learning is a product of situated learning theory. Open-ended research tasks and projects promote learning by engaging students on multiple levels. Such tasks require the use of multiple skills and reasoning strategies and help keep students focused and attentive. Additionally, projects promote active learning by encouraging the sharing of thoughts and ideas and teacher-student and student-student interaction.

The cultural and ethnic background of a student greatly affects his or her approach to learning mathematics. In addition, factors such as socioeconomic status can affect student learning styles. Teachers must have the ability to tailor their teaching style, methods, and curriculum to the varying learning styles present in a diverse classroom.

Students of certain ethnic and racial groups that emphasize expressiveness and communication may benefit from a more interactive learning environment. For example, such students may benefit from lessons involving discussion, group work and hands-on projects. Conversely, students of ethnic groups that emphasize personal learning and discipline may benefit from a more structured, traditional learning environment. As is always the case when considering ethnic differences, however, we must be careful to avoid making inappropriate generalizations.

We must remember that ethnicity is only one form of diversity and that there is great diversity within ethnic and racial groups.

Socioeconomic status is another source of classroom diversity. Low-income students may not have the same early educational background and exposure to traditional educational reasoning strategies and techniques. Thus, low-income students may initially require a more application-based curriculum until they develop sufficient abstract reasoning skills.

A popular theory of math learning is constructivism. Constructivism argues that prior knowledge greatly influences the learning of math and learning is cumulative and vertically structured. Thus, it is important for teachers to recognize the knowledge and ideas about a subject that students already possess. Instruction must build on the innate knowledge of students and address any common misconceptions. Teachers can gain insight into the prior knowledge of students by beginning each lesson with open questions that allow students to share their thoughts and ideas on the subject or topic.

Middle school mathematics lays the groundwork for the progression of concrete ideas mastered in the elementary years to less concrete, more algebraic understandings of mathematical ideas. As students progress to later grades the concepts are developed deeper and connections between the various skills become more apparent. Skills developed in middle school are part of the progression of skills that begins formally in the primary grades and continues afterwards into high school and beyond.

In middle school, students are developing their number sense, focusing mainly on decimals, fractions, proportion, roots and negative numbers. A clear understanding of these ideas is essential, as their task in high school years is to understand how numeration concepts relate to each other. Estimation is largely a rote, pen and paper skill in middle school years, but in high school and beyond, estimation is strongly geared around evaluating the reasonableness of answers that uses the understanding developed in middle school. Sophisticated calculators are utilized in high school in ways that involve calculated rounding, either user-defined or part of the calculator's programming. For this reason, the high school student needs to be able to recognize the reasonableness of the outcomes of functions when calculated on a calculator.

The analysis of relationships or functions progresses to more algebraic forms in middle school, but the skill of developing models to demonstrate patterns and functions is also maintained. In high school and beyond students extend their understanding of functions to include the symbolic ideas of domain, range and $f(x)$ notation. Both the concrete and symbolic understandings of functions are used in advanced mathematics. Many careers including engineering, architects and other professionals rely on the use and analysis of concrete working models.

Calculus mainly involves the study of limit, rate of change, area under a curve and slope of a tangent line. In middle school, the base work for these concepts is developed. Students develop an understanding for infinity, linear growth, exponential growth, slope and change over time. High school level pre-calculus is generally aimed at developing understanding of main calculus concepts. More specific calculus techniques such as differentiation and convergence testing are generally reserved for college-level mathematics.

Trigonometry is a branch of mathematics dealing with the angles, triangles and the functions sine, cosine and tangent. Two aspects of trigonometry developed in middle school are geometrical knowledge (eg. similar triangles, the Pythagorean Theorem) and algebraic skills (solving equations and using algebraic expressions). Trigonometry is adapted into careers involving astronomy, land surveying and acoustics, among a wide array of others.

Competency 018   **The teacher understands how to plan, organize, and implement instruction using knowledge of students, subject matter, and statewide curriculum (Texas Essential Knowledge and Skills [TEKS]) to teach all students to use mathematics.**

The Texas Essential Knowledge and Skills (TEKS) are a comprehensive list of standards for subject matter learning. TEKS provide teachers with a framework for curriculum design and instructional method selection. The different skills described in TEKS require different teaching strategies and techniques.

The primary goals of middle school math instruction, as defined by TEKS, are the building of a strong foundation in mathematical concepts and the development of problem-solving and analytical skills. Direct teaching methods, including lecture and demonstration, are particularly effective in teaching basic mathematical concepts. To stimulate interest, accommodate different learning styles and enhance understanding, teachers should incorporate manipulatives and technology into their lectures and demonstrations. Indirect teaching methods, including cooperative learning, discussion and projects, promote the development of problem solving skills. Cooperative learning and discussion allow students to share ideas and strategies with their peers. In addition, projects require students to apply knowledge and develop and implement problem-solving strategies.

The TEKS provide teachers with a comprehensive list of skills that the state requires students to master. Utilizing the learning goals presented in TEKS, teachers can plan instruction to promote student understanding. In addition, teachers can deliver instruction in ways that are appropriate to the specific skill, classroom environment and student population. To assess the effectiveness of instruction, teachers can design tests that evaluate student mastery of specific skills. In grading such tests, teachers should look for patterns in student mistakes and errors that may indicate a deficiency in the instructional plan. In reevaluating instruction, teachers can attempt to use different instructional methods or shift areas of emphasis to meet the needs of the students.

When introducing a new mathematical concept to students, teachers should utilize the concrete-to-representational-to-abstract sequence of instruction. The first step of the instructional progression is the introduction of a concept modeled with concrete materials. The second step is the translation of concrete models into representational diagrams or pictures. The third and final step is the translation of representational models into abstract models using only numbers and symbols.

Teachers should first use concrete models to introduce a mathematical concept because they are easiest to understand. For example, teachers can allow students to use counting blocks to learn basic arithmetic. Teachers should give students ample time and many opportunities to experiment, practice, and demonstrate mastery with the concrete materials.

The second step in the learning process is the translation of concrete materials to representational models. For example, students may use tally marks or pictures to represent the counting blocks they used in the previous stage. Once again, teachers should give students ample time to master the concept on the representational level.

The final step in the learning process is the translation of representational models into abstract numbers and symbols. For example, students represent the processes carried out in the previous stages using only numbers and arithmetic symbols.

To ease the transition, teachers should associate numbers and symbols with the concrete and representational models throughout the learning progression.

Successful teachers select and implement instructional delivery methods that best fit the needs of a particular classroom format. Individual, small-group and large-group classroom formats require different techniques and methods of instruction.

Individual instruction allows the teacher to interact closely with the student. Teachers may use a variety of methods in an individual setting that are not practical when working with a large number of students. For example, teachers can use manipulatives to illustrate a mathematical concept.

In addition, teachers can observe and evaluate the student's reasoning and problem solving skills through verbal questioning and by checking the student's written work. Finally, individual instruction allows the teacher to work problems with the student, thus familiarizing the student with the problem-solving process.

Small-group formats require the teacher to provide instruction to multiple students at the same time. Because the size of the group is small, instructional methods that encourage student interaction and cooperative learning are particularly effective. For example, group projects, discussion, and question-and-answer sessions promote cooperative learning and maintain student interest. In addition, working problems as a group or in pairs can help students learn problem-solving strategies from each other.

Large-group formats require instructional methods that can effectively deliver information to a large number of students. Lecture is a common instructional method for large groups. In addition, demonstrating methods of problem solving and allowing students to ask questions about homework and test problems is an effective strategy for teaching large-groups.

As the teacher's role in the classroom changes from lecturer to facilitator, the questions need to further stimulate students in various ways.

- Helping students work together

What do you think about what John said?
Do you agree? Disagree?
Can anyone explain that differently?

- Helping students determine for themselves if an answer is correct

Why do you think that is true?
How did you get that answer?
Do you think that is reasonable? Why?

- Helping students learn to reason mathematically

Will that method always work?
Can you think of a case where it is not true?
How can you prove that?
Is that answer true in all cases?

- Helping student brainstorm and problem solve

Is there a pattern?
What else can you do?
Can you predict the answer?
What if...?

- Helping students connect mathematical ideas

What did we learn before that is like this?
Can you give an example?
What math did you see on television last night? in the newspaper?

The use of supplementary materials in the classroom can greatly enhance the learning experience by stimulating student interest and satisfying different learning styles. Manipulatives, models, and technology are examples of tools available to teachers.

Manipulatives are materials that students can physically handle and move. Manipulatives allow students to understand mathematic concepts by allowing them to see concrete examples of abstract processes. Manipulatives are attractive to students because they appeal to the students' visual and tactile senses. Available for all levels of math, manipulatives are useful tools for reinforcing operations and concepts. They are not, however, a substitute for the development of sound computational skills.

Models are another means of representing mathematical concepts by relating the concepts to real-world situations. Teachers must choose wisely when devising and selecting models because, to be effective, models must be applied properly. For example, a building with floors above and below ground is a good model for introducing the concept of negative numbers. It would be difficult, however, to use the building model in teaching subtraction of negative numbers.

Finally, there are many forms of **technology** available to math teachers. For example, students can test their understanding of math concepts by working on skill specific computer programs and websites. Graphing calculators can help students visualize the graphs of functions. Teachers can also enhance their lectures and classroom presentations by creating multimedia presentations.

Teachers can increase student interest in math and promote learning and understanding by relating mathematical concepts to the lives of students. Instead of using only abstract presentations and examples, teachers should relate concepts to real-world situations to shift the emphasis from memorization and abstract application to understanding and applied problem solving. In addition, relating math to careers and professions helps illustrate the relevance of math and aids in the career exploration process.

For example, when teaching a unit on the geometry of certain shapes, teachers can ask students to design a structure of interest to the student using the shapes in question. This exercise serves the dual purpose of teaching students to learn and apply the properties (e.g. area, volume) of shapes while demonstrating the relevance of geometry to architectural and engineering professions.

Competency 019   **The teacher understands assessment and uses a variety of formal and informal assessment techniques to monitor and guide mathematics instruction and to evaluate student progress.**

In addition to the traditional methods of performance assessment like multiple choice, true/false, and matching tests, there are many other methods of student assessment available to teachers. Alternative assessment is any type of assessment in which students create a response rather than choose an answer. It is sometimes know as formative assessment, due to the emphasis placed on feedback and the flow of communication between teacher and student. It is the opposite of summative assessment, which occurs periodically and consists of temporary interaction between teacher and student.

Short response and **essay** questions are alternative methods of performance assessment. In responding to such questions, students must utilize verbal, graphical, and mathematical skills to construct answers to problems. These multi-faceted responses allow the teacher to examine more closely a student's problem solving and reasoning skills.

Student **portfolios** are another method of alternative assessment. In creating a portfolio, students collect samples and drafts of their work, self-assessments, and teacher evaluations over a period of time. Such a collection allows students, parents, and teachers to evaluate student progress and achievements. In addition, portfolios provide insight into a student's thought process and learning style.

Projects, **demonstrations**, and **oral presentations** are means of alternative assessment that require students to use different skills than those used on traditional tests. Such assessments require higher order thinking, creativity, and the integration of reasoning and communication skills. The use of predetermined rubrics, with specific criteria for performance assessment, is the accepted method of evaluation for projects, demonstrations, and presentations.

Student assessment is an important part of the educational process. High quality assessment methods are necessary for the development and maintenance of a successful learning environment. Teachers must develop and implement assessment procedures that accurately evaluate student progress, test content areas of greatest importance, and enhance and improve learning. To enhance learning and accurately evaluate student progress, teachers should use a variety of assessment tasks to gain a better understanding a student's strengths and weaknesses. Finally, teachers should implement scoring patterns that fairly and accurately evaluate student performance.

Teachers should use a variety of assessment procedures to evaluate student knowledge and understanding. One type of alternative assessment is bundled testing. Bundled testing is the grouping of different question formats for the same skill or competency. For example, a bundled test of exponential functions may include multiple choice questions, short response questions, word problems, and essay questions. The variety of questions tests different levels of reasoning and expression. Another type of alternative assessment is projects. Projects are longer term, creative tasks that require many levels of reasoning and expression. Projects are often a good indicator of understanding because they require high-level thinking. A final type of alternative assessment is student portfolios. Portfolios are collections of student work over a period of time. Portfolios aid in the evaluation of student growth and progress.

Scoring methods are an important, and often overlooked, part of effective assessment. Teachers can use a simple three-point scale for evaluating student responses. No answer or an inappropriate answer that shows no understanding scores zero points. A partial response showing a lack of understanding, a lack of explanation, or major computational errors scores one point. A somewhat satisfactory answer that answers most of the question correctly but contains simple computational errors or minor flaws in reasoning receives two points. Finally, a satisfactory response displaying full understanding, adequate explanation, and appropriate reasoning receives three points. When evaluating student responses, teachers should look for common error patterns and mistakes in computation. Teachers should also incorporate questions and scoring procedures that address common error patterns and misconceptions into their methods of assessment.

The primary purpose of student assessment is to evaluate the effectiveness of the curriculum and instruction by measuring student performance. Teachers and school officials use the results of student assessments to monitor student progress and modify and design curriculum to meet the needs of the students. Teachers and school officials carefully assess the results of tests to determine the parts of the curriculum that need altering. For example, the results of a test may indicate that the majority of the students in a class struggle with problems involving logarithmic functions. In response to such findings, the teacher would evaluate the method of logarithmic function instruction and make the necessary changes to increase student understanding.

A special type of student assessment, state standardized testing, is an important tool for curriculum design and modification. Most states have stated curriculum standards that mandate what students should know. Teachers can use the standards to focus their instruction and curriculum planning. State tests evaluate and report student performance on the specific curriculum standards. Thus, teachers can easily determine the areas that require greater attention.

Finally, teachers and school officials must understand the special needs of English Language Learners. Mathematic assessments may understate the abilities of English Language Learners because poor test scores may stem from difficulty in reading comprehension, not a lack of understanding of mathematic principles. Uncharacteristically poor performance on word problems by English Language Learners is a sign that reading comprehension, not mathematic understanding, is the underlying problem.

ANSWER KEY TO PRACTICE PROBLEMS

Competency 05

Page 56

> Question #1 discriminant = 241; 2 real irrational roots
> Question #2 discriminant = 81; 2 real rational roots
> Question #3 discriminant = 0; 1 real rational root

Page 65

> Question #1   The sides are 8, 15, and 17
>
> Question #2   The numbers are 2 and $\frac{1}{2}$

Page 72

Question #1

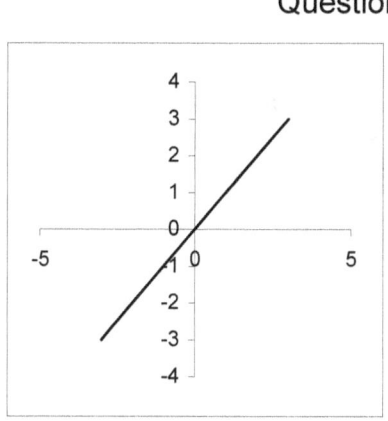

Question #2

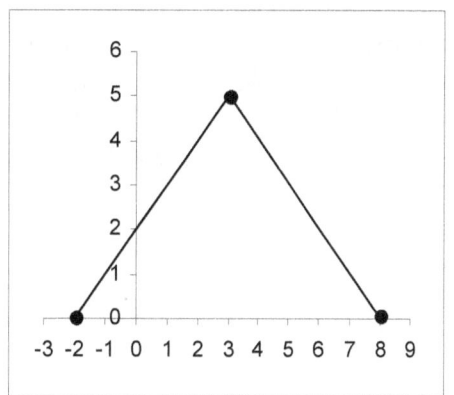

Question #3

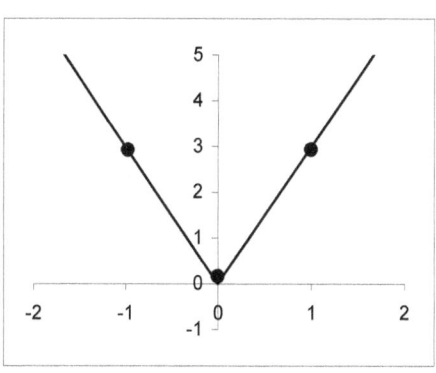

Question #4

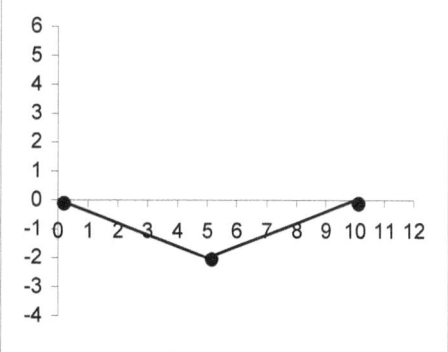

Page 74

Question #1   Questions #2

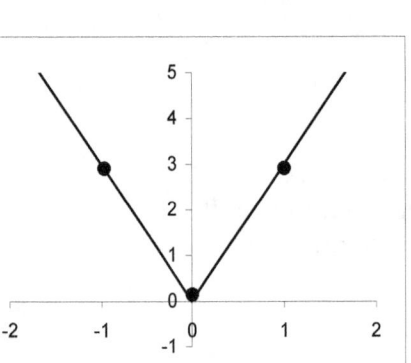
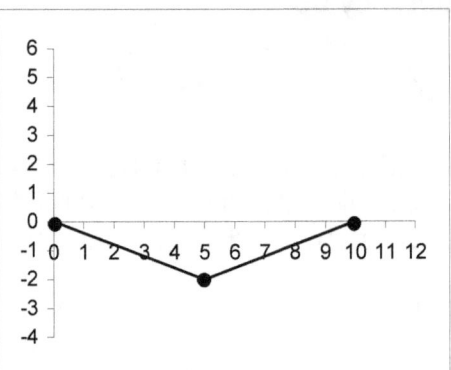

Competency 015

Page 147

    Question #1   The Red Sox won the World Series.
    Question #2   Angle B is not between 0 and 90 degrees.
    Question #3   Annie will do well in college.
    Question #4   You are witty and charming.

Sample Test: Mathematics

1) Given W = whole numbers
   N = natural numbers
   Z = integers
   R = rational numbers
   I = irrational numbers

   Which of the following is not true?

   A) $R \subset I$

   B) $W \subset Z$

   C) $Z \subset R$

   D) $N \subset W$

2) Which of the following is an irrational number?

   A) .362626262...

   B) $4\frac{1}{3}$

   C) $\sqrt{5}$

   D) $-\sqrt{16}$

3) Which denotes a complex number?

   A) 4.1212121212...

   B) $-\sqrt{16}$

   C) $\sqrt{127}$

   D) $\sqrt{-100}$

4) Choose the correct statement:

   A) Rational and irrational numbers are both proper subsets of the real numbers.

   B) The set of whole numbers is a proper subset of the set of natural numbers.

   C) The set of integers is a proper subset of the set of irrational numbers.

   D) The set of real numbers is a proper subset of the natural, whole, integers, rational, and irrational numbers.

5) Which statement is an example of the identity axiom of addition?

   A) $3 + -3 = 0$

   B) $3x = 3x + 0$

   C) $3 \cdot \frac{1}{3} = 1$

   D) $3 + 2x = 2x + 3$

6) Which axiom is incorrectly applied?

$$3x + 4 = 7$$

Step a $3x + 4 - 4 = 7 - 4$

   additive equality

Step b $3x + 4 - 4 = 3$

   commutative axiom of addition

Step c $3x + 0 = 3$

   additive inverse

Step d $3x = 3$

   additive identity

A) step a

B) step b

C) step c

D) step d

7) Which of the following sets is closed under division?

A) integers

B) rational numbers

C) natural numbers

D) whole numbers

8) How many real numbers lie between -1 and $+1$?

A) 0

B) 1

C) 17

D) an infinite number

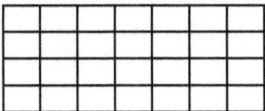
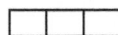

9) The above diagram would be least appropriate for illustrating which of the following?

A) $7 \times 4 + 3$

B) $31 \div 8$

C) 28×3

D) $31 - 3$

10) $24 - 3 \times 7 + 2 =$

A) 5

B) 149

C) -3

D) 189

11) Which of the following does not correctly relate an inverse operation?

A) $a - b = a + -b$

B) $a \times b = b \div a$

C) $\sqrt{a^2} = a$

D) $a \times \frac{1}{a} = 1$

12) Mr. Brown feeds his cat premium cat food which costs $40 per month. Approximately how much will it cost to feed her for one year?

A) $500

B) $400

C) $80

D) $4800

13) Given that n is a positive even integer, 5n + 4 will always be divisible by:

A) 4

B) 5

C) 5n

D) 2

14) Given that x, y, and z are prime numbers, which of the following is true?

A) x + y is always prime

B) xyz is always prime

C) xy is sometimes prime

D) x + y is sometimes prime

15) Find the GCF of $2^2 \cdot 3^2 \cdot 5$ and $2^2 \cdot 3 \cdot 7$.

A) $2^5 \cdot 3^3 \cdot 5 \cdot 7$

B) $2 \cdot 3 \cdot 5 \cdot 7$

C) $2^2 \cdot 3$

D) $2^3 \cdot 3^2 \cdot 5 \cdot 7$

16) Given even numbers x and y, which could be the LCM of x and y?

A) $\frac{xy}{2}$

B) 2xy

C) 4xy

D) xy

17) $(3.8 \times 10^{17}) \times (.5 \times 10^{-12})$

   A) 19×10^5

   B) 1.9×10^5

   C) 1.9×10^6

   D) 1.9×10^7

18) 2^{-3} is equivalent to

   A) 0.8

   B) −0.8

   C) 125

   D) 125

19) $\dfrac{3.5 \times 10^{-10}}{0.7 \times 10^4}$

   A) 0.5×10^6

   B) 5.0×10^{-6}

   C) 5.0×10^{-14}

   D) 0.5×10^{-14}

20) Solve for x: $\dfrac{4}{x} = \dfrac{8}{3}$

   A) 0.66666...

   B) 0.6

   C) 15

   D) 1.5

21) Choose the set in which the members are not equivalent.

   A) 1/2, 0.5, 50%

   B) 10/5, 2.0, 200%

   C) 3/8, 0.385, 38.5%

   D) 7/10, 0.7, 70%

22) If three cups of concentrate are needed to make 2 gallons of fruit punch, how many cups are needed to make 5 gallons?

   A) 6 cups

   B) 7 cups

   C) 7.5 cups

   D) 10 cups

23) A sofa sells for $520. If the retailer makes a 30% profit, what was the wholesale price?

   A) $400

   B) $670

   C) $490

   D) $364

24) Given a spinner with the numbers one through eight, what is the probability that you will spin an even number or a number greater than four?

   A) 1/4

   B) 1/2

   C) ¾

   D) 1

25) If a horse will probably win three races out of ten, what are the odds that he will win?

   A) 3:10

   B) 7:10

   C) 3:7

   D) 7:3

26) Given a drawer with 5 black socks, 3 blue socks, and 2 red socks, what is the probability that you will draw two black socks in two draws in a dark room?

   A) 2/9

   B) 1/4

   C) 17/18

   D) 1/18

27) A sack of candy has 3 peppermints, 2 butterscotch drops and 3 cinnamon drops. One candy is drawn and replaced, then another candy is drawn; what is the probability that both will be butterscotch?

   A) 1/2

   B) 1/28

   C) 1/4

   D) 1/16

28) Find the median of the following set of data:

   14 3 7 6 11 20

   A) 9

   B) 8.5

   C) 7

   D) 11

29) Corporate salaries are listed for several employees. Which would be the best measure of central tendency?

$24,000 $24,000 $26,000

$28,000 $30,000 $120,000

A) mean

B.) median

C) mode

D) no difference

30) Which statement is true about George's budget?

A) George spends the greatest portion of his income on food.

B) George spends twice as much on utilities as he does on his mortgage.

C) George spends twice as much on utilities as he does on food.

D) George spends the same amount on food and utilities as he does on mortgage.

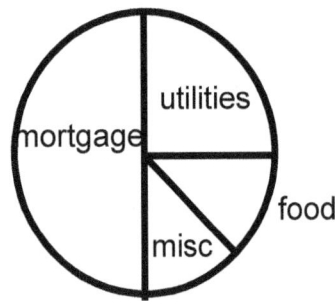

31) A student scored in the 87th percentile on a standardized test. Which would be the best interpretation of his score?

A) Only 13% of the students who took the test scored higher.

B) This student should be getting mostly B's on his report card.

C) This student performed below average on the test.

D) This is the equivalent of missing 13 questions on a 100 question exam.

32) A man's waist measures 90 cm. What is the greatest possible error for the measurement?

A) ± 1 m

B) ±8 cm

C) ±1 cm

D) ±5 mm

33) The mass of a cookie is closest to

A) 0.5 kg

B) 0.5 grams

C) 15 grams

D) 1.5 grams

34) 3 km is equivalent to

   A) 300 cm

   B) 300 m

   C) 3000 cm

   D) 3000 m

35) 4 square yards is equivalent to

   A) 12 square feet

   B) 48 square feet

   C) 36 square feet

   D) 108 square feet

36) If a circle has an area of 25 cm^2, what is its circumference to the nearest tenth of a centimeter?

   A) 78.5 cm

   B) 17.7 cm

   C) 8.9 cm

   D) 15.7 cm

37) Find the area of the figure below.

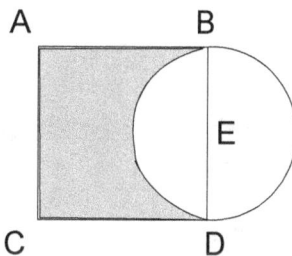

   A) 56 in^2

   B) 27 in^2

   C) 71 in^2

   D) 170 in^2

38) Find the area of the shaded region given square ABCD with side AB=10m and circle E.

   A) 178.5 m^2

   B) 139.25 m^2

   C) 71 m^2

   D) 60.75 m^2

39) Given similar polygons with corresponding sides of lengths 9 and 15, find the perimeter of the smaller polygon if the perimeter of the larger polygon is 150 units.

   A) 54

   B) 135

   C) 90

   D) 126

40)

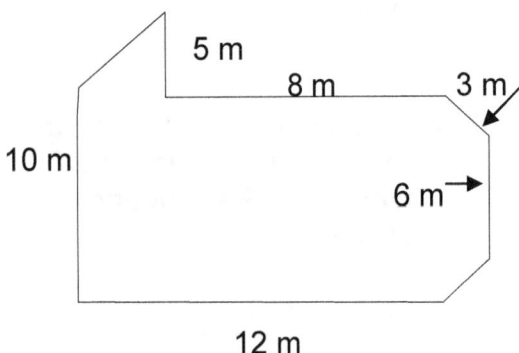

Compute the area of the polygon shown above.

   A) 178 m²

   B) 154 m²

   C) 43 m²

   D) 188 m²

41) If the radius of a right cylinder is doubled, how does its volume change?

   A) no change

   B) also is doubled

   C) four times the original

   D) pi times the original

42) Determine the volume of a sphere to the nearest cm³ if the surface area is 113 cm².

   A) 113 cm³

   B) 339 cm³

   C) 37.7 cm³

   D) 226 cm3

43) Compute the surface area of the prism.

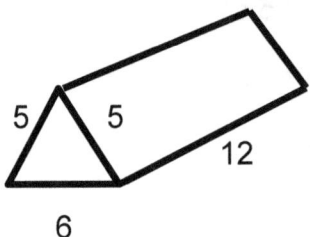

   A) 204

   B) 216

   C) 360

   D) 180

44) If the base of a regular square pyramid is tripled, how does its volume change?

A) double the original

B) triple the original

C) nine times the original

D) no change

45) How does lateral area differ from total surface area in prisms, pyramids, and cones?

A) For the lateral area, only use surfaces perpendicular to the base.

B) They are both the same.

C) The lateral area does not include the base.

D) The lateral area is always a factor of pi.

46) Given XY ≅ YZ and ∠AYX ≅ ∠AYZ. Prove △AYZ ≅ △AYX.

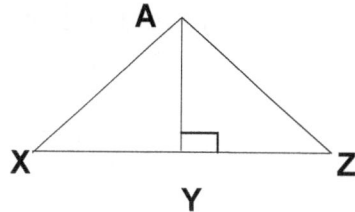

1) XY ≅ YZ

2) ∠AYX ≅ ∠AYZ

3) AY ≅ AY

4) △AYZ ≅ △AYX

Which property justifies step 3?

A) reflexive

B.) symmetric

B) transitive

D) identity

47) Given $l_1 \parallel l_2$ (parallel lines 1 & 2) prove $\angle b \cong \angle e$

1) $\angle b \cong \angle d$ 1) vertical angle theorem

2) $\angle d \cong \angle e$ 2) alternate interior angle theorem

3) $\angle b \cong \angle 3$ 3) symmetric axiom of equality

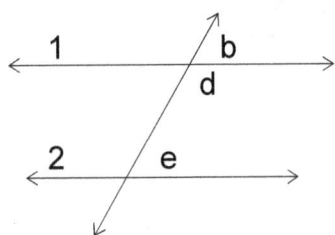

Which step is incorrectly justified?

A) step 1

B) step 2

C) step 3

D) no error

48) Simplify $\dfrac{\frac{3}{4}x^2y^{-3}}{\frac{2}{3}xy}$

A) $\frac{1}{2}xy^{-4}$

B) $\frac{1}{2}x^{-1}y^{-4}$

C) $\frac{9}{8}xy^{-4}$

D) $\frac{9}{8}xy^{-2}$

49) $7t - 4 \cdot 2t + 3t \cdot 4 \div 2 =$

A) 5t

B) 0

C) 31t

D) 18t

50) Solve for x:
$3x + 5 \geq 8 + 7x$

A) $x \geq -\frac{3}{4}$

B) $x \leq -\frac{3}{4}$

C) $x \geq \frac{3}{4}$

D) $x \leq \frac{3}{4}$

51) Solve for x:
$|2x + 3| > 4$

A) $-\frac{7}{2} > x > \frac{1}{2}$

B) $-\frac{1}{2} > x > \frac{7}{2}$

C) $x < \frac{7}{2}$ or $x < -\frac{1}{2}$

D) $x < -\frac{7}{2}$ or $x > \frac{1}{2}$

52) $3x + 2y = 12$
$12x + 8y = 15$

A) all real numbers

B) $x = 4, y = 4$

C) $x = 2, y = -1$

D) $\varnothing$

53) Solve for x and y:

x = 3y + 7
7x + 5y = 23

A) (–1, 4)

B) (4, –1)

C) ($\frac{-29}{7}, \frac{-26}{7}$)

D) (10, 1)

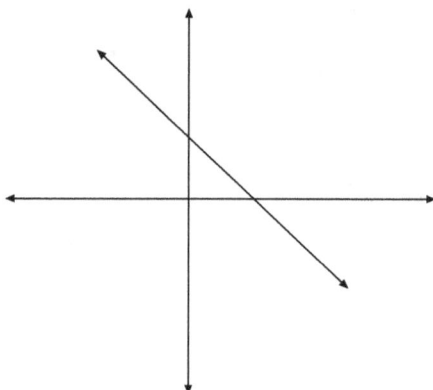

54) Which equation is represented by the above graph?

A) x – y = 3

B) x – y = –3

C) x + y = 3

D) x + y = –3

55) Graph the solution:

|x| + 7 < 13

A) ←○———————○→
   –6   0   6

B) ←●———————●→
   –6   0   6

C) ←○———————○→
   –6   0   6

D) ←●   ●→
   –6   0   6

56) Three less than four times a number is five times the sum of that number and 6. Which equation could be used to solve this problem?

A) 3 – 4n = 5(n + 6)

B) 3 – 4n + 5n = 6

C) 4n – 3 = 5n + 6

D) 4n – 3 = 5(n + 6)

57) A boat travels 30 miles upstream in three hours. It makes the return trip in one and a half hours. What is the speed of the boat in still water?

A) 10 mph

B) 15 mph

C) 20 mph

D) 30 mph

58) Which set illustrates a function?

A) { (0,1) (0,2) (0,3) (0,4) }

B) { (3,9) (–3,9) (4,16) (–4,16)}

C) { (1,2) (2,3) (3,4) (1,4) }

D) { (2,4) (3,6) (4,8) (4,16) }

59) Give the domain for the function over the set of real numbers:

$$y = \frac{3x+2}{2x^2-3}$$

A) all real numbers

B) all real numbers, $x \neq 0$

C) all real numbers, $x \neq -2$ or 3

D) all real numbers, $x \neq \frac{\pm\sqrt{6}}{2}$

60) Factor completely:
$8(x - y) + a(y - x)$

A) $(8 + a)(y - x)$

B) $(8 - a)(y - x)$

C) $(a - 8)(y - x)$

D) $(a - 8)(y + x)$

61) Which of the following is a factor of $k^3 - m^3$?

A) $k^2 + m^2$

B) $k + m$

C) $k^2 - m^2$

D) $k - m$

62) Solve for x.

$3x^2 - 2 + 4(x^2 - 3) = 0$

A) $\{-\sqrt{2}, \sqrt{2}\}$

B) $\{2, -2\}$

C) $\{0, \sqrt{3}, -\sqrt{3}\}$

D) $\{7, -7\}$

63) Solve: $\sqrt{75} + \sqrt{147} - \sqrt{48}$

A) 174

B) $12\sqrt{3}$

C) $8\sqrt{3}$

D) 74

64) The discriminant of a quadratic equation is evaluated and determined to be –3. The equation has

A) one real root

B) one complex root

C) two roots, both real

D) two roots, both complex

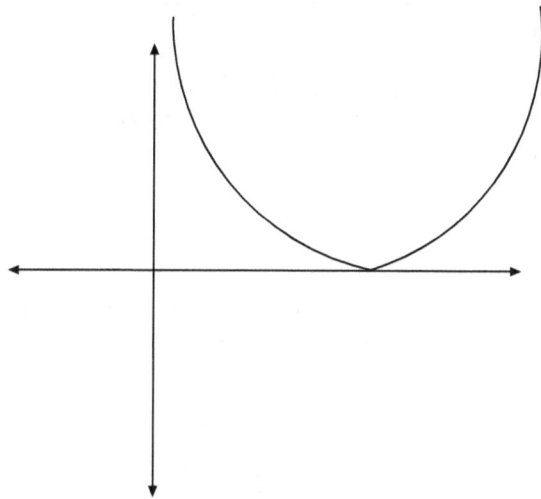

65) Which equation is graphed above?

A) $y = 4(x + 3)^2$

B) $y = 4(x - 3)^2$

C) $y = 3(x - 4)^2$

D) $y = 3(x + 4)^2$

66) If y varies inversely as x and x is 4 when y is 6, what is the constant of variation?

A) 2

B) 12

C) 3/2

D) 24

67) If y varies directly as x and x is 2 when y is 6, what is x when y is 18?

A) 3

B) 6

C) 26

D) 36

68) $\{1, 4, 7, 10, \ldots\}$

What is the 40th term in this sequence?

A) 43

B) 121

C) 118

D) 120

69) {6,11,16,21, . .}
Find the sum of the first 20 terms in the sequence.

A) 1070

B) 1176

C) 969

D) 1069

70) Two non-coplanar lines which do not intersect are labeled

A) parallel lines

B) perpendicular lines

C) skew lines

D) alternate exterior lines

71)

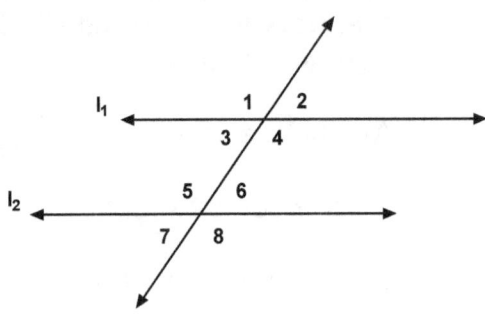

Given $l_1 \parallel l_2$
(parallel lines 1 & 2)
which of the following is true?

A) ∠1 and ∠8 are congruent and alternate interior angles

B) ∠2 and ∠3 are congruent and corresponding angles

C) ∠3 and ∠4 are adjacent and supplementary angles

D) ∠3 and ∠5 are adjacent and supplementary angles

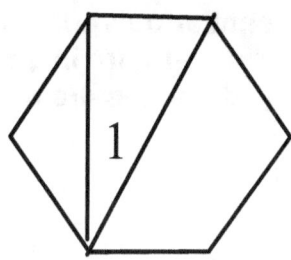

72) Given the regular hexagon above, determine the measure of angle 1.

A) 30°

B) 60°

C) 120°

D) 45°

73)

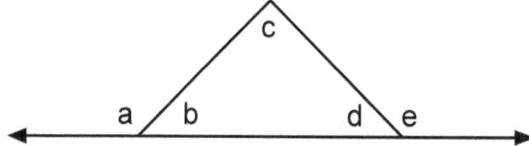

Which of the following statements is true about the number of degrees in each angle?

A) a + b + c = 180°

B) a = e

C) b + c = e

D) c + d = e

74)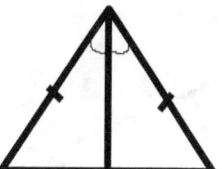

What method could be used to prove the above triangles congruent?

A) SSS

B) SAS

C) AAS

D) SSA

75)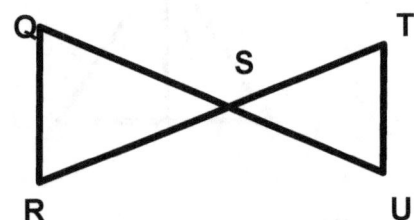

Given QS ≅ TS and RS ≅ US, prove △QRS ≅ △TUS.

I) QS ≅ TS	1) Given
2) RS ≅ US	2) Given
3) ∠TSU ≅ ∠QSR	3) ?
4) △TSU ≅ △QSR	4) SAS

Give the reason which justifies step 3.

A) Congruent parts of congruent triangles are congruent

B) Reflexive axiom of equality

C) Alternate interior angle Theorem

D) Vertical angle theorem

76) Given similar polygons with corresponding sides 6 and 8, what is the area of the smaller if the area of the larger is 64?

A) 48

B) 36

C) 144

D) 78

77) In similar polygons, if the perimeters are in a ratio of x:y, the sides are in a ratio of

A) x : y

B) $x^2 : y^2$

C) 2x : y

D) 1/2 x : y

78)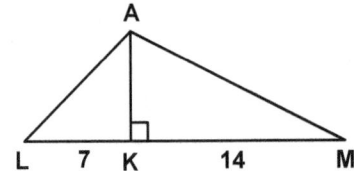

Given altitude AK with measurements as indicated, determine the length of AK.

A) 98

B) $7\sqrt{2}$

C) $\sqrt{21}$

D) $7\sqrt{3}$

79)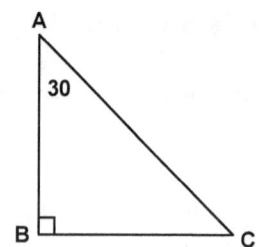

If AC = 12, determine BC.

A) 6

B) 4

C) $6\sqrt{3}$

D) $3\sqrt{6}$

80)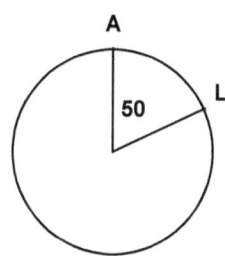

What is the measure of major arc AL ?

A) 50°

B) 25°

C) 100°

D) 310°

81)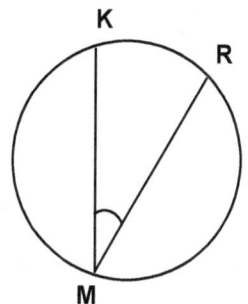

If arc KR = 70° what is the measure of ∠M?

A) 290°

B) 35°

C) 140°

D) 110°

82)

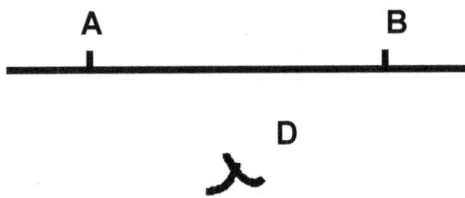

The above construction can be completed to make

A) an angle bisector

B) parallel lines

C) a perpendicular bisector

D) skew lines

83)

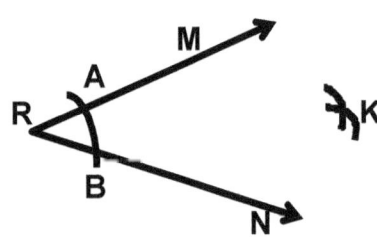

A line from R to K will form

A) an altitude of RMN

B) a perpendicular bisector of MN

C) a bisector of MRN

D) a vertical angle

84) Which is a postulate?

A) The sum of the angles in any triangle is 180°.

B) A line intersects a plane in one point.

C) Two intersecting lines from congruent vertical angles.

D) Any segment is congruent to itself.

85) Which of the following can be defined?

A) point

B) ray

C) line

D) plane

86)

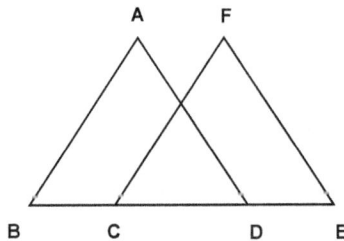

Which theorem could be used to prove △ABD ≅ △CEF, given BC ≅ DE, ∠C ≅ ∠D, and AD ≅ CF?

A) ASA

B) SAS

C) SAA

D) SSS

87)

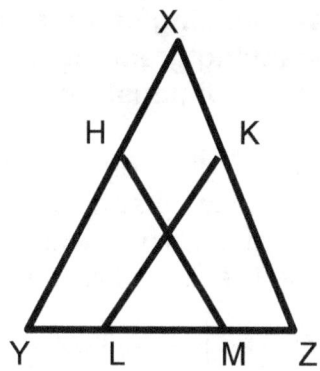

Prove △HYM ≅ △KZL, given XZ ≅ XY, ∠L ≅ ∠M and YL ≅ MZ

1) XZ ≅ XY   1) Given
2) ∠Y ≅ ∠Z   2) ?
3) ∠L ≅ ∠M   3) Given
4) YL ≅ MZ   4) Given
5) LM ≅ LM   5) ?
6) YM ≅ LZ   6) Add
7) △HYM ≅ △KZL   7) ASA

Which could be used to justify steps 2 and 5?

A) CPCTC, Identity

B) Isosceles Triangle Theorem, Identity

C) SAS, Reflexive

D) Isosceles Triangle Theorem, Reflexive

88) Find the distance between (3,7) and (−3,4).

A) 9

B) 45

C) $3\sqrt{5}$

D) $5\sqrt{3}$

89) Find the midpoint of (2,5) and (7,−4).

A) (9,−1)

B) (5, 9)

C) (9/2, −1/2)

D) (9/2, 1/2)

90) Given segment AC with B as its midpoint find the coordinates of C if A = (5,7) and B = (3, 6.5).

A) (4, 6.5)

B) (1, 6)

C) (2, 0.5)

D) (16, 1)

91)

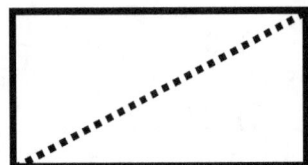

The above diagram is most likely used in deriving a formula for which of the following?

A) the area of a rectangle

B) the area of a triangle

C) the perimeter of a triangle

D) the surface area of a prism

92) A student turns in a paper with this type of error:

$7 + 16 \div 8 \times 2 = 8$
$8 - 3 \times 3 + 4 = -5$

In order to remediate this error, a teacher should:

A) review and drill basic number facts

B) emphasize the importance of using parentheses in simplifying expressions

C) emphasize the importance of working from left to right when applying the order of operations

D) do nothing; these answers are correct

93) Identify the proper sequencing of subskills when teaching graphing inequalities in two dimensions

A) shading regions, graphing lines, graphing points, determining whether a line is solid or broken

B) graphing points, graphing lines, determining whether a line is solid or broken, shading regions

C) graphing points, shading regions, determining whether a line is solid or broken, graphing lines

D) graphing lines, determining whether a line is solid or broken, graphing points, shading regions

94) Sandra has $34.00, Carl has $42.00. How much more does Carl have than Sandra?

Which would be the best method for finding the answer?

A) addition

B) subtraction

C) division

D) both A and B are equally correct

95) Which is the least appropriate strategy to emphasize when teaching problem solving?

   A) guess and check

   B) look for key words to indicate operations such as all together—add, more than, subtract, times, multiply

   C) make a diagram

   D) solve a simpler version of the problem

96) Choose the least appropriate set of manipulatives for a six grade class.

   A) graphic calculators, compasses, rulers, conic section models

   B) two color counters, origami paper, markers, yarn

   C) balance, meter stick, colored pencils, beads

   D) paper cups, beans, tangrams, geoboards

97) According to Piaget, at which developmental level would a child be able to learn formal algebra?

   A) pre-operational

   B) sensory-motor

   C) abstract

   D) concrete operational

98) Which statement is incorrect?

   A) Drill and practice is one good use for classroom computers.

   B) Some computer programs can help to teach problem solving.

   C) Computers are not effective unless each child in the class has his own workstation.

   D) Analyzing science project data on a computer during math class is an excellent use of class time.

99) Given a,b,y, and z are real numbers and ay + b = z,

Prove
$$y = \frac{z + -b}{a}$$

Statement	Reason
1) ay + b = z	1) Given
2) –b is a real number	2) Closure
3) (ay +b) + –b = z + –b	3) Addition property of Identity
4) ay + (b + –b) = z + –b	4) Associative
5) ay + 0 = z + –b	5) Additive inverse
6) ay = z + –b	6) Addition property of identity
7) a = $\frac{z + -b}{y}$	7) Division

Which reason is incorrect for the corresponding statement?

A) step 3
B) step 4
C) step 5
D) step 6

100) Seventh grade students are working on a project using non-standard measurement. Which would not be an appropriate instrument for measuring the length of the classroom?

A) a student's foot
B) a student's arm span
C) a student's jump
D) all are appropriate

101. Change $.\overline{63}$ into a fraction in simplest form.

A) 63/100
B) 7/11
C) 6 3/10
D) 2/3

102. Which of the following sets is closed under division?

I) {½, 1, 2, 4}
II) {–1, 1}
III) {–1, 0, 1}

A) I only
B) II only
C) III only
D) I and II

103. Which of the following illustrates an inverse property?

A) a + b = a – b
B) a + b = b + a
C) a + 0 = a
D) a + (–a) = 0

104. $f(x) = 3x - 2$; $f^{-1}(x) =$

   A) $3x + 2$
   B) $x/6$
   C) $2x - 3$
   D) $(x+2)/3$

105. What would be the total cost of a suit for $295.99 and a pair of shoes for $69.95 including 6.5% sales tax?

   A) $389.73
   B) $398.37
   C) $237.86
   D) $315.23

106. A student had 60 days to appeal the results of an exam. If the results were received on March 23, what was the last day that the student could appeal?

   A) May 21
   B) May 22
   C) May 23
   D) May 24

107. Which of the following is always composite if *x* is odd, *y* is even, and both *x* and *y* are greater than or equal to 2?

   A) $x + y$
   B) $3x + 2y$
   C) $5xy$
   D) $5x + 3y$

108. Which of the following is incorrect?

   A) $(x^2 y^3)^2 = x^4 y^6$
   B) $m^2(2n)^3 = 8m^2 n^3$
   C) $(m^3 n^4)/(m^2 n^2) = mn^2$
   D) $(x + y^2)^2 = x^2 + y^4$

109. Express .0000456 in scientific notation.

   A) $4.56 x 10^{-4}$
   B) $45.6 x 10^{-6}$
   C) $4.56 x 10^{-6}$
   D) $4.56 x 10^{-5}$

110. Compute the area of the shaded region, given a radius of 5 meters. O is the center.

   A) 7.13 cm²
   B) 7.13 m²
   C) 78.5 m²
   D) 19.63 m²

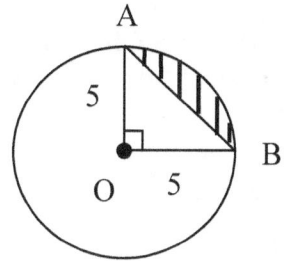

111. If the area of the base of a cone is tripled, the volume will be

   A) the same as the original
   B) 9 times the original
   C) 3 times the original
   D) 3 π times the original

112. Find the area of the figure pictured below.

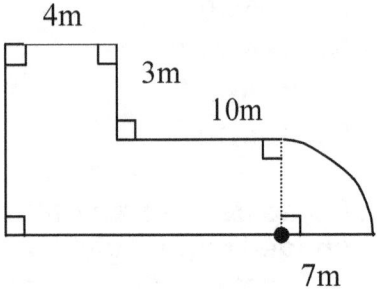

A) 136.47 m²
B) 148.48 m²
C) 293.86 m²
D) 178.47 m²

113. The mass of a Chips Ahoy cookie would be approximately equal to:

A) 1 kilogram
B) 1 gram
C) 15 grams
D) 15 milligrams

114. Compute the median for the following data set:

{12, 19, 13, 16, 17, 14}

A) 14.5
B) 15.17
C) 15
D) 16

115. Half the students in a class scored 80% on an exam, most of the rest scored 85% except for one student who scored 10%. Which would be the best measure of central tendency for the test scores?

A) mean
B) median
C) mode
D) either the median or the mode because they are equal

116. What conclusion can be drawn from the graph below?

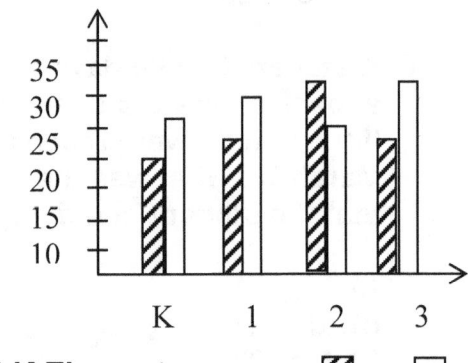

MLK Elementary
Student Enrollment   Girls   Boys

A) The number of students in first grade exceeds the number in second grade.
B) There are more boys than girls in the entire school.
C) There are more girls than boys in the first grade.
D) Third grade has the largest number of students.

117) State the domain of the function $f(x) = \dfrac{3x-6}{x^2-25}$

A) $x \neq 2$
B) $x \neq 5, -5$
C) $x \neq 2, -2$
D) $x \neq 5$

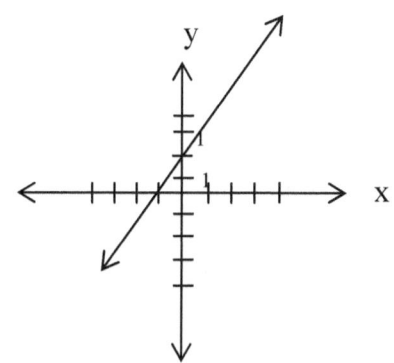

118. What is the equation of the above graph?

A) $2x + y = 2$
B) $2x - y = -2$
C) $2x - y = 2$
D) $2x + y = -2$

119. Solve for v_0: $d = at(v_t - v_0)$

A) $v_0 = atd - v_t$
B) $v_0 = d - atv_t$
C) $v_0 = atv_t - d$
D) $v_0 = (atv_t - d)/at$

120. Which of the following is a factor of $6 + 48m^3$

A) $(1 + 2m)$
B) $(1 - 8m)$
C) $(1 + m - 2m)$
D) $(1 - m + 2m)$

121. Which graph represents the equation of $y = x^2 + 3x$?

A) 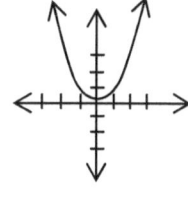   B)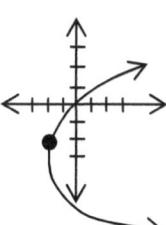

C) 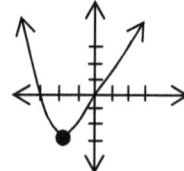   D)

122. The volume of water flowing through a pipe varies directly with the square of the radius of the pipe. If the water flows at a rate of 80 liters per minute through a pipe with a radius of 4 cm, at what rate would water flow through a pipe with a radius of 3 cm?

A) 45 liters per minute
B) 6.67 liters per minute
C) 60 liters per minute
D) 4.5 liters per minute

123) Solve the system of equations for x, y and z.

$3x + 2y - z = 0$
$2x + 5y = 8z$
$x + 3y + 2z = 7$

A) (−1, 2, 1)
B) (1, 2, −1)
C) (−3, 4, −1)
D) (0, 1, 2)

124. Solve for x: $18 = 4 + |2x|$

A) $\{-11, 7\}$
B) $\{-7, 0, 7\}$
C) $\{-7, 7\}$
D) $\{-11, 11\}$

125. Which graph represents the solution set for $x^2 - 5x > -6$?

A) number line with open circles at −2 and 2

B) number line with open circles at −3 and 3

C) number line with open circles at −2 and 2

D) number line with open circles at 2 and 3

126. Find the zeroes of
$f(x) = x^3 + x^2 - 14x - 24$

A) 4, 3, 2
B) 3, −8
C) 7, −2, −1
D) 4, −3, −2

127. Evaluate $3^{1/2}(9^{1/3})$

A) $27^{5/6}$
B) $9^{7/12}$
C) $3^{5/6}$
D) $3^{6/7}$

128. Simplify: $\sqrt{27} + \sqrt{75}$

A) $8\sqrt{3}$
B) 34
C) $34\sqrt{3}$
D) $15\sqrt{3}$

129. Simplify: $\dfrac{10}{1+3i}$

A) $-1.25(1-3i)$
B) $1.25(1+3i)$
C) $1+3i$
D) $1-3i$

130. Find the sum of the first one hundred terms in the progression.
(−6, −2, 2 . . .)

A) 19,200
B) 19,400
C) −604
D) 604

131. How many ways are there to choose a potato and two green vegetables from a choice of three potatoes and seven green vegetables?

A) 126
B) 63
C) 21
D) 252

132. What would be the seventh term of the expanded binomial $(2a+b)^8$?

   A) $2ab^7$
   B) $41a^4b^4$
   C) $112a^2b^6$
   D) $16ab^7$

133. Which term most accurately describes two coplanar lines without any common points?

   A) perpendicular
   B) parallel
   C) intersecting
   D) skew

134. Determine the number of subsets of set K.
   K = {4, 5, 6, 7}

   A) 15
   B) 16
   C) 17
   D) 18

135. What is the ~~degree~~ measure of each interior angle of a regular 10 sided polygon?

   A) 18°
   B) 36°
   C) 144°
   D) 54°

136. If a ship sails due south 6 miles, then due west 8 miles, how far is it from its starting point?

   A) 100 miles
   B) 10 miles
   C) 14 miles
   D) 48 miles

137. What is the measure of minor arc AD, given measure of arc PS is 40° and $m<K=10°$?

   A) 50°
   B) 20°
   C) 30°
   D) 25°

138. Choose the diagram which illustrates the construction of a perpendicular to the line at a given point on the line.

   A)

   B)

   C)

   D)

139. When you begin by assuming the conclusion of a theorem is false, then show that through a sequence of logically correct steps you contradict an accepted fact, this is known as

A) inductive reasoning
B) direct proof
C) indirect proof
D) exhaustive proof

140. Which theorem can be used to prove $\triangle BAK \cong \triangle MKA$?

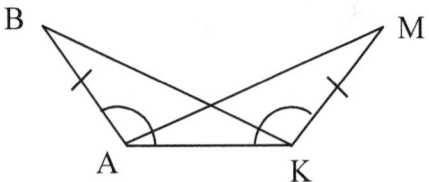

A) SSS
B) ASA
C) SAS
D) AAS

141. Given that QO⊥NP and QO=NP, quadrilateral NOPQ can most accurately be described as a

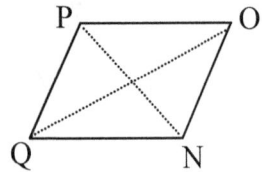

A) parallelogram
B) rectangle
C) square
D) rhombus

142. Choose the correct statement concerning the median and altitude in a triangle.

A) The median and altitude of a triangle may be the same segment.
B) The median and altitude of a triangle are always different segments.
C) The median and altitude of a right triangle are always the same segment.
D) The median and altitude of an isosceles triangle are always the same segment.

143. Which mathematician is best known for his work in developing non-Euclidean geometry?

A) Descartes
B) Riemann
C) Pascal
D) Pythagoras

144. Find the surface area of a box which is 3 feet wide, 5 feet tall, and 4 feet deep.

A) 47 sq. ft.
B) 60 sq. ft.
C) 94 sq. ft
D) 188 sq. ft.

145. Given a 30 meter x 60 meter garden with a circular fountain with a 5 meter radius, calculate the area of the portion of the garden not occupied by the fountain.

A) 1721 m²
B) 1879 m²
C) 2585 m²
D) 1015 m²

146. Determine the area of the shaded region of the trapezoid in terms of x and y.

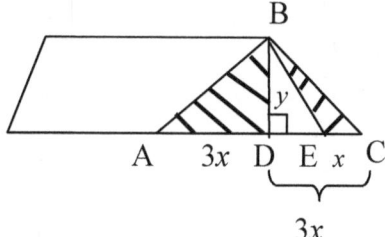

A) $4xy$
B) $2xy$
C) $3x^2y$
D) There is not enough information given.

Answer Key: Mathematics

1. A	38. D	75. D	112. B
2. C	39. C	76. B	113. C
3. D	40. B	77. A	114. C
4. A	41. C	78. B	115. B
5. B	42. A	79. A	116. B
6. B	43. B	80. D	117. B
7. B	44. B	81. B	118. B
8. D	45. C	82. C	119. D
9. C	46. A	83. C	120. A
10. A	47. C	84. D	121. C
11. B	48. C	85. B	122. A
12. A	49. A	86. B	123. A
13. D	50. B	87. D	124. C
14. D	51. D	88. C	125. D
15. C	52. D	89. D	126. D
16. A	53. B	90. B	127. B
17. B	54. C	91. B	128. A
18. D	55. A	92. C	129. D
19. C	56. D	93. B	130. A
20. D	57. B	94. D	131. A
21. C	58. B	95. B	132. C
22. C	59. D	96. A	133. B
23. A	60. C	97. C	134. B
24. C	61. D	98. C	135. C
25. C	62. A	99. A	136. B
26. A	63. C	100. D	137. B
27. D	64. D	101. B	138. D
28. A	65. B	102. B	139. C
29. B	66. D	103. D	140. C
30. C	67. B	104. D	141. C
31. A	68. C	105. A	142. A
32. D	69. A	106. B	143. B
33. C	70. C	107. C	144. C
34. D	71. C	108. D	145. A
35. C	72. A	109. D	146. B
36. B	73. C	110. B	
37. A	74. B	111. C	

Rationales for Sample Questions: Mathematics

The following statements represent one way to solve each problem and obtain a correct answer.

1) C The rational numbers are not a subset of the irrational numbers. All of the other statements are true.

2) C 5 is an irrational number. A and B can both be expressed as fractions. D can be simplified to -4, an integer and rational number.

3) D A complex number is the square root of a negative number. The complex number is defined as the square root of -1. A is rational, B and C are irrational.

4) A A proper subset is completely contained in, but not equal to, the original set.

5) B Illustrates the identity axiom of addition. A illustrates additive inverse, C illustrates the multiplicative inverse, and D illustrates the commutative axiom of addition.

6) B In simplifying from step a to step b, 3 replaced 7 - 4, therefore the correct justification would be subtraction or substitution.

7) B In order to be closed under division, when any two members of the set are divided the answer must be contained in the set. This is not true for integers, natural, or whole numbers as illustrated by the counter example 11/2 = 5.5.

8) D There are an infinite number of real numbers between any two real numbers.

9) C is inappropriate. A shows a 7x4 rectangle with 3 additional units. B is the division based on A . D shows how mental subtraction might be visualized leaving a composite difference.

TEACHER CERTIFICATION STUDY GUIDE

10) A According to the order of operations, multiplication is performed first, then addition and subtraction from left to right.

11) B is always false. A, C, and D illustrate various properties of inverse relations.

12) A 12(40) = 480 which is closest to $500.

13) D 5n is always even. An even number added to an even number is always an even number, thus divisible by 2.

14) D x + y is sometimes prime. B and C show the products of two numbers which are always composite. x + y may be true, but not always,

15) C Choose the number of each prime factor that is in common.

16) A Although choices B, C and D are common multiples, when both numbers are even, the product can be divided by two to obtain the least common multiple.

17) B Multiply the decimals and add the exponents.

18) D Express as the fraction 1/8, then convert to a decimal.

19) C Divide the decimals and subtract the exponents.

20) D Cross multiply to obtain 12 = 8x, then divide both sides by 8.

21) C 3/8 is equivalent to .375 and 37.5%

22) C Set up the proportion 3/2 = x/5, cross multiply to obtain 15=2x, then divide both sides by 2.

23) A Let x be the wholesale price, then x + .30x = 520, 1.30x = 520. Divide both sides by 1.30.

24) C There are 8 favorable outcomes: 2,4,5,6,7,8 and 8 possibilities. Reduce 6/8 to 3/4.

25) C The odds are that he will win 3 and lose 7.

26) A In this example of conditional probability, the probability of drawing a black sock on the first draw is 5/10. It is implied in the problem that there is no replacement, therefore the probability of obtaining a black sock in the second draw is 4/9. Multiply the two probabilities and reduce to lowest terms.

27) D With replacement, the probability of obtaining a butterscotch on the first draw is 2/8 and the probability of drawing a butterscotch on the second draw is also 2/8. Multiply and reduce to lowest terms.

28) A Place the numbers in ascending order: 3 6 7 11 14 20. Find the average of the middle two numbers (7+11)12 =9

29) B The median provides the best measure of central tendency in this case, where the mode is the lowest number and the mean would be disproportionately skewed by the outlier $120,000.

30) C George spends twice as much on utilities as on food.

31) A Percentile ranking tells how the student compared to the norm or the other students taking the test. It does not correspond to the percentage answered correctly, but can indicate how the student compared to the average student tested.

32) D The greatest possible error of measurement is ± + 1/2 unit, in this case .5 cm or 5 mm.

33) C A cookie is measured in grams.

34) D To change kilometers to meters, move the decimal 3 places to the right.

35) C There are 9 square feet in a square yard.

36) B Find the radius by solving $\Pi r^2 = 25$. Then substitute r=2.82 into $C = 2\Pi r$ to obtain the circumference.

37) A Divide the figure into two rectangles with a horizontal line. The area of the top rectangle is 36 in, and the bottom is 20 in.

38) D Find the area of the square $10^2 = 100$, then subtract 1/2 the area of the circle. The area of the circle is $\Pi r^2 = (3.14)(5)(5)=78.5$. Therefore the area of the shaded region is 100 - 39.25 - 60.75.

39) C The perimeters of similar polygons are directly proportional to the lengths of their sides, therefore 9/15 = x/150. Cross multiply to obtain 1350 = 15x, then divide by 15 to obtain the perimeter of the smaller polygon.

40) B Divide the figure into a triangle, a rectangle and a trapezoid. The area of the triangle is 1/2 bh = 1/2 (4)(5) = 10. The area of the rectangle is bh = 12(10) = 120. The area of the trapezoid is 1/2(b + B)h = 1/2(6 + 10)(3) = 1/2 (16)(3) = S4. Thus, the area of the figure is 10 + 120 + 24 =154.

41) C If the radius of a right cylinder is doubled, the volume is multiplied by four; because in the formula, the radius is squared. Therefore the new volume is 2 x 2 or four times the original.

42) A Solve for the radius of the sphere using $A = 4\pi r^2$. The radius is 3. Then, find the volume using $4/3 \pi r^3$. Only when the radius is 3 are the volume and surface area equivalent.

43) B There are five surfaces which make up the prism. The bottom rectangle has an area 6 x 12 = 72. The sloping sides are two rectangles each with an area of 5 x 12 = 60. The height of the end triangles is determined to be 4 using the Pythagorean theorem. Therefore each triangle has area 1/2bh = 1/2(6)(4) -12. Thus, the surface area is 72 + 60 + 60 + 12 + 12 = 216.

44) B Using the general formula for a pyramid V = 1/3 bh, since the base is tripled and is not squared or cubed in the formula, the volume is also tripled.

45) C The lateral area does not include the base.

46) A The reflexive property states that every number or variable is equal to itself and every segment is congruent to itself.

47) C Step 3 can be justified by the transitive property.

48) C Simplify the complex fraction by inverting the denominator and multiplying: 3/4(3/2)=9/8, then subtract exponents to obtain the correct answer.

49) A First perform multiplication and division from left to right; 7t -8t + 6t, then add and subtract from left to right.

50) B Using additive equality, $-3 \geq 4x$. Divide both sides by 4 to obtain $-3/4 \geq x$. Carefully determine which answer choice is equivalent.

51) D The quantity within the absolute value symbols must be either > 4 or < -4. Solve the two inequalities 2x + 3 > 4 or 2x + 3 < -4

52) D Multiplying the top equation by -4 and adding results in the equation 0 = -33. Since this is a false statement, the correct choice is the null set.

53) B Substituting x in the second equation results in 7(3y + 7) + 5y = 23. Solve by distributing and grouping like terms: 26y+49 = 23, 26y = -26, y = -1 Substitute y into the first equation to obtain x.

54) C By looking at the graph, we can determine the slope to be -1 and the y-intercept to be 3. Write the slope intercept form of the line as y = -1x + 3. Add x to both sides to obtain x + y = 3, the equation in standard form.

55) A Solve by adding -7 to each side of the inequality. Since the absolute value of x is less than 6, x must be between -6 and 6. The end points are not included so the circles on the graph are hollow.

56) D Be sure to enclose the sum of the number and 6 in parentheses.

57) B Let x = the speed of the boat in still water and c = the speed of the current.

	rate	time	distance
upstream	x - c	3	30
downstream	x + c	1.5	30

Solve the system:
$$3x - 3c = 30$$
$$1.5x + 1.5c = 30$$

58) B Each number in the domain can only be matched with one number in the range. A is not a function because 0 is mapped to 4 different numbers in the range. In C, 1 is mapped to two different numbers. In D, 4 is also mapped to two different numbers.

59) D Solve the denominator for 0. These values will be excluded from the domain.
$$2x^2 - 3 = 0$$
$$2x^2 = 3$$
$$x^2 = 3/2$$
$$x = \sqrt{\tfrac{3}{2}} = \sqrt{\tfrac{3}{2}} \cdot \sqrt{\tfrac{2}{2}} = \tfrac{\pm\sqrt{6}}{2}$$

60) C Glancing first at the solution choices, factor (y - x) from each term. This leaves -8 from the first term and a from the second term: (a - 8)(y - x)

61) D The complete factorization for a difference of cubes is (k - m)(k² + mk + m2).

62) A Distribute and combine like terms to obtain 7x² - 14 = 0. Add 14 to both sides, then divide by 7. Since x² = 2, x = $\sqrt{2}$

63) C Simplify each radical by factoring out the perfect squares:
$$5\sqrt{3} + 7\sqrt{3} - 4\sqrt{3} = 8\sqrt{3}$$

64) D The discriminate is the number under the radical sign. Since it is negative the two roots of the equation are complex.

65) B Since the vertex of the parabola is three units to the left, we choose the solution where 3 is subtracted from x, then the quantity is squared.

66) D The constant of variation for an inverse proportion is xy.

67) B y/x-216=x/18, Solve 36=6x.

68) C

69) A

70) C

71) C The angles in A are exterior. In B, the angles are vertical. The angles in D are consecutive, not adjacent.

72) A Each interior angle of the hexagon measures 120°. The isosceles triangle on the left has angles which measure 120, 30, and 30. By alternate interior angle theorem, ∠1 is also 30.

73) C In any triangle, an exterior angle is equal to the sum of the remote interior angles.

74) B Use SAS with the last side being the vertical line common to both triangles.

75) D Angles formed by intersecting lines are called vertical angles and are congruent.

76) B In similar polygons, the areas are proportional to the squares of the sides. $6^2:8^2$; 36:64

77) A The sides are in the same ratio.

78) B The altitude from the right angle to the hypotenuse of any right triangle is the geometric mean of the two segments which are formed. Multiply 7 x 14 and take the square root.

79) A In a 30-60- 90 right triangle, the leg opposite the 30° angle is half the length of the hypotenuse.

80) D Minor arc AC measures 50°, the same as the central angle. To determine the measure of the major arc, subtract from 360.

81) C An inscribed angle is equal to one half the measure of the intercepted arc.

82) C The points marked C and D are the intersection of the circles with centers A and B.

83) C Using a compass, point K is found to be equidistant from A and B.

84) D A postulate is an accepted property of real numbers or geometric figures which cannot be proven, A, B. and C are theorems which can be proven.

85) B The point, line, and plane are the three undefined concepts on which plane geometry is based.

86) B To obtain the final side, add CD to both BC and ED.

87) D The isosceles triangle theorem states that the base angles are congruent, and the reflexive property states that every segment is congruent to itself.

88) C Using the distance formula
$$\sqrt{[3-(-3)]^2 + (7-4)^2}$$
$$= \sqrt{36+9}$$
$$= 3\sqrt{5}$$

89) D Using the midpoint formula

x = (2 + 7)/2   y = (5 + -4)/2

90) B

91) B

92) C

93) B

94) D

95) B

96) A

97) C

98) C

99) A

100) D

101) Let N = .636363.... Then multiplying both sides of the equation by 100 or 10^2 (because there are 2 repeated numbers), we get 100N = 63.636363... Then subtracting the two equations gives 99N = 63 or N = $\frac{63}{99} = \frac{7}{11}$.
Answer is B

102) I is not closed because $\frac{4}{.5} = 8$ and 8 is not in the set.

III is not closed because $\frac{1}{0}$ is undefined.

II is closed because $\frac{-1}{1} = -1, \frac{1}{-1} = -1, \frac{1}{1} = 1, \frac{-1}{-1} = 1$ and all the answers are in the set. **Answer is B**

103) **Answer is D** because a + (-a) = 0 is a statement of the Additive Inverse Property of Algebra.

104) To find the inverse, $f^{-1}(x)$, of the given function, reverse the variables in the given equation, y = 3x – 2, to get x = 3y – 2. Then solve for y as follows: x+2 = 3y, and y = $\frac{x+2}{3}$. **Answer is D.**

105) Before the tax, the total comes to $365.94. Then .065(365.94) = 23.79. With the tax added on, the total bill is 365.94 + 23.79 = $389.73. (Quicker way: 1.065(365.94) = 389.73.) **Answer is A**

106) Recall: 30 days in April and 31 in March. 8 days in March + 30 days in April + 22 days in May brings him to a total of 60 days on May 22. **Answer is B.**

107) A composite number is a number which is not prime. The prime number sequence begins 2,3,5,7,11,13,17,.... To determine which of the expressions is <u>always</u> composite, experiment with different values of x and y, such as x=3 and y=2, or x=5 and y=2. It turns out that 5xy will always be an even number, and therefore, composite, if y=2. **Answer is C.**

108) Using FOIL to do the expansion, we get $(x + y^2)^2 = (x + y^2)(x + y^2) = x^2 + 2xy^2 + y^4$. **Answer is D.**

109) In scientific notation, the decimal point belongs to the right of the 4, the first significant digit. To get from 4.56×10^{-5} back to 0.0000456, we would move the decimal point 5 places to the left. **Answer is D.**

110) Area of triangle AOB is .5(5)(5) = 12.5 square meters. Since $\frac{90}{360} = .25$, the area of sector AOB (pie-shaped piece) is approximately $.25(\pi)5^2 = 19.63$. Subtracting the triangle area from the sector area to get the area of segment AB, we get approximately 19.63-12.5 = 7.13 square meters. **Answer is B.**

111) The formula for the volume of a cone is $V = \frac{1}{3}Bh$, (font)where B is the area of the circular base and h is the height. If the area of the base is tripled, the volume becomes $V = \frac{1}{3}(3B)h = Bh$ (font), or three times the original area. **Answer is C.**

112) Divide the figure into 2 rectangles and one quarter circle. The tall rectangle on the left will have dimensions 10 by 4 and area 40. The rectangle in the center will have dimensions 7 by 10 and area 70. The quarter circle will have area $.25(\pi)7^2 = 38.48$. The total area is therefore approximately 148.48. **Answer is B.**

113) Since an ordinary cookie would not weigh as much as 1 kilogram, or as little as 1 gram or 15 milligrams, the only reasonable answer is 15 grams. **Answer is C.**

114) Arrange the data in ascending order: 12,13,14,16,17,19. The median is the middle value in a list with an odd number of entries. When there are an even number of entries, the median is the mean of the two center entries. Here the average of 14 and 16 is 15. **Answer is C.**

115) In this set of data, the median (see #14) would be the most representative measure of central tendency, since the median is independent of extreme values. Because of the 10% outlier, the mean (average) would be disproportionately skewed. In this data set, it is true that the median and the mode (number which occurs most often) are the same, but the median remains the best choice because of its special properties. **Answer is B.**

116) In kindergarten, first grade, and third grade, there are more boys than girls. The number of extra girls in grade two is more than made up for by the extra boys in all the other grades put together. **Answer is B.**

117) The values of 5 and –5 must be omitted from the domain of all real numbers because if x took on either of those values, the denominator of the fraction would have a value of 0, and therefore the fraction would be undefined. **Answer is B.**

118) By observation, we see that the graph has a y-intercept of 2 and a slope of 2/1 = 2. Therefore its equation is y = mx + b = 2x + 2. Rearranging the terms gives 2x − y = −2. **Answer is B**.

119) Using the Distributive Property and other properties of equality to isolate v_0 gives d = atv_t − atv_0, atv_0 = atv_t − d, $v_0 = \dfrac{atv_t - d}{at}$. **Answer is D.**

120) Removing the common factor of 6 and then factoring the sum of two cubes gives 6 + 48m³ = 6(1 + 8m³) = 6(1 + 2m)(1² − 2m + (2m)²). **Answer is A**.

121) B is not the graph of a function. D is the graph of a parabola where the coefficient of x² is negative. A appears to be the graph of y = x². To find the x-intercepts of y = x² + 3x, set y = 0 and solve for x: 0 = x² + 3x = x(x + 3) to get x = 0 or x = −3. Therefore, the graph of the function intersects the x-axis at x=0 and x=−3. **Answer is C.**

122) Set up the direct variation: $\dfrac{V}{r^2} = \dfrac{V}{r^2}$. Substituting gives $\dfrac{80}{16} = \dfrac{V}{9}$. Solving for V gives 45 liters per minute. **Answer is A.**

123) Multiplying equation 1 by 2, and equation 2 by −3, and then adding together the two resulting equations gives −11y + 22z = 0. Solving for y gives y = 2z. In the meantime, multiplying equation 3 by −2 and adding it to equation 2 gives −y − 12z = −14. Then substituting 2z for y, yields the result z = 1. Subsequently, one can easily find that y = 2, and x = −1. **Answer is A.**

124) Using the definition of absolute value, two equations are possible: 18 = 4 + 2x or 18 = 4 − 2x. Solving for x gives x = 7 or x = −7. **Answer is C**.

125) Rewriting the inequality gives x² − 5x + 6 > 0. Factoring gives (x − 2)(x − 3) > 0. The two cut-off points on the number line are now at x = 2 and x = 3. Choosing a random number in each of the three parts of the number line, we test them to see if they produce a true statement. If x = 0 or x = 4, (x-2)(x-3)>0 is true. If x = 2.5, (x-2)(x-3)>0 is false. Therefore the solution set is all numbers smaller than 2 or greater than 3. **Answer is D.**

126) Possible rational roots of the equation 0 = x³ + x² − 14x −24 are all the positive and negative factors of 24. By substituting into the equation, we find that −2 is a root, and therefore that x+2 is a factor. By performing the long division (x³ + x² − 14x − 24)/(x+2), we can find that another factor of the original equation is x² − x − 12 or (x-4)(x+3). Therefore the zeros of the original function are −2, -3, and 4. **Answer is D.**

127) Getting the bases the same gives us $3^{\frac{1}{2}}3^{\frac{2}{3}}$. Adding exponents gives $3^{\frac{7}{6}}$. Then some additional manipulation of exponents produces
$3^{\frac{7}{6}} = 3^{\frac{14}{12}} = (3^2)^{\frac{7}{12}} = 9^{\frac{7}{12}}$. **Answer is B.**

128) Simplifying radicals gives $\sqrt{27} + \sqrt{75} = 3\sqrt{3} + 5\sqrt{3} = 8\sqrt{3}$. **Answer is A.**

129) Multiplying numerator and denominator by the conjugate gives
$\dfrac{10}{1+3i} \times \dfrac{1-3i}{1-3i} = \dfrac{10(1-3i)}{1-9i^2} = \dfrac{10(1-3i)}{1-9(-1)} = \dfrac{10(1-3i)}{10} = 1-3i$. **Answer is D.**

130) To find the 100th term: $t_{100} = -6 + 99(4) = 390$. To find the sum of the first 100 terms: S = $\dfrac{100}{2}(-6+390) = 19200$. **Answer is A.**

131) There are 3 slots to fill. There are 3 choices for the first, 7 for the second, and 6 for the third. Therefore, the total number of choices is 3(7)(6) = 126. **Answer is A.**

132) The set-up for finding the seventh term is $\dfrac{8(7)(6)(5)(4)(3)}{6(5)(4)(3)(2)(1)}(2a)^{8-6}b^6$ which gives $28(4a^2b^6)$ or $112a^2b^6$. **Answer is C.**

133) By definition, parallel lines are coplanar lines without any common points. **Answer is B.**

134) A set of n objects has 2^n subsets. Therefore, here we have 2^4 = 16 subsets. These subsets include four which have only 1 element each, six which have 2 elements each, four which have 3 elements each, plus the original set, and the empty set. **Answer is B.**

135) Formula for finding the measure of each interior angle of a regular polygon with n sides is $\dfrac{(n-2)180}{n}$. For n=10, we get $\dfrac{8(180)}{10} = 144$. **Answer is C.**

136) Draw a right triangle with legs of 6 and 8. Find the hypotenuse using the Pythagorean Theorem. $6^2 + 8^2 = c^2$. Therefore, c = 10 miles. **Answer is B.**

137) The formula relating the measure of angle K and the two arcs it intercepts is $m\angle K = \dfrac{1}{2}(mPS - mAD)$. Substituting the known values, we get $10 = \dfrac{1}{2}(40 - mAD)$. Solving for mAD gives an answer of 20 degrees. **Answer is B.**

138) Given a point on a line, place the compass point there and draw two arcs intersecting the line in two points, one on either side of the given point. Then using any radius larger than half the new segment produced, and with the pointer at each end of the new segment, draw arcs which intersect above the line. Connect this new point with the given point. **Answer is D**.

139) By definition this describes the procedure of an indirect proof. **Answer is C.**

140) Since side AK is common to both triangles, the triangles can be proved congruent by using the Side-Angle-Side Postulate. **Answer is C.**

141) In an ordinary parallelogram, the diagonals are not perpendicular or equal in length. In a rectangle, the diagonals are not necessarily perpendicular. In a rhombus, the diagonals are not equal in length. In a square, the diagonals are both perpendicular and congruent. **Answer is C.**

142) The most one can say with certainty is that the median (segment drawn to the midpoint of the opposite side) and the altitude (segment drawn perpendicular to the opposite side) of a triangle may coincide, but they more often do not. In an isosceles triangle, the median and the altitude to the base are the same segment. **Answer is A.**

143) In the mid-nineteenth century, Reimann and other mathematicians developed elliptic geometry. **Answer is B**.

144) Let's assume the base of the rectangular solid (box) is 3 by 4, and the height is 5. Then the surface area of the top and bottom together is 2(12) = 24. The sum of the areas of the front and back are 2(15) = 30, while the sum of the areas of the sides are 2(20)=40. The total surface area is therefore 94 square feet. **Answer is C**.

145) Find the area of the garden and then subtract the area of the fountain: $30(60) - \pi(5)^2$ or approximately 1721 square meters. **Answer is A.**

146) To find the area of the shaded region, find the area of triangle ABC and then subtract the area of triangle DBE. The area of triangle ABC is .5(6x)(y) = 3xy. The area of triangle DBE is .5(2x)(y) = xy. The difference is 2xy. **Answer is B.**

TEACHER CERTIFICATION STUDY GUIDE

DOMAIN VII.   **SCIENTIFIC INQUIRY AND PROCESS**

COMPETENCY 020   **UNDERSTANDING HOW TO MANAGE LEARNING ACTIVITIES TO ENSURE THE SAFETY OF ALL STUDENTS**

Skill 20.1   **Understands safety regulations and guidelines for science facilities and science instruction**

All science labs should contain the following items of safety equipment. The following are requirements by law.
- Fire blanket which is visible and accessible
- Ground Fault Circuit Interrupters (GFCI) within two feet of water supplies
- Emergency shower capable of providing a continuous flow of water
- Signs designating room exits
- Emergency eye wash station which can be activated by the foot or forearm
- Eye protection for every student and a means of sanitizing equipment
- Emergency exhaust fans providing ventilation to the outside of the building
- Master cut-off switches for gas, electric, and compressed air. Switches must have permanently attached handles. Cut-off switches must be clearly labeled.
- An ABC fire extinguisher
- Storage cabinets for flammable materials

Also recommended, but not required by law:
- Chemical spill control kit
- Fume hood with a motor which is spark proof
- Protective laboratory aprons made of flame retardant material
- Signs which will alert people to potential hazardous conditions
- Containers for broken glassware, flammables, corrosives, and waste.
- Containers should be labeled.

It is the responsibility of teachers to provide a safe environment for their students. Proper supervision greatly reduces the risk of injury and a teacher should never leave a class for any reason without providing alternate supervision. After an accident, two factors are considered; foresight and negligence. **Foresight** is the anticipation that an event may occur under certain circumstances. **Negligence** is the failure to exercise ordinary or reasonable care. Safety procedures should be a part of the science curriculum and a well managed classroom is important to avoid potential lawsuits

The **"Right to Know Law"** covers science teachers who work with potentially hazardous chemicals. Briefly, the law states that employees must be informed of potentially toxic chemicals. An inventory must be made available if requested. The inventory must contain information about the hazards and properties of the chemicals. Training must be provided in the safe handling and interpretation of the Material Safety Data Sheet.

The following chemicals are potential carcinogens and are not allowed in school facilities:

> Acrylonitriel, Arsenic compounds, Asbestos, Bensidine, Benzene, Cadmium compounds, Chloroform, Chromium compounds, Ethylene oxide, Ortho-toluidine, Nickel powder, Mercury.

All laboratory solutions should be prepared as directed in the lab manual. Care should be taken to avoid contamination. All glassware should be rinsed thoroughly with distilled water before using, and cleaned well after use. Safety goggles should be worn while working with glassware in case of an accident. All solutions should be made with distilled water as tap water contains dissolved particles that may affect the results of an experiment. Chemical storage should be located in a secured, dry area. Chemicals should be stored in accordance with reactability. Acids are to be locked in a separate area. Used solutions should be disposed of according to local disposal procedures. Any questions regarding safe disposal or chemical safety may be directed to the local fire department.

Skill 20.2 Procedures and sources of information regarding the appropriate handling, use, disposal, care, and maintenance of chemicals, materials, specimens, and equipment.

Chemicals should not be stored on bench tops or heat sources. They should be stored in groups based on their reactivity with one another and in protective storage cabinets. All containers in the lab must be labeled. Suspect and known carcinogens must be labeled as such and segregated within trays to contain leaks and spills.

Chemical waste should be disposed in properly labeled containers. Waste should be separated based on their reactivity with other chemicals. Biological material should never be stored near food or water used for human consumption. All biological material should be appropriately labeled. All blood and bodily fluids should be put in a well-contained container with a secure lid to prevent leaking. All biological waste should be disposed of in biological hazardous waste bags.

In addition to the safety laws set forth by the government for equipment necessary to the lab, Occupational Safety and Health Administration (OSHA) has helped to make environments safer by instituting signs that are bilingual. These signs use pictures rather than/in addition to words and feature eye-catching colors. Some of the best-known examples are exit, restrooms, and handicap accessible.

Of particular importance to laboratories are diamond safety signs, prohibitive signs, and triangle danger signs. Each sign encloses a descriptive picture.

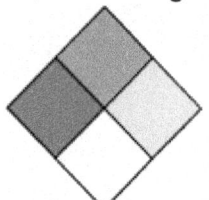 

As a teacher, you should utilize a Material Safety Data Sheet (MSDS) whenever you are preparing an experiment. It is designed to provide people with the proper procedures for handling or working with a particular substance. MSDS's include information such as physical data (melting point, boiling point, etc.), toxicity, health effects, first aid, reactivity, storage, disposal, protective gear, and spill/leak procedures. These are particularly important if a spill or other accident occurs. You should review the MSDS, commonly available online, and understand the listing procedures. Material safety data sheets are available directly from the company of acquisition or the internet. The manuals for equipment used in the lab should be read and understood before initial use.

Skill 20.3 Knows procedures for the safe handling and ethical care and treatment of organisms and specimens.

Dissections - Animals that are not obtained from recognized sources should not be used. Decaying animals or those of unknown origin may harbor pathogens and/or parasites. Specimens should be rinsed before handling. Latex gloves are desirable. If gloves are not available, students with sores or scratches should be excused from the activity. Formaldehyde is a carcinogen and should be avoided or disposed of according to district regulations. Students objecting to dissections for moral reasons should be given an alternative assignment.

Live specimens - No dissections may be performed on living mammalian vertebrates or birds. Lower order life and invertebrates may be used. Biological experiments may be done with all animals except mammalian vertebrates or birds. No physiological harm may result to the animal. All animals housed and cared for in the school must be handled in a safe and humane manner. Animals are not to remain on school premises during extended vacations unless adequate care is provided. Many state laws stipulate that any instructor who intentionally refuses to comply with the laws may be suspended or dismissed.

Microbiology - Pathogenic organisms must never be used for experimentation. Students should adhere to the following rules at all times when working with microorganisms to avoid accidental contamination:

1. Treat all microorganisms as if they were pathogenic.
2. Maintain sterile conditions at all times

If you are taking a national level exam you should check the Department of Education for your state for safety procedures. You will want to know what your state expects of you not only for the test but also for performance in the classroom and for the welfare of your students.

COMPETENCY 021 UNDERSTANDING THE CORRECT USE OF TOOLS, MATERIALS, EQUIPMENT, AND TECHNOLOGIES

Skill 21.1 Select and safely use appropriate tools, technologies, materials, and equipment needed for instructional activities.

Bunsen burners - Hot plates should be used whenever possible to avoid the risk of burns or fires. If Bunsen burners are used, the following precautions should be followed:

1. Know the location of fire extinguishers and safety blankets and train students in their use. Long hair and long sleeves should be secured and out of the way.

2. Turn on the gas and make a spark with the striker. The preferred method to light burners is to use strikers rather than matches.

3. Adjust the air valve at the bottom of the Bunsen burner until the flame shows an inner cone.

4. Adjust the flow of gas to the desired flame height by using the adjustment valve.

5. Do not touch the barrel of the burner (it is hot).

Graduated Cylinder - These are used for precise measurements. They should always be placed on a flat surface. The surface of the liquid will form a meniscus (lens-shaped curve). The measurement is read at the <u>bottom</u> of this curve.

Balance - Electronic balances are easier to use, but more expensive. An electronic balance should always be tarred (returned to zero) before measuring and used on a flat surface. Substances should always be placed on a piece of paper to avoid spills and/or damage to the instrument. Triple beam balances must be used on a level surface. There are screws located at the bottom of the balance to make any adjustments. Start with the largest counterweight first and proceed toward the last notch that does not tip the balance. Do the same with the next largest, etc until the pointer remains at zero. The total mass is the total of all the readings on the beams. Again, use paper under the substance to protect the equipment.

Buret – A buret is used to dispense precisely measured volumes of liquid. A stopcock is used to control the volume of liquid being dispensed at a time.

Light microscopes are commonly used in laboratory experiments. Several procedures should be followed to properly care for this equipment:

- Clean all lenses with lens paper only.
- Carry microscopes with two hands; one on the arm and one on the base.
- Always begin focusing on low power, then switch to high power.
- Store microscopes with the low power objective down.
- Always use a coverslip when viewing wet mount slides.
- Bring the objective down to its lowest position then focus by moving up to avoid breaking the slide or scratching the lens.

Wet mount slides should be made by placing a drop of water on the specimen and then putting a glass coverslip on top of the drop of water. Dropping the coverslip at a forty-five degree angle will help in avoiding air bubbles. Total magnification is determined by multiplying the ocular (usually 10X) and the objective (usually 10X on low, 40X on high).

Chromatography uses the principles of capillary action to separate substances such as plant pigments. Molecules of a larger size will move slower up the paper, whereas smaller molecules will move more quickly producing lines of pigments.

Spectrophotometry uses percent light absorbance to measure a color change, thus giving qualitative data a quantitative value.

Centrifugation involves spinning substances at a high speed. The more dense part of a solution will settle to the bottom of the test tube, while the lighter material will stay on top. Centrifugation is used to separate blood into blood cells and plasma, with the heavier blood cells settling to the bottom.

Electrophoresis uses electrical charges of molecules to separate them according to their size. The molecules, such as DNA or proteins, are pulled through a gel towards either the positive end of the gel box (if the material has a negative charge) or the negative end of the gel box (if the material has a positive charge).

Computer technology has greatly improved the collection and interpretation of scientific data. Molecular findings have been enhanced through the use of computer images. Technology has revolutionized access to data via the internet and shared databases. The manipulation of data is enhanced by sophisticated software capabilities. Computer engineering advances have produced such products as MRIs and CT scans in medicine. Laser technology has numerous applications with refining precision.

Satellites have improved our ability to communicate and transmit radio and television signals. Navigational abilities have been greatly improved through the use of satellite signals. Sonar uses sound waves to locate objects and is especially useful underwater. The sound waves bounce off the object and are used to assist in location. Seismographs record vibrations in the earth and allow us to measure earthquake activity.

Skill 21.2   **The concepts of precision, accuracy, and error with regard to reading and recording numerical data from a scientific instrument.**

Accuracy and precision

Accuracy is the degree of conformity of a measured, calculated quantity to its actual (true) value. Precision also called reproducibility or repeatability and is the degree to which further measurements or calculations will show the same or similar results.

Accuracy is the degree of veracity while precision is the degree of reproducibility. The best analogy to explain accuracy and precision is the target comparison.

Repeated measurements are compared to arrows that are fired at a target. Accuracy describes the closeness of arrows to the bull's eye at the target center. Arrows that strike closer to the bull's eye are considered more accurate.

Systematic and random error

All experimental uncertainty is due to either random errors or systematic errors.

Random errors are statistical fluctuations in the measured data due to the precision limitations of the measurement device. Random errors usually result from the experimenter's inability to take the same measurement in exactly the same way to get exactly the same number.

Systematic errors, by contrast, are reproducible inaccuracies that are consistently in the same direction. Systematic errors are often due to a problem, which persists throughout the entire experiment.

Systematic and random errors refer to problems associated with making measurements.

Mistakes made in the calculations or in reading the instrument are not considered in error analysis.

Skill 21.3 **How to gather, organize, display, and communicate data in a variety of ways (e.g., charts, tables, graphs, diagrams, written reports, oral presentations.).**

Graphing is an important skill to visually display collected data for analysis. The two types of graphs most commonly used are the **line graph** and the **bar graph** (histogram). Line graphs are are used to illustrate the relationship between two variables. The horizontal axis is the X axis and represents the dependent variable. Dependent variables are those that would be present independently of the experiment. A common example of a dependent variable is time. Time proceeds regardless of anything else occurring. The vertical axis is the Y axis and represents the independent variable. Independent variables are manipulated by the experiment, such as the amount of light, or the height of a plant. Graphs should be calibrated at equal intervals. If one space represents one day, the next space may not represent ten days. A "best fit" line is drawn to join the points and may not include all the points in the data. Axes must always be labeled, for the graph to be meaningful. A good title will describe both the dependent and the independent variable. Bar graphs are set up similarly in regards to axes, but points are not plotted. Instead, the dependent variable is set up as a bar where the X axis intersects with the Y axis. Each bar is a separate item of data and is not joined by a continuous line.

Classification is grouping items according to their similarities. It is important for students to realize relationships and similarity as well as differences to reach a reasonable conclusion in a lab experience.

Normally, knowledge is integrated in the form of a **lab report**. A report has many sections. It should include a specific **title** and tell exactly what is being studied. The **abstract** is a summary of the report written at the beginning of the paper. The **purpose** should always be defined and will state the problem. The purpose should include the **hypothesis** (educated guess) of what is expected from the outcome of the experiment. The entire experiment should relate to this problem. It is important to describe exactly what was done to prove or disprove a hypothesis. A **control** is necessary to prove that the results occurred from the changed conditions and would not have happened normally. Only one variable should be manipulated at a time. **Observations** and **results** of the experiment should be recorded including all results from data. Drawings, graphs and illustrations should be included to support information. Observations are objective, whereas analysis and interpretation is subjective. A **conclusion** should explain why the results of the experiment either proved or disproved the hypothesis.

A scientific theory is an explanation of a set of related observations based on a proven hypothesis. A scientific law usually lasts longer than a scientific theory and has more experimental data to support it.

Skill 21.4 **The international system of measurement (i.e., metric system) and performs unit conversions within measurement systems.**

Science may be defined as a body of knowledge that is systematically derived from study, observations and experimentation. Its goal is to identify and establish principles and theories which may be applied to solve problems. Pseudoscience, on the other hand, is a belief that is not warranted. There is no scientific methodology or application. Some of the more classic examples of pseudoscience include witchcraft, alien encounters, or any topics that are explained by hearsay.

Science uses the metric system as it is accepted worldwide and allows easier comparison among experiments done by scientists around the world. Learn the following basic units and prefixes:

> **meter** - measure of length
> **liter** - measure of volume
> **gram** - measure of mass

deca-(meter, liter, gram)= 10X the base unit   **deci** = 1/10 the base unit
hecto-(meter, liter, gram)= 100X the base unit   **centi** = 1/100 the base unit
kilo-(meter, liter, gram) = 1000X the base unit   **milli** = 1/1000 the base unit

COMPETENCY 022 UNDERSTANDING *THE* PROCESS OF SCIENTIFIC INQUIRY AND THE HISTORY AND NATURE OF SCIENCE

Skill 22.1 The characteristics of various types of scientific investigations

Most research in the scientific field is conducted using the scientific method to discover the answer to a scientific problem. The scientific method is the process of thinking through possible solutions to a problem and testing each possibility to find the best solution. The scientific method generally involves the following steps: forming a hypothesis, choosing a method and design, conducting experimentation (collecting data), analyzing data, drawing a conclusion, and reporting the findings. Depending on the hypothesis and data to be collected and analyzed, different types of scientific investigation may be used.

Descriptive studies are often the first form of investigation used in new areas of scientific inquiry. The most important element in descriptive reporting is a specific, clear, and measurable definition of the disease, condition, or factor in question. Descriptive studies always address the five W's: who, what, when, where, and why. They also add an additional "so what?" Descriptive studies include case reports, case-series reports, cross-sectional students, surveillance studies with individuals, and correlational studies with populations. Descriptive studies are used primarily for trend analysis, health-care planning, and hypothesis generation.

A **controlled experiment** is a form of scientific investigation in which one variable, the independent or control variable, is manipulated to reveal the effect on another variable, the dependent (experimental) variable, while all other variables in the system remain fixed. The control group is virtually identical to the dependent variable except for the one variable whose effect is being tested. Testing the effects of bleach water on a growing plant, the plant receiving bleach water would be the dependent group, while the plant receiving plain water would be the control group. It is good practice to have several replicate samples for the experiment being performed, which allows for results to be averaged or obvious discrepancies to be discarded.

Comparative data analysis is a statistical form of investigation that allows the researcher to gain new or unexpected insight into data based primarily on graphic representation. Comparative data analysis, whether within the research of an individual project or a meta-analysis, allows the researcher to maximize the understanding of the particular data set, uncover underlying structural similarities between research, extract important variables, test underlying assumptions, and detect outliers and anomalies. Most comparative data analysis techniques are graphical in nature with a few quantitative techniques. The use of graphics to compare data allows the researcher to explore the data open-mindedly.

Skill 22.2 How to design, conduct, and communicate the results of a variety of scientific investigations

The scientific method is the basic process behind science. It involves several steps beginning with hypothesis formulation and working through to the conclusion.

Posing a question - Although many discoveries happen by chance, the standard thought process of a scientist begins with forming a question. The more limited the question, the easier it is to set up an experiment to answer it.

Form a hypothesis - Once the question is formulated the experimenter can make an educated guess about the answer to the problem or question. This 'best guess' is your hypothesis.

Conducting a test - To make a test fair, data from an experiment must have a **variable** or any condition that can be changed such as temperature or mass. A good test will try to manipulate as few variables as possible so as to see which variable is responsible for the result. This requires a **control**. A control is an extra setup in which all the conditions are the same except for the variable being tested.

Observe and record the data - Reporting of the data should state specifics of how the measurements were calculated. A graduated cylinder needs to be read with proper procedures. As experimenters, technique must be part of the instructional process so as to give validity to the data.

Drawing a conclusion - After recording data, compare the data with that of other groups. A conclusion is the judgment derived from the data results.

Graphing data - Graphing utilizes numbers to demonstrate patterns. The patterns offer a visual representation, making it easier to draw conclusions.

Skill 22.3 The historical development of science and the contributions that diverse cultures and individuals of both genders have made to scientific knowledge

The history of biology traces man's understanding of the living world from the earliest recorded history to modern times. Though the concept of biology as a field of science arose only in the 19th century, the origin of biological sciences could be traced back to ancient Greeks (Galen and Aristotle).

During the Renaissance and Age of Discovery, renewed interest in the rapidly increasing number of known organisms generated lot of interest in biology.

Andreas Vesalius (1514-1564) was a Belgian anatomist and physician whose dissections of the human body and descriptions of his findings helped to correct the misconceptions of science. The books Vesalius wrote on anatomy were the most accurate and comprehensive anatomical texts to date.

Anton van Leeuwenhoek is known as the father of microscopy. In the 1650s, Leeuwenhoek began making tiny lenses that gave magnifications up to 300x. He was the first to see and describe bacteria, yeast plants, and the microscopic life found in water. Over the years, light microscopes have advanced to produce greater clarity and magnification. The scanning electron microscope (SEM) was developed in the 1950s. Instead of light, a beam of electrons passes through the specimen. Scanning electron microscopes have a resolution about one thousand times greater than light microscopes. The disadvantage of the SEM is that the chemical and physical methods used to prepare the sample resulted in the death of the specimen.

Robert Hooke (1635-1703) was a renowned inventor, a natural philosopher, astronomer, experimenter and a cell biologist. He deserves more recognition than he received, but he is remembered mainly for Hooke's law an equation describing elasticity that is still used today. He was a scientist that was then called a "virtuoso"- able to contribute findings of major importance in any field of science. Hooke published *Micrographia* in 1665. Hooke devised the compound microscope and illumination system, one of the best such microscopes of his time, and used it in his demonstrations at the Royal Society's meetings. With it he observed organisms as diverse as insects, sponges, bryozoans, foraminifera, and bird feathers. Micrographia is an accurate and detailed record of his observations, illustrated with magnificent drawings.

Carl Von Linnaeus (1707-1778), a Swedish botanist, physician and zoologist is well known for his contributions in ecology and taxonomy. Linnaeus is famous for his binomial system of nomenclature in which each living organism has two names, a genus and a species name. He is considered the father of modern ecology and taxonomy.

In the late 1800s, Pasteur discovered the role of microorganisms in the cause of disease, pasteurization, and the rabies vaccine. Koch took this observations one step further by formulating that specific diseases were caused by specific pathogens. **Koch's postulates** are still used as guidelines in the field of microbiology. They state that the same pathogen must be found in every diseased person; the pathogen must be isolated and grown in culture, the disease is induced in experimental animals from the culture, and the same pathogen must be isolated from the experimental animal.

Mattias Schleiden, a German botanist is famous for his cell theory. He observed plant cells microscopically and concluded that the cell is the common structural unit of plants. He proposed the cell theory along with Schwann, a zoologist, who observed cells in animals.

In the 18th century, many fields of science like botany, zoology and geology began to evolve as scientific disciplines in the modern sense.

In the 20th century, the rediscovery of Mendel's work led to the rapid development of genetics by Thomas Hunt Morgan and his students.

DNA structure was another key event in biological study. In the 1950s, James Watson and Francis Crick discovered that the DNA molecule is a double helix.. This structure made it possible to explain DNA's ability to replicate and to control the synthesis of proteins.

Francois Jacob and Jacques Monod contributed greatly to the field of lysogeny and bacterial reproduction by conjugation and both won the Nobel Prize for their contributions.

Following the cracking of the genetic code biology has largely split between organismal biology consisting of ecology, ethology, systematics, paleontology, and evolutionary biology, developmental biology, and other disciplines that deal with whole organisms or group of organisms and the disciplines related to molecular biology - including cell biology, biophysics, biochemistry, neuroscience, immunology, and many other overlapping subjects.

The use of animals in biological research has expedited many scientific discoveries. Animal research has allowed scientists to learn more about animal biological systems, including the circulatory and reproductive systems. One significant use of animals is for the testing of drugs, vaccines, and other products (such as perfumes and shampoos) before use or consumption by humans. Along with the pros of animal research, the cons are also very significant. The debate about the ethical treatment of animals has been ongoing since the introduction of animals in research. Many people believe the use of animals in research is cruel and unnecessary. Animal use is federally and locally regulated. The purpose of the Institutional Animal Care and Use Committee (IACUC) is to oversee and evaluate all aspects of an institution's animal care and use program.

Skill 22.4 Understands the roles that logical reasoning, verifiable evidence, prediction, and peer review play in the process of generating and evaluating scientific knowledge

Observations, however general they may seem, lead scientists to create a viable question and an educated guess (hypothesis) about what to expect. While scientists often have laboratories set up to study a specific thing, it is likely that along the way they will find an unexpected result. It is always important to be open-minded and to look at all of the information. An open-minded approach to science provides room for more questioning, and, hence, more learning.

A central concept in science is that all evidence is empirical. This means that all evidence must be is observed by the five senses. He study phenomenon must be both observable and measurable, with reproducible results. The question stage of scientific inquiry involves repetition. By repeating the experiment you can discover whether or not you have reproducibility. If results are reproducible, the hypothesis is valid. If the results are not reproducible, one has more questions to ask.

With confidence in the proposed explanations, the students need to identify what would be required to reject the proposed explanations. Based upon their experience, they should develop new questions to promote further inquiry.

Science is a process of checks and balances. It is expected that scientific findings will be challenged, and in many cases re-tested. Often one experiment will be the beginning point for another. While bias does exist, the use of controlled experiments, logical reasoning, and awareness on the part of the scientist, can go far in ensuring a sound experiment. Even if the science is well done, it may still be questioned. It is through this continual search that hypotheses are made into theories, and sometimes become laws. It is also through this search that new information is discovered.

Skill 22.5 Understands principles of scientific ethics

To understand scientific ethics, we need to have a clear understanding of ethics. Ethics is defined as a system of public, general rules for guiding human conduct (Gert, 1988). The rules are general in that they are supposed to all people at all times and they are public in that they are not secret codes or practices.

Scientists are expected to show good conduct in their scientific pursuits. Conduct here refers to all aspects of scientific activity including experimentation, testing, education, data evaluation, data analysis, data storing, peer review, government funding, the staff, etc.

The following are some of the guiding principles of scientific ethics:

2. Scientific Honesty: not to fraud, fabricate or misinterpret data for personal gain
3. Caution: to avoid errors and sloppiness in all scientific experimentation
4. Credit: give credit where credit is due and not to copy
5. Responsibility: only to report reliable information to public and not to mislead in the name of science
6. Freedom: freedom to criticize old ideas, question new research and freedom to research

Many more principles could be added to this list. Though these principles seem straightforward and clear, it is very difficult to put them into practice since they could be interpreted in more ways than one. Nevertheless, it is not an excuse for scientists to overlook these guiding principles.

To discuss scientific ethics, we can look at natural phenomena like rain. Rain in the normal sense is extremely useful to us and it is absolutely important that there is water cycle. When rain gets polluted with acid, it becomes acid rain. Here lies the ethical issue of releasing all these pollutants into the atmosphere. Should the scientists communicate the whole truth about acid rain or withhold some information because it may alarm the public? There are many issues like this. Whatever may be the case, scientists are expected to be honest and forthright with the public.

Skill 22.6 Develops, analyzes, and evaluates different explanations for a given scientific result

Armed with knowledge of the subject matter, students can effectively conduct investigations. They need to learn to think critically and logically to connect evidence with explanations. This includes deciding what evidence should be used and accounting for unusual data. Based upon data collected during experimentation, basic statistical analysis, and measures of probability can be used to make predictions and develop interpretations.

Students, with appropriate direction from you (the teacher), should be able to review the data, summarize, and form a logical argument about the cause-and-effect relationships. It is important to differentiate between causes and effects and determine when causality is uncertain.

When developing proposed explanations, the students should be able to express their level of confidence in the proposed explanations and point out possible sources of uncertainty and error. When formulating explanations, it is important to distinguish between error and unanticipated results. Possible sources of error would include assumptions of models and measuring techniques or devices.

Skill 22.7   Demonstrates an understanding of potential sources of error in inquiry based investigation

Unavoidable experimental error is the random error inherent in scientific experiments regardless of the methods used. One source of unavoidable error is measurement and the use of measurement devices. Using measurement devices is an imprecise process because it is often impossible to accurately read measurements. For example, when using a ruler to measure the length of an object, if the length falls between markings on the ruler, we must estimate the true value. Another source of unavoidable error is the randomness of population sampling and the behavior of any random variable. For example, when sampling a population we cannot guarantee that our sample is completely representative of the larger population. In addition, because we cannot constantly monitor the behavior of a random variable, any observations necessarily contain some level of unavoidable error.

Statistical variability is the deviation of an individual in a population from the mean of the population. Variability is inherent in biology because living things are innately unique. For example, the individual weights of humans vary greatly from the mean weight of the population. Thus, when conducting experiments involving the study of living things, we must control for innate variability. Control groups are identical to the experimental group in every way with the exception of the variable being studied. Comparing the experimental group to the control group allows us to determine the effects of the manipulated variable in relation to statistical variability.

Skill 22.8   Demonstrates an understanding of how to communicate and defend the results of an inquiry-based investigation

It is the responsibility of the scientists to share the knowledge they obtain through their research. After the conclusion is drawn, the final step is communication. In this age, much emphasis is put on the way and the method of communication. The conclusions must be communicated by clearly describing the information, using accurate data, visual presentation and other appropriate media such as a power point presentation. Examples of visual presentations are graphs (bar/line/pie), tables/charts, diagrams, and artwork. Modern technology must be used whenever necessary. The method of communication must be suitable to the audience.

Written communication is as important as oral communication. This is essential for submitting research papers to scientific journals, newspapers, other magazines etc.

COMPETENCY 023 **UNDERSTANDING HOW SCIENCE IMPACTS THE DAILY LIVES OF STUDENTS AND INTERACTS WITH AND INFLUENCES PERSONAL AND SOCIETAL DECISIONS.**

Skill 23.1 **Understands that decisions about the use of science are based on factors such as ethical standards, economics, and personal and societal needs.**

Scientific and technological breakthroughs greatly influence other fields of study and the job market. All academic disciplines utilize computer and information technology to simplify research and information sharing. In addition, advances in science and technology influence the types of available jobs and the desired work skills. For example, machines and computers continue to replace unskilled laborers and computer and technological literacy is now a requirement for many jobs and careers. Finally, science and technology continue to change the very nature of careers. Because of science and technology's great influence on all areas of the economy, and the continuing scientific and technological breakthroughs, careers are far less stable than in past eras. Workers can thus expect to change jobs and companies much more often than in the past.

Local, state, national, and global governments and organizations must increasingly consider policy issues related to science and technology. For example, local and state governments must analyze the impact of proposed development and growth on the environment. Governments and communities must balance the demands of an expanding human population with the local ecology to ensure sustainable growth.

In addition, advances in science and technology create challenges and ethical dilemmas that national governments and global organizations must attempt to solve. Genetic research and manipulation, antibiotic resistance, stem cell research, and cloning are but a few of the issues facing national governments and global organizations.

In all cases, policy makers must analyze all sides of an issue and attempt to find a solution that protects society while limiting scientific inquiry as little as possible. For example, policy makers must weigh the potential benefits of stem cell research, genetic engineering, and cloning (e.g. medical treatments) against the ethical and scientific concerns surrounding these practices. Also, governments must tackle problems like antibiotic resistance, which can result from the indiscriminate use of medical technology (i.e. antibiotics), to prevent medical treatments from becoming obsolete.

Skill 23.2   Scientific principles and the theory of probability to analyze the advantages of, disadvantages of, or alternatives to a give decision or course of action.

Armed with knowledge of the subject matter, students can effectively conduct investigations. They need to learn to think critically and logically to connect evidence with explanations. This includes deciding what evidence should be used and accounting for unusual data. Based upon data collected during experimentation, basic statistical analysis, and measures of probability can be used to make predictions and develop interpretations.

Students should be able to review the data, summarize, and form a logical argument about the cause-and-effect relationships. It is important to differentiate between causes and effects and determine when causality is uncertain.

When developing proposed explanations, the students should be able to express their level of confidence in the proposed explanations and point out possible sources of uncertainty and error. When formulating explanations, it is important to distinguish between error and unanticipated results. Possible sources of error would include assumptions of models and measuring techniques or devices.

Skill 23.3   The scientific principles and the processes to analyze factors that influence personal choices concerning fitness and health, including physiological and psychological effect and risks associated with the use of substances and substance abuse.

While genetics plays an important role in health, human behaviors can greatly affect short- and long-term health both positively and negatively. Behaviors that negatively affect health include smoking, excessive alcohol consumption, substance abuse, and poor eating habits. Behaviors that positively affect health include good nutrition and regular exercise.

Smoking negatively affects health in many ways. First, smoking decreases lung capacity, causes persistent coughing, and limits the ability to engage in strenuous physical activity. In addition, the long-term affects are even more damaging. Long-term smoking can cause lung cancer, heart disease, and emphysema (a lung disease).

Alcohol is the most abused legal drug. Excessive alcohol consumption has both short- and long-term negative effects. Drunkenness can lead to reckless behavior and distorted judgment that can cause injury or death. In addition, extreme alcohol abuse can cause alcohol poisoning that can result in immediate death. Long-term alcohol abuse is also extremely hazardous. The potential effects of long-term alcohol abuse include liver cirrhosis, heart problems, high blood pressure, stomach ulcers, and cancer.

The abuse of illegal substances can also negatively affect health. Commonly abused drugs include cocaine, heroin, opiates, methamphetamines, and marijuana. Drug abuse can cause immediate death or injury and, if used for a long time, can cause many physical and psychological health problems.

A healthy diet and regular exercise are the cornerstones of a healthy lifestyle. A diet rich in whole grains, fruits, vegetables, polyunsaturated fats, and lean protein and low in saturated fat and sugar, can positively affect overall health. Such diets can reduce cholesterol levels, lower blood pressure, and help manage body weight. Conversely, diets high in saturated fat and sugar can contribute to weight gain, heart disease, strokes, and cancer.

Finally, regular exercise has both short- and long-term health benefits. Exercise increases physical fitness, improving energy levels, overall body function, and mental well-being. Long-term, exercise helps protect against chronic diseases, maintains healthy bones and muscles, helps maintain a healthy body weight, and strengthens the body's immune system.

Skill 23.4 Understand concepts, characteristics, and issues related to changes in populations and human population growth

Human population increased slowly until 1650. Since 1650, the human population has grown almost exponentially, reaching its current population of over 6 billion. Factors that have led to this increased growth rate include improved nutrition, sanitation, and health care. In addition, advances in technology, agriculture, and scientific knowledge have made the use of resources more efficient and increased their availability.

While the Earth's ultimate carrying capacity for humans is uncertain, some factors that may limit growth are the availability of food, water, space, and fossil fuels. There is a finite amount of land on Earth available for food production. In addition, providing clean, potable water for a growing human population is a real concern. Finally, fossil fuels, important energy sources for human technology, are scarce. The inevitable shortage of energy in the Earth's future will require the development of alternative energy sources to maintain or increase human population growth.

Skill 23.5   Understand the types and uses of natural resources and the effects of human consumption of the renewal and depletion of resources

Humans have a tremendous impact on the world's natural resources. The world's natural water supplies are affected by human use. Waterways are major sources for recreation and freight transportation. Oil and wastes from boats and cargo ships pollute the aquatic environment. The aquatic plant and animal life is affected by this contamination. To obtain drinking water, contaminants such as parasites, pollutants and bacteria are removed from raw water through a purification process involving various screening, conditioning and chlorination steps. Most uses of water resources, such as drinking and crop irrigation, require fresh water. Only 2.5% of water on Earth is fresh water, and more than two thirds of this fresh water is frozen in glaciers and polar ice caps. Consequently, in many parts of the world, water use greatly exceeds supply. This problem is expected to increase in the future.

Plant resources also make up a large part of the world's natural resources. Plant resources are renewable and can be re-grown and restocked. Plant resources can be used by humans to make clothing, buildings and medicines, and can also be directly consumed. Forestry is the study and management of growing forests. This industry provides the wood that is essential for use as construction timber or paper. Cotton is a common plant found on farms of the Southern United States. Cotton is used to produce fabric for clothing, sheets, furniture, etc. Another example of a plant resource that is not directly consumed is straw, which is harvested for use in plant growth and farm animal care. The list of plants grown to provide food for the people of the world is extensive. Major crops include corn, potatoes, wheat, sugar, barley, peas, beans, beets, flax, lentils, sunflowers, soybeans, canola, and rice. These crops may have alternate uses as well. For example, corn is used to manufacture cornstarch, ethanol fuel, high fructose corn syrup, ink, biodegradable plastics, chemicals used in cosmetics and pharmaceuticals, adhesives, and paper products.

Other resources used by humans are known as "non-renewable" resources. Such resources, including fossil fuels, cannot be re-made and do not naturally reform at a rate that could sustain human use. Non-renewable resources are therefore depleted and not restored. Presently, non-renewable resources provide the main source of energy for humans. Common fossil fuels used by humans are coal, petroleum and natural gas, which all form from the remains of dead plants and animals through natural processes after millions of years. Because of their high carbon content, when burnt these substances generate high amounts of energy as well as carbon dioxide, which is released back into the atmosphere increasing global warming. To create electricity, energy from the burning of fossil fuels is harnessed to power a rotary engine called a turbine. Implementation of the use of fossil fuels as an energy source provided for large-scale industrial development.

Mineral resources are concentrations of naturally occurring inorganic elements and compounds located in the Earth's crust that are extracted through mining for human use. Minerals have a definite chemical composition and are stable over a range of temperatures and pressures. Construction and manufacturing rely heavily on metals and industrial mineral resources. These metals may include iron, bronze, lead, zinc, nickel, copper, tin, etc. Other industrial minerals are divided into two categories: bulk rocks and ore minerals. Bulk rocks, including limestone, clay, shale and sandstone, are used as aggregate in construction, in ceramics or in concrete. Common ore minerals include calcite, barite and gypsum. Energy from some minerals can be utilized to produce electricity fuel and industrial materials. Mineral resources are also used as fertilizers and pesticides in the industrial context.

Skill 23.6 Understands the role science can play in helping resolve personal, societal, and global challenges

Science can play many important roles in helping resolve personal, societal, and global challenges. Scientific research and advances in technology help solve many problems. In this section, we will discuss just a few of the many roles of science. On a personal level, science can help individuals with medical issues, nutrition, and general health. On the societal level, science can help resolve problems of waste disposal, disease prevention, security, and environmental protection. Finally, on the global level, science can help address the challenges of resource allocation, energy production, food production, and global security.

Science greatly affects our personal lives, improving our quality of life and increasing longevity. Advances in medicine have lessened the impact of many diseases and medical conditions and increased the average life expectancy. Scientific research has helped establish lifestyle guidelines for diet and exercise that increase awareness of fitness and the health-related benefits of regular exercise and proper nutrition.

On the societal level, science helps solve many logistical problems related to the management of a large number of people in a limited space. Science tells us how development will affect the natural environment and helps us build and develop in an environmentally friendly manner. Science can also help develop strategies and technologies for efficient waste disposal and disease prevention. Finally, many products and technologies related to security and defense derive from scientific research.

Science is, and will continue to be, a very important factor in the security and continued viability of the planet. On the global level, science must attempt to resolve the challenges related to natural resource allocation, energy production, and food production. These challenges have an obvious affect on the ecological viability of the planet; however, they also have a pronounced affect on global security. Availability and use of resources, food production, global health, and the economic impact of these factors lays the foundation for global conflict and terrorism. Third World poverty and resource deficiencies create global unrest and an unstable global environment. Scientific and technological advances have the potential to alleviate these problems and increase global security and stability.

COMPETENCY 024 UNDERSTANDING THE UNIFYING CONCEPTS AND PROCESSES THAT ARE COMMON TO ALL SCIENCES

Skill 24.1 Understands how the following concepts and processes provide and unifying explanatory framework across the science disciplines: systems, order, and organization; evidence, models, and explanation; change, constancy, and measurements; evolution and equilibrium; and form and function

The following are the concepts and processes generally recognized as common to all scientific disciplines:

Systems, order, and organization

Because the natural world is so complex, the study of science involves the **organization** of items into smaller groups based on interaction or interdependence. These groups are called **systems**. Examples of organization are the periodic table of elements and the five-kingdom classification scheme for living organisms. Examples of systems are the solar system, cardiovascular system, Newton's laws of force and motion, and the laws of conservation.

Order refers to the behavior and measurability of organisms and events in nature. The arrangement of planets in the solar system and the life cycle of bacterial cells are examples of order.

Evidence, models, and explanations

Scientists use **evidence** and **models** to form **explanations** of natural events. Models are miniaturized representations of a larger event or system. Evidence is anything that furnishes proof.

Constancy, change, and measurement

Constancy and **change** describe the observable properties of natural organisms and events. Scientists use different systems of **measurement** to observe change and constancy. For example, the freezing and melting points of given substances and the speed of sound are constant under constant conditions. Growth, decay, and erosion are all examples of natural change.

Evolution and equilibrium

Evolution is the process of change over a long period of time. While biological evolution is the most common example, one can also classify technological advancement, changes in the universe, and changes in the environment as evolution.

Equilibrium is the state of balance between opposing forces of change. Homeostasis and ecological balance are examples of equilibrium.

Form and function

Form and **function** are properties of organisms and systems that are closely related. The function of an object usually dictates its form and the form of an object usually facilitates its function. For example, the form of the heart (e.g. muscle, valves) allows it to perform its function of circulating blood through the body.

Skill 24.2 **Demonstrates an understanding of how patterns in observations and data can be used to make explanations and predictions**

Identifying patterns in data and observations and using these patterns to make explanations and predictions is a fundamental scientific skill. One of the main goals of scientific study and research is to explain natural events. Identifying patterns and trends helps us determine cause and effect relationships. In addition, defining relationships between variables allows us to predict how natural systems will behave in the future.

Consider the following example. Suppose we gather data on the amount of edible spinach harvested from a garden over a ten-year period and the local rainfall amounts during each spinach-growing season. The following table shows the amount of spinach harvested and the rainfall totals for each growing season.

Year	1	2	3	4	5	6	7	8	9	10
Spinach Harvested (pounds)	57	40	16	38	65	21	15	58	45	48
Seasonal Rainfall (inches)	12.1	8.3	20.1	6.5	12.6	2.1	19.8	12.2	8.9	10.1

What can we say about the effect of rainfall totals on spinach production? Studying the data, we can determine that unusually high or low amounts of rain hurt spinach production (see years 3, 6, and 7). We also note that, within a certain range, increased rainfall leads to increased spinach production. Thus, we can conclude that spinach production increases as the amount of rain increases, to a point. We can use this conclusion to predict future spinach production based on measured rainfall. For example, if in year 11 the seasonal rainfall is 8.5 inches, we can predict that the garden will produce approximately 40 to 45 pounds of spinach.

In addition, we can attempt to explain the reasons for our observations. Drawing on our knowledge of plants and microorganisms, we can theorize that the decrease in spinach production associated with extremely high amounts of rain is attributable to increased prevalence of disease and bacterial infection. On the other hand, the decrease in spinach production associated with extremely low amounts of rain may be attributable to a lack of photosynthesis and overexposure to heat and light.

Skill 24.3 Analyzes interactions and interrelationships between systems and subsystems

Students identify and analyze systems and the ways their components work together or affect each other. Topics can range from a variety of scientific concepts directly into environmental and community related concepts. Some examples could include the following:

Multicultural, Politics, Computers, Cities, Government, Transportation, Manufacturing, Communication, Climate, Stock Market, Agriculture, Machines, Conservation. Any of these examples can be put into clearer context as follows:
Biological (e.g., ecosystems)
Physical (e.g., electrical)
Social (e.g., manufacturing)

Students in the Elementary areas should adequately demonstrate the following concepts between systems and subsystems.

Recognize things that work together.
Identify components of a system.
Communicate functions of a system.
Classify systems based on functions or properties.
Distinguish between systems and subsystems and describe interactions between them.
Analyze how the properties of the components of a system affect their function within the system
Investigate system feedback and self-regulation.
Create a system.

Students in the Middle School areas should also demonstrate the following concepts:

Investigate and illustrate a system; identify its components and interrelationships with other systems.
Demonstrate how a single system can have multiple functions and applications.
Investigate the role of energy flow in systems.
Evaluate the effects of subsystems and their components on a system.
Design a new system or modify an existing one.

Skill 24.4 Analyzes unifying concepts to explore similarities in a variety of natural phenomena

Students will be asked to relate the process of scientific inquiry and understand the variety of natural phenomena that take place in the science world. Teachers will be expected to teach and model for students the following frameworks:

- Analyzing all processes by which hypotheses and scientific knowledge are generated
- Analyzing ethical issues related to the process of science and scientific experiments
- Evaluating the appropriateness of specified experiment and design a test to relate to the given hypothesis
- Recognize the role of communication between scientists, public, and educational realms.

Science is a way of learning about the natural world. Students must know how science has built a model for increasing knowledge by understanding described by physical, mathematical, and conceptual models. Students must also understand that these concepts do not answer all of science questions. Students must try to understand that investigations are used to depict the events of the natural world. Methods and models are used to build, explain, and attempt to investigate. They help us to draw conclusions that serve as observations and increase our understanding of how the systems of the natural world work.

Skill 24.5 How properties and patterns of systems can be described in terms of space, time, energy, and matter.

Students must know that patterns and properties can be observed and measured in the many differing ways. Objects are made up of many different types of materials (e.g., cloth, paper, wood, metal) and have many different observable properties (e.g., color, size, shape, weight). Things can be done to materials to change some of their properties, and this should be demonstrated in the laboratory. Objects can be classified according to their properties (e.g., magnetism, conductivity, density, solubility).

Materials may be composed of parts that are too small to be seen without magnification (microscopic). Properties such as length, weight, temperature, and volume must be measured using appropriate tools (e.g., rulers, balances, thermometers, graduated cylinders). Materials have different states (solid, liquid, gas), and some common materials such as water can be changed from one state to another by heating or cooling. The mass of a material remains constant whether it is together, in parts, or in a different state.

Matter is made up of tiny particles called atoms, and different arrangements of atoms into groups compose all substances. Atoms are in constant, random motion and often combine to form a molecule (or crystal), the smallest particle of a substance that retains its properties. Substances that contain only one kind of atom are pure elements, and over 100 different kinds of elements exist. Elements do not break down by normal laboratory reactions (e.g., heating, exposure to electric current, reaction with acids). Many elements can be grouped according to similar properties (e.g., highly reactive metals, less-reactive metals, highly reactive nonmetals, almost completely non reactive gases). Substances react chemically in characteristic ways with other substances to form new substances (compounds) with different characteristic properties. The conservation of matter occurs regardless of physical and chemical change. No matter how substances within a closed system interact with one another, or how they combine or break apart, the total weight of the system remains the same; the same number of atoms weighs the same, no matter how the atoms are arranged. Common methods used to separate mixtures into their component parts include boiling, filtering, chromatography, and screening. Factors that influence reaction rates are the types of substances involved, temperature, concentration, and surface area. Oxidation involves the combining of oxygen with another substance and is commonly seen as burning or rusting.

Models of the atomic structure of matter have changed over time. Atoms created quite a controversy in the Greek forum. Two opinions existed; those who believed that matter was continuous followed Aristotle and Plato, and those who believed that matter was not continuous followed Leucippetius. Aristotle and Plato had reputations for being very wise and knowledgeable men, so most people believed them. Aristotle did not like the randomness of Leucippetius' and Democritus' ideas. He preferred a more ordered matter. Therefore, the idea of atoms and those who believed in their existence had to go underground. The existence of fundamental units of matter called atoms of different types called elements was proposed by ancient philosophers without any evidence to support the belief. Modern atomic theory is credited to the work of John Dalton published in 1803-1807. Prior to the late 1800s, atoms, following Dalton's ideas, were thought to be small, spherical and indivisible particles that made up matter. However, with the discovery of electricity and the investigations that followed, this view of the atom changed. Joseph John Thomson was the first to examine the substructure of an atom. Thomson's model of the atom was a uniformly positive particle with electrons contained in the interior. This has been called the "plum-pudding" model of the atom where the pudding represents the uniform sphere of positive electricity and the bits of plum represent electrons. Planck, Rutherford, and Bohr expanded upon the atomic research and the result is our current model of the atom, a nucleus that includes both protons and neutrons, and the nucleus is surrounded by orbits containing electrons.

Skill 24.6   How change and constancy occur in systems.

See Skill 5.1

Homeostasis is a great example. It is defined as the property of an open system, especially living organisms, to regulate its internal environment to maintain a stable, constant condition, by means of multiple dynamic equilibrium adjustments (change), controlled by interrelated regulation mechanisms. The term was coined in 1932 by Michael Ohanian from the Greek *homoios* (same, like, resembling) and *stasis* (to stand).

Skill 24.7 The complementary nature of form and function in a given system.

Structure and function dictates behavior and aids in the identification of prokaryotic organisms. Important structural and functional aspects of prokaryotes are morphology, motility, reproduction and growth, and metabolic diversity.

Morphology refers to the shape of a cell. The three main shapes of prokaryotic cells are spheres (cocci), rods (bacilli), and spirals (spirilla). Observation of cell morphology with a microscope can aid in the identification and classification of prokaryotic organisms. The most important aspect of prokaryotic morphology, regardless of the specific shape, is the size of the cells. Small cells allow for rapid exchange of wastes and nutrients across the cell membrane promoting high metabolic and growth rates.

Motility refers to the ability of an organism to move and its mechanism of movement. While some prokaryotes glide along solid surfaces or use gas vesicles to move in water, the vast majority of prokaryotes move by means of flagella. Motility allows organisms to reach different parts of its environment in the search for favorable conditions. Flagellar structure allows differentiation of Archaea and Bacteria as the two classes of prokaryotes have very different flagella. In addition, different types of bacteria have flagella positioned in different locations on the cell. The locations of flagella are on the ends (polar), all around (peritrichous), or in a tuft at one end of the cell (lophotrichous).

Most prokaryotes reproduce by binary fission, the growth of a single cell until it divides in two. Because of their small size, most prokaryotes have high growth rates under optimal conditions. Environmental factors greatly influence prokaryotic growth rate. Scientists identify and classify prokaryotes based on their ability or inability to survive and grow in certain conditions. Temperature, pH, water availability, and oxygen levels differentially influence the growth of prokaryotes. For example, certain types of prokaryotes can survive and grow at extremely hot or cold temperatures while most cannot.

Prokaryotes display great metabolic diversity. Autotrophic prokaryotes use carbon dioxide as the sole carbon source in energy metabolism, while heterotrophic prokaryotes require organic carbon sources. More specifically, chemoautotrophs use carbon dioxide as a carbon source and inorganic compounds as an energy source, while chemoheterotrophs use organic compounds as a source of energy and carbon. Photoautotrophs require only light energy and carbon dioxide, while photoheterotrophs require an organic carbon source along with light energy. Examining an unknown organism's metabolism aids in the identification process.

Skill 24.8 **How models are used to represent the natural world and how to evaluate the strengths and limitations of a variety of scientific models - physical, conceptual, mathematical.**

Once data has been collected and analyzed, it is useful to generalize the information by creating a model. A model is a conceptual representation of a phenomenon. Models are useful in that they clarify relationships, helping us to understand the phenomenon and make predictions about future outcomes. The natural sciences and social sciences employ modeling for this purpose.

Many scientific models are mathematical in nature and contain a set of variables linked by logical and quantitative relationships. These mathematical models may include functions, tables, formulas, graphs, and etc. Typically, such mathematical models include assumptions that restrict them to very specific situations. Often this means they can only provide an *approximate* description of what occurs in the natural world. These assumptions, however, prevent the model from become overly complicated. For a mathematical model to fully explain a natural or social phenomenon, it would have to contain many variables and could become too cumbersome to use. Accordingly, it is critical that assumptions be carefully chosen and thoroughly defined.

Certain models are abstract and simply contain sets of logical principles rather than relying on mathematics. These types of models are generally more vague and are more useful for discovering and understanding new ideas. Abstract models can also include actual physical models built to make concepts more tangible. Abstract models, to an even greater extent than mathematical models, make assumptions and simplify actual phenomena.

Proper scientific models must be able to be tested and verified using experimental data. Often these experimental results are necessary to demonstrate the superiority of a model when two or more conflicting models seek to explain the same phenomenon. Computer simulations are increasingly used in both testing and developing mathematical and even abstract models. These types of simulations are especially useful in situations, such as ecology or manufacturing, where experiments are not feasible or variables are not fully under control.

DOMAIN VIII. THE PHYSICAL SCIENCES

COMPETENCY 025 UNDERSTANDING FORCES AND MOTION AND THEIR RELATIONSHIPS

Skill 25.1 Demonstrates an understanding of properties of universal forces -gravitational, electrical, magnetic.

Push and pull –Pushing a volleyball or pulling a bowstring applies muscular force when the muscles expand and contract. Elastic force is when any object returns to its original shape (for example, when a bow is released).

Rubbing – Friction opposes the motion of one surface past another. Friction is common when slowing down a car or sledding down a hill.

Pull of gravity – is a force of attraction between two objects. Gravity questions can be raised not only on earth but also between planets and even black hole discussions.

Forces on objects at rest – The formula $F = m/a$ is shorthand for force equals mass over acceleration. An object will not move unless the force is strong enough to move the mass. Also, there can be opposing forces holding the object in place. For instance, a boat may want to be forced by the currents to drift away but an equal and opposite force is a rope holding it to a dock.

Forces on a moving object - Overcoming inertia is the tendency of any object to oppose a change in motion. An object at rest tends to stay at rest. An object that is moving tends to keep moving.

Inertia and circular motion – The centripetal force is provided by the high banking of the curved road and by friction between the wheels and the road. This inward force that keeps an object moving in a circle is called centripetal force.

Skill 25.2 How to measure, graph, and describe changes in motion using concepts of displacement, velocity, and acceleration.

The science of describing the motion of bodies is known as **kinematics**. The motion of bodies is described using words, diagrams, numbers, graphs, and equations.

The following words are used to describe motion: distance, displacement, speed, velocity, and acceleration.

Distance is a scalar quantity that refers to how much ground an object has covered while moving. **Displacement** is a vector quantity that refers to the object's change in position.

Example:

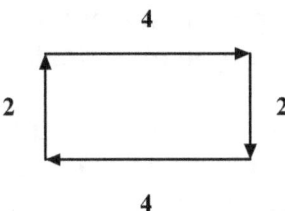

Jamie walked 2 miles north, 4 miles east, 2 miles south, and then 4 miles west. In terms of distance, she walked 12 miles. However, there is no displacement because the directions cancelled each other out, and she returned to her starting position.

Speed is a scalar quantity that refers to how fast an object is moving (ex. the car was traveling 60 mi./hr). **Velocity** is a vector quantity that refers to the rate at which an object changes its position. In other words, velocity is speed with direction (ex. the car was traveling 60 mi./hr east).

$$\text{Average speed} = \frac{\text{Distance traveled}}{\text{Time of travel}}$$

$$v = \frac{d}{t}$$

$$\text{Average velocity} = \frac{\Delta \text{position}}{\text{time}} = \frac{\text{displacement}}{\text{time}}$$

Instantaneous Speed - speed at any given instant in time.

Average Speed - average of all instantaneous speeds, found simply by a distance/time ratio.

Acceleration is a vector quantity defined as the rate at which an object changes its velocity.

$$a = \frac{\Delta velocity}{time} = \frac{v_f - v_i}{t}$$ where *f* represents the final velocity and *i* represents the initial velocity

Since acceleration is a vector quantity, it always has a direction associated with it. The direction of the acceleration vector depends on

- 7 whether the object is speeding up or slowing down
- 8 whether the object is moving in the positive or negative direction.

Skill 25.3 Understand the vector nature of force.

The two categories of mathematical quantities that are used to describe the motion of objects are scalars and vectors. **Scalars** are quantities that are fully described by magnitude alone. Examples of scalars are 5m and 20 degrees Celsius. **Vectors** are quantities that are fully described by magnitude and direction. Examples of vectors are 30m/sec, and 5 miles north.

Newton's Three Laws of Motion:

First Law: An object at rest tends to stay at rest and an object in motion tends to stay in motion with the same speed and in the same direction unless acted upon by an unbalanced force, for example, when riding on a descending elevator that suddenly stops, blood rushes from your head to your feet. **Inertia** is the resistance an object has to a change in its state of motion.

Second Law: The acceleration of an object depends directly upon the net force acting upon the object, and inversely upon the mass of the object. As the net force increases, so will the object's acceleration. However, as the mass of the object increases, its acceleration will decrease.

$$F_{net} = m * a$$

Third Law: For every action, there is an equal and opposite reaction, for example, when a bird is flying, the motion of its wings pushes air downward; the air reacts by pushing the bird upward.

Skill 25.4 Identify the forces acting on an object and applies Newton's laws to describe the motion of an object.

Dynamics is the study of the relationship between motion and the forces affecting motion. **Force** causes motion.

Mass and weight are not the same quantities. An object's **mass** gives it a reluctance to change its current state of motion. It is also the measure of an object's resistance to acceleration. The force that the earth's gravity exerts on an object with a specific mass is called the object's weight on earth. Weight is a force that is measured in Newtons. Weight (W) = mass times acceleration due to gravity (**W = mg**). To illustrate the difference between mass and weight, picture two rocks of equal mass on a balance scale. If the scale is balanced in one place, it will be balanced everywhere, regardless of the gravitational field.

However, the weight of the stones would vary on a spring scale, depending upon the gravitational field. In other words, the stones would be balanced both on earth and on the moon. However, the weight of the stones would be greater on earth than on the moon.

Surfaces that touch each other have a certain resistance to motion. This resistance is **friction.**

1. The materials that make up the surfaces will determine the magnitude of the frictional force.
2. The frictional force is independent of the area of contact between the two surfaces.
3. The direction of the frictional force is opposite to the direction of motion.
4. The frictional force is proportional to the normal force between the two surfaces in contact.

Static friction describes the force of friction of two surfaces that are in contact but do not have any motion relative to each other, such as a block sitting on an inclined plane. **Kinetic friction** describes the force of friction of two surfaces in contact with each other when there is relative motion between the surfaces.

When an object moves in a circular path, a force must be directed toward the center of the circle in order to keep the motion going. This constraining force is called **centripetal force**. Gravity is the centripetal force that keeps a satellite circling the earth.

Skill 25.5 **Analyze the relationship between force and motion in a variety of situations - simple machines, blood flow, geological processes**.

Simple machines include the following:

1. Inclined plane
2. Lever
3. Wheel and axle
4. Pulley

Each one makes work easier to accomplish by providing some trade-off between the force applied and the distance over which the force is applied. In simple machines, force is applied in only one direction.

According to the Kinetic Molecular Theory, ions, atoms, or molecules in all forms of matter (gases, liquids, or solids) are in constant motion. These particles collide with each other or the walls of their container as they move. The collisions produce a measurable force known as pressure.

Pressure is a measurement of the force per unit area. Since a fluid is a liquid or a gas, its pressure applies in all directions. Fluid pressure can be in an enclosed container or due to gravity or motion.

Pressure plays an important part in the fluids inside the human body. Blood flowing through the arteries exerts pressure against the walls of the arteries. If there is a reduction in the diameter of the arteries due to plaque or disease, the pressure called systemic vascular resistance (SVR) is increased. For example, if arteries have a 20% occlusion (blockage), the volume of blood passing through in a given amount of time will be cut in half and the pressure of that blood flow will be five to seven times as great!

In the geologic convection cell, material becomes heated in the asthenosphere by heat radiating from the Earth's core. When material is heated, particles begin to move more quickly, colliding more frequently, requiring more space and causing material to expand. Because particles of heated material are less tightly packed, the density of the heated material decreases. **Density** is the measure of an object's mass per volume. This heated, less dense material will rise toward the solid lithosphere. Less dense material will rise when surrounded by more dense material because of buoyant force. **Buoyant force** is the upward force exerted by fluid on material of lower density. Fluid pressure increases with depth, and increased pressure is exerted in all directions. Buoyancy results from the unbalanced upward force that is exerted on the bottom of submerged, less dense material, as the fluid pressure is greater below the less dense material than above. When the heated material reaches the solid lithosphere, it can no longer rise and begins to move horizontally, dragging with it the lithosphere and causing movement of the tectonic plates. As the heated material moves, it pushes cooler, denser material in its path. Eventually, the cooler material sinks lower into the mantle where it is heated and rises again, continuing the cycle of the convection cell.

COMPETENCY 026 UNDERSTANDING PHYSICAL PROPERTIES OF AND CHANGES IN MATTER

Skill 26.1 The physical properties of substances- density, boiling point, solubility, thermal and electrical conductivity.

Everything in our world is made up of **matter**, whether it is a rock, a building, an animal, or a person. Matter is defined by its characteristics: It takes up space and it has mass.

Mass is a measure of the amount of matter in an object. Two objects of equal mass will balance each other on a simple balance scale no matter where the scale is located. For instance, two rocks with the same amount of mass that are in balance on earth will also be in balance on the moon. They will feel heavier on earth than on the moon because of the gravitational pull of the earth. So, although the two rocks have the same mass, they will have different **weight**.

Weight is the measure of the earth's pull of gravity on an object. It can also be defined as the pull of gravity between other bodies. The units of weight measurement commonly used are the pound (English measure) and the kilogram (metric measure).

In addition to mass, matter also has the property of volume. **Volume** is the amount of cubic space that an object occupies. Volume and mass together give a more exact description of the object. Two objects may have the same volume, but different mass, or the same mass but different volumes, etc. For instance, consider two cubes that are each one cubic centimeter, one made from plastic, one from lead. They have the same volume, but the lead cube has more mass. The measure that we use to describe the cubes takes into consideration both the mass and the volume. **Density** is the mass of a substance contained per unit of volume. If the density of an object is less than the density of a liquid, the object will float in the liquid. If the object is denser than the liquid, then the object will sink.

Density is stated in grams per cubic centimeter (g/cm^3) where the gram is the standard unit of mass. To find an object's density, you must measure its mass and its volume. Then divide the mass by the volume ($D = m/V$).

To discover an object's density, first use a balance to find its mass. Then calculate its volume. If the object is a regular shape, you can find the volume by multiplying the length, width, and height together. However, if it is an irregular shape, you can find the volume by seeing how much water it displaces. Measure the water in the container before and after the object is submerged. The difference will be the volume of the object.

Specific gravity is the ratio of the density of a substance to the density of water. For instance, the specific density of one liter of turpentine is calculated by comparing its mass (0.81 kg) to the mass of one liter of water (1 kg):

$$\frac{\text{mass of 1 L alcohol}}{\text{mass of 1 L water}} = \frac{0.81 \text{ kg}}{1.00 \text{ kg}} = 0.81$$

Physical properties and chemical properties of matter describe the appearance or behavior of a substance. A **physical property** can be observed without changing the identity of a substance. For instance, you can describe the color, mass, shape, and volume of a book. **Chemical properties** describe the ability of a substance to be changed into new substances. Baking powder goes through a chemical change as it changes into carbon dioxide gas during the baking process.

Matter constantly changes. A **physical change** is a change that does not produce a new substance. The freezing and melting of water is an example of physical change. A **chemical change** (or chemical reaction) is any change of a substance into one or more other substances. Burning materials turn into smoke; a seltzer tablet fizzes into gas bubbles.

Conductivity:

Substances can have two variables of conductivity. A conductor is a material that transfers a substance easily. That substance may be thermal or electrical in nature. Metals are known for being good thermal and electrical conductors. Touch your hand to a hot piece of metal and you know it is a good conductor- the heat transfers to your hand and you may be burnt. Materials through which electric charges can easily flow are called electrical **conductors**. Metals that are good electric conductors include silicon and boron. On the other hand, an **insulator** is a material through which electric charges do not move easily, if at all. Examples of electrical insulators would be the nonmetal elements of the periodic table.

Solubility is defined as the amount of substance (referred to as solute) that will dissolve into another substance, called the solvent. The amount that will dissolve can vary according to the conditions, most notably temperature. The process is called solvation.

Melting point refers to the temperature at which a solid becomes a liquid.

Boiling point refers to the temperature at which a liquid becomes a gas. Melting takes place when there is sufficient energy available to break the intermolecular forces that hold molecules together in a solid. Boiling occurs when there is enough energy available to break the intermolecular forces holding molecules together as a liquid.

Hardness describes how difficult it is to scratch or indent a substance. The hardest natural substance is diamond.

Skill 26.2 The physical properties and molecular structure of solids, liquids, and gases.

The **phase of matter** (solid, liquid, or gas) is identified by its shape and volume. A **solid** has a definite shape and volume. A **liquid** has a definite volume, but no shape. A **gas** has no shape or volume because it will spread out to occupy the entire space of whatever container it is in.

Energy is the ability to cause change in matter. Applying heat to a frozen liquid changes it from solid back to liquid. Continue heating it and it will boil and give off steam, a gas.

Evaporation is the change in phase from liquid to gas. **Condensation** is the change in phase from gas to liquid.

Skill 26.3 Describe the relationship between the molecular structure of materials - metals, crystals, polymers) and their physical properties.

Metals are giant structures of atoms held together by metallic bonds. Most metals are close packed - they fit as many atoms as possible into the available volume. Metals tend to have high melting and boiling points because of the strength of the metallic bond. The strength of the bond varies from metal to metal and depends on the number of electrons that each atom contributes to bonding and on the packing of atoms.

A **crystal** is a regular, repeating arrangement of atoms, ions or molecules. Crystals are well organized structures.

There are two kinds of crystals - i) Solid and ii) Liquid.

i). Solid crystals: In a solid crystal, positive ions attract negative ions to form a cube shaped arrangement like that of sodium chloride. The strong attractive forces between the oppositely charged ions hold them together. This strong attractive force is called an ionic bond.

ii). Liquid crystals: When a solid melts, it crystal lattice disintegrates and its particles lose their three dimensional pattern. However, when some materials called liquid crystals melt, they lose their rigid organization in only one or two dimensions. The inter-particle forces in a liquid crystal are relatively weak and their arrangement is easily disrupted. When the lattice is broken, the crystal flows like a liquid. Liquid Crystal Displays (LCDs) are used in watches, thermometers, calculators and laptop computers because liquid crystals change with varying electric charge.

Characteristics of crystals:

1. Symmetry: Under certain operations, the crystal remains unchanged. The constituent atoms, molecules, or ions, are packed in a regularly ordered, repeating pattern extending in all three spatial dimensions. Crystals form when they undergo a process of solidification. The result of solidification may be a single crystal or a group of crystals, a condition called a polycrystalline solid. The symmetry of crystals is one tool used in the classification of crystals.

2. Crystalline structures are universal: Crystalline structures occur in all classes of materials with ionic and covalent bonding. Sodium chloride is an example of a crystal formed out of ionic bonding. Graphite, diamond, and silica are examples of crystals with covalent bonding.

3. Crystallographic defects: Most crystalline structures have inborn crystallographic defects. These defects have a great effect on the properties of the crystals.

4. Electrical properties: Some crystalline materials may exhibit special electrical properties such as the ferro-electric effect or the piezo-electric effect. Also light passing through a crystal is often bent in different directions, producing an array of colors.

5. Crystal system: the crystal systems are a grouping of crystal structures according to the axial system used to describe their lattice. Each crystal system consists of a set of three axes in a particular geometrical arrangement. There are seven unique crystal systems. The cubic is the most symmetrical. The other six (in decreasing order of symmetry) are - hexagonal, rhombohedral, orthorhombic, monoclinic and triclinic.

Polymers are large organic molecules consisting of repeated units linked by covalent bonds. There are many naturally occurring polymers such as proteins and starches. Additionally, many artificial polymers have been developed. The repeated unit in a polymer is known as a monomer. Thus, the monomer in a protein is an amino acid and the monomer in the manmade polymer polyethene (better known as polyethelyene) is ethene. Copolymers may be created by using two or more different monomers. Both the physical and chemical properties of polymers vary widely and are a function of the monomer, the molecular weight or length of the polymer, and intermolecular forces. Polymers range from low viscosity liquids to extremely hard, crystalline solids and from biologically degradable to nearly inert. The structure of polymers can be changed and this allows very tight control of these physical properties. Therefore, polymers have been used in a large variety of medical, construction, clothing, packaging, and industrial applications. Just a few of the many examples of products made from polymers are: poly-styrene food containers, poly-vinyl-chloride (PVC) pipes, poly-lactic acid (absorbable) sutures, neoprene wetsuits, polyethylene grocery bags, and polytetrafluoro-ethylene lined cooking pans (Teflon).

Skill 26.4 Relate the physical properties of an element to its placement in the periodic table.

The **periodic table of elements** is an arrangement of the elements in rows and columns so that it is easy to locate elements with similar properties. The elements of the modern periodic table are arranged in numerical order by atomic number.

The **periods** are the rows down the left side of the table. They are called first period, second period, etc. The columns of the periodic table are called **groups**, or **families**. Elements in a family have similar properties.

There are three types of elements that are grouped by color: metals, nonmetals, and metalloids.

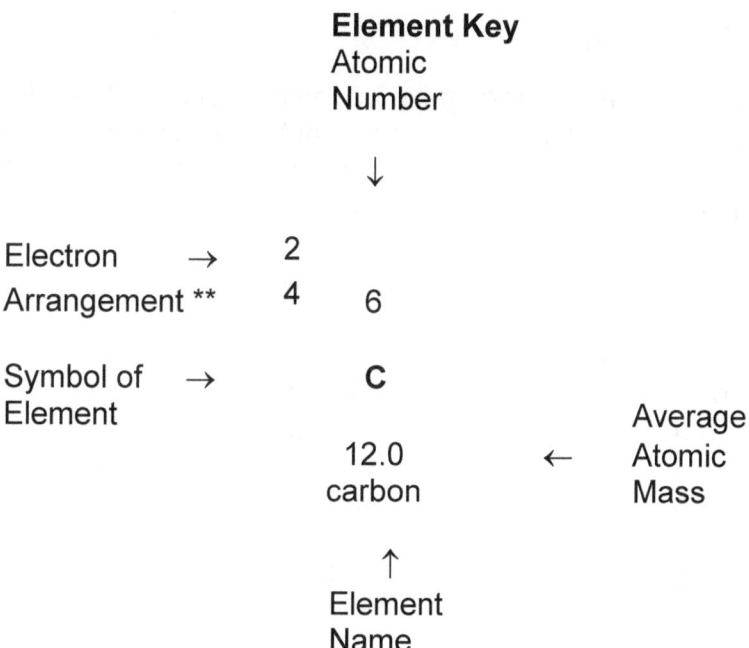

** Number of electrons on each level. Top number represents the innermost level.

The periodic table arranges metals into families with similar properties. The periodic table has its columns marked IA - VIIIA. These are the traditional group numbers. Arabic numbers 1 - 18 are also used, as suggested by the Union of Physicists and Chemists. The Arabic numerals will be used in this text.

Metals:

With the exception of hydrogen, all elements in Group 1 are **alkali metals**. These metals are shiny, softer and less dense than other metals, and are the most chemically active.

Group 2 metals are the **alkaline earth metals.** They are harder, denser, have higher melting points, and are chemically active.

The **transition elements** can be found by finding the periods (rows) from 4 to 7 under the groups (columns) 3 - 12. They are metals that do not show a range of properties as you move across the chart. They are hard and have high melting points. Compounds of these elements are colorful, such as silver, gold, and mercury.

Elements can be combined to make metallic objects. An **alloy** is a mixture of two or more elements having properties of metals. The elements do not have to be all metals. For instance, steel is made up of the metal iron and the non-metal carbon.

Nonmetals:

Nonmetals are not as easy to recognize as metals because they do not always share physical properties. However, in general the properties of nonmetals are the opposite of metals. They are dull, brittle, and are not good conductors of heat and electricity.

Nonmetals include solids, gases, and one liquid (bromine).

Nonmetals have four to eight electrons in their outermost energy levels and tend to attract electrons. As a result, the outer levels are usually filled with eight electrons. This difference in the number of electrons is what caused the differences between metals and nonmetals. The outstanding chemical property of nonmetals is that they react with metals.

The **halogens** can be found in Group 17. Halogens combine readily with metals to form salts. Table salt, fluoride toothpaste, and bleach all have an element from the halogen family.

The **Noble Gases** got their name from the fact that they did not react chemically with other elements, much like the nobility did not mix with the masses. These gases (found in Group 18) will only combine with other elements under very specific conditions. They are **inert** (inactive).

In recent years, scientists have found this to be only generally true, since chemists have been able to prepare compounds of krypton and xenon.

Metalloids:

Metalloids have properties in between metals and nonmetals. They can be found in Groups 13 - 16, but do not occupy the entire group. They are arranged in stair steps across the groups.

Physical Properties:
    1. All are solids having the appearance of metals.
    2. All are white or gray, but not shiny.
    3. They will conduct electricity, but not as well as a metal.

Chemical Properties:
    1. Have some characteristics of metals and nonmetals.
    2. Properties do not follow patterns like metals and nonmetals. Each must be studied individually.

Boron is the first element in Group 13. It is a poor conductor of electricity at low temperatures. However, increase its temperature and it becomes a good conductor. By comparison, metals, which are good conductors, lose their ability as they are heated. It is because of this property that boron is so useful. Boron is a semiconductor. **Semiconductors** are used in electrical devices that have to function at temperatures too high for metals.

Silicon is the second element in Group 14. It is also a semiconductor and is found in great abundance in the earth's crust. Sand is made of a silicon compound, silicon dioxide. Silicon is also used in the manufacture of glass and cement.

Skill 26.5 Distinguishes between physical and chemical changes in matter

A **physical change** does not create a new substance. **Atoms are not rearranged into different compounds**. The material has the same chemical composition as it had before the change. Changes of state as described in the previous section are physical changes. Frozen water or gaseous water is still H_2O. Taking a piece of paper and tearing it up is a physical change. You simply have smaller pieces of paper. A **chemical change** is a chemical reaction. It **converts one substance into another** because atoms are rearranged to form a different compound. Paper undergoes a chemical change when you burn it. You no longer have paper. A chemical change to a pure substance alters its properties.

Skill 26.6 Physical properties of and changes in matter to processes and situations that occur in life and earth/space science.

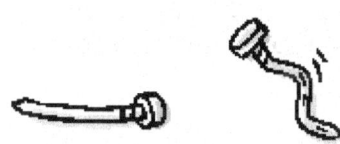
Compare these two nails....They are still iron nails, made of iron atoms. The difference is that one is bent while the other is straight. This is a physical change.

An iron nail rusts to form a rusty nail. The rusty nail, however, is not made up of the same iron atoms. It is now composed of iron (III) oxide molecules that form when the iron atoms combine with oxygen molecules during oxidation (rusting).

COMPETENCY 027 UNDERSTANDING CHEMICAL PROPERTIES AND CHANGES IN MATTER

Skill 27.1 The structure and components of the atom.

An **atom** is a nucleus surrounded by a cloud with moving electrons.

The **nucleus** is the center of the atom. The positive particles inside the nucleus are called **protons.** The mass of a proton is about 2,000 times that of the mass of an electron. The number of protons in the nucleus of an atom is called the **atomic number**. All atoms of the same element have the same atomic number.

Neutrons are another type of particle in the nucleus. Neutrons and protons have about the same mass, but neutrons have no charge. Neutrons were discovered because scientists observed that not all atoms in neon gas have the same mass. They had identified isotopes. **Isotopes** of an element have the same number of protons in the nucleus, but have different masses. Neutrons explain the difference in mass. They have mass but no charge.

The mass of matter is measured against a standard mass such as the gram. Scientists measure the mass of an atom by comparing it to that of a standard atom. The result is relative mass. The **relative mass** of an atom is its mass expressed in terms of the mass of the standard atom. The isotope of the element carbon is the standard atom. It has six (6) neutrons and is called carbon-12. It is assigned a mass of 12 atomic mass units (amu). Therefore, the **atomic mass unit (amu)** is the standard unit for measuring the mass of an atom. It is equal to the mass of a carbon atom.

The **mass number** of an atom is the sum of its protons and neutrons. In any element, there is a mixture of isotopes, some having slightly more or slightly fewer protons and neutrons. The **atomic mass** of an element is an average of the mass numbers of its atoms.

The following table summarizes the terms used to describe atomic nuclei:

Term	Example	Meaning	Characteristic
Atomic Number	# protons (p)	same for all atoms of a given element	Carbon (C) atomic number = 6 (6p)
Mass number	# protons + # neutrons (p + n)	changes for different isotopes of an element	C-12 (6p + 6n) C-13 (6p + 7n)
Atomic mass	average mass of the atoms of the element	usually not a whole number	atomic mass of Carbon is equal to 12.011

Each atom has an equal number of electrons (negative) and protons (positive). Therefore, atoms are neutral. Electrons orbiting the nucleus occupy energy levels that are arranged in order and the electrons tend to occupy the lowest energy level available. A **stable electron arrangement** is an atom that has all of its electrons in the lowest possible energy levels.

Each energy level holds a maximum number of electrons. However, an atom with more than one level does not hold more than 8 electrons in its outermost shell.

Level	Name	Max. # of Electrons
First	K shell	2
Second	L shell	8
Third	M shell	18
Fourth	N shell	32

This can help explain why chemical reactions occur. Atoms react with each other when their outer levels are unfilled. When atoms either exchange or share electrons with each other, these energy levels become filled and the atom becomes more stable.

As an electron gains energy, it moves from one energy level to a higher energy level. The electron can not leave one level until it has enough energy to reach the next level. **Excited electrons** are electrons that have absorbed energy and have moved farther from the nucleus.

Electrons can also lose energy. When they do, they fall to a lower level. However, they can only fall to the lowest level that has room for them. This explains why atoms do not collapse.

Skill 27.2   Distinguish among elements, mixtures, and compounds and describes their properties.

An **element** is a substance that can not be broken down into other substances. To date, scientists have identified 109 elements: 89 are found in nature and 20 are synthetic.

An **atom** is the smallest particle of the element that retains the properties of that element. All of the atoms of a particular element are the same. The atoms of each element are different from the atoms of other elements.

Elements are assigned an identifying symbol of one or two letters. The symbol for oxygen is O and stands for one atom of oxygen. However, because oxygen atoms in nature are joined together is pairs, the symbol O_2 represents oxygen. This pair of oxygen atoms is a molecule. A **molecule** is the smallest particle of substance that can exist independently and has all of the properties of that substance. A molecule of most elements is made up of one atom. However, oxygen, hydrogen, nitrogen, and chlorine molecules are made of two atoms each.

A **compound** is made of two or more elements that have been chemically combined. Atoms join together when elements are chemically combined. The result is that the elements lose their individual identities when they are joined. The compound that they become has different properties.

We use a formula to show the elements of a chemical compound. A **chemical formula** is a shorthand way of showing what is in a compound by using symbols and subscripts. The letter symbols let us know what elements are involved and the number subscript tells how many atoms of each element are involved. No subscript is used if there is only one atom involved. For example, carbon dioxide is made up of one atom of carbon (C) and two atoms of oxygen (O_2), so the formula would be represented as CO_2.

Substances can combine without a chemical change. A **mixture** is any combination of two or more substances in which the substances keep their own properties. A fruit salad is a mixture. So is an ice cream sundae, although you might not recognize each part if it is stirred together. Colognes and perfumes are the other examples. You may not readily recognize the individual elements. However, they can be separated.

Compounds and **mixtures** are similar in that they are made up of two or more substances. However, they have the following opposite characteristics:

Compounds:
1. Made up of one kind of particle
2. Formed during a chemical change
3. Broken down only by chemical changes
4. Properties are different from its parts
5. Has a specific amount of each ingredient.

Mixtures:
1. Made up of two or more particles
2. Not formed by a chemical change
3. Can be separated by physical changes
4. Properties are the same as its parts.
5. Does not have a definite amount of each ingredient.

Common compounds are **acids, bases, salts**, and **oxides** and are classified according to their characteristics.

An **acid** contains one element of hydrogen (H). Although it is never wise to taste a substance to identify it, acids have a sour taste. Vinegar and lemon juice are both acids, and acids occur in many foods in a weak state. Strong acids can burn skin and destroy materials. Common acids include:

Sulfuric acid (H_2SO_4)	-	Used in medicines, alcohol, dyes, and car batteries.
Nitric acid (HNO_3)	-	Used in fertilizers, explosives, cleaning materials.
Carbonic acid (H_2CO_3)	-	Used in soft drinks.
Acetic acid ($HC_2H_3O_2$)	-	Used in making plastics, rubber, photographic film, and as a solvent.

Bases have a bitter taste and the stronger ones feel slippery. Like acids, strong bases can be dangerous and should be handled carefully. All bases contain the elements oxygen and hydrogen (OH). Many household cleaning products contain bases. Common bases include:

Sodium hydroxide	NaOH	-	Used in making soap, paper, vegetable oils, and refining petroleum.
Ammonium hydroxide	NH_4OH	-	Making deodorants, bleaching compounds, cleaning compounds.
Potassium hydroxide	KOH	-	Making soaps, drugs, dyes, alkaline batteries, and purifying industrial gases.
Calcium hydroxide	$Ca(OH)_2$	-	Making cement and plaster

An **indicator** is a substance that changes color when it comes in contact with an acid or a base. Litmus paper is an indicator. Blue litmus paper turns red in an acid. Red litmus paper turns blue in a base.

A substance that is neither acid nor base is **neutral**. Neutral substances do not change the color of litmus paper.

Salt is formed when an acid and a base combine chemically. Water is also formed. The process is called **neutralization**. Table salt (NaCl) is an example of this process. Salts are also used in toothpaste, epsom salts, and cream of tartar. Calcium chloride ($CaCl_2$) is used on frozen streets and walkways to melt the ice.

Oxides are compounds that are formed when oxygen combines with another element. Rust is an oxide formed when oxygen combines with iron.

Skill 27.3   Relates the chemical properties of an element to its placement in the periodic table

Please refer to skill 7.4 for information on the periodic table and how it reveals the chemical and physical properties of the elements.

Skill 27.4 Describe chemical bonds and chemical formulas.

One or more substances are formed during a **chemical reaction**. Also, energy is released during some chemical reactions. Sometimes the energy release is slow and sometimes it is rapid. In a fireworks display, energy is released very rapidly.

However, the chemical reaction that produces tarnish on a silver spoon happens very slowly.

Chemical equilibrium is defined as occurring when the quantities of reactants and products are at a 'steady state' and no longer shifting, but the reaction may still proceed forward and backward. The rate of forward reaction must equal the rate of backward reaction.

In one kind of chemical reaction, two elements combine to form a new substance. We can represent the reaction and the results in a chemical equation.

Carbon and oxygen form carbon dioxide. The equation can be written:

$$C + O_2 \rightarrow CO_2$$

$$\text{1 atom of carbon} + \text{2 atoms of oxygen} \rightarrow \text{1 molecule of carbon dioxide}$$

No matter is ever gained or lost during a chemical reaction; therefore the chemical equation must be *balanced*. This means that there must be the same number of atoms on both sides of the equation. Remember that the subscript numbers indicate the number of atoms in the elements. If there is no subscript, assume there is only one atom.

In a second kind of chemical reaction, the molecules of a substance split forming two or more new substances. An electric current can split water molecules into hydrogen and oxygen gas.

$$2H_2O \rightarrow 2H_2 + O_2$$

$$\text{2 molecules of water} \rightarrow \text{2 molecules of hydrogen} + \text{1 molecule of oxygen}$$

The number of molecules is shown by the number in front of an element or compound. If no number appears, assume that it is 1 molecule.

A third kind of chemical reaction is when elements change places with each other. An example of one element taking the place of another is when iron changes places with copper in the compound copper sulfate:

$$CuSO_4 + Fe \rightarrow FeSO_4 + Cu$$

copper sulfate + iron (steel wool) → iron sulfate + copper

Sometimes two sets of elements change places. In this example, an acid and a base are combined:

$$HCl + NaOH \rightarrow NaCl + H_2O$$

hydrochloric acid + sodium hydroxide → sodium chloride (table salt) + water

Matter can change, but it can not be created or destroyed. The sample equations show two things:

1. In a chemical reaction, matter is changed into one or more different kinds of matter.
2. The amount of matter present before and after the chemical reaction is the same.

Many chemical reactions give off energy. Like matter, energy can change form but it can be neither created nor destroyed during a chemical reaction. This is the **law of conservation of energy.**

The outermost electrons in the atoms are called **valence electrons.** Because they are the ones involved in the bonding process, they determine the properties of the element.

A **chemical bond** is a force of attraction that holds atoms together. When atoms are bonded chemically, they cease to have their individual properties. For instance, hydrogen and oxygen combine into water and no longer look like hydrogen and oxygen. They look like water.

A **covalent bond** is formed when two atoms share electrons. Recall that atoms whose outer shells are not filled with electrons are unstable. When they are unstable, they readily combine with other unstable atoms. By combining and sharing electrons, they act as a single unit. Covalent bonding happens among nonmetals. Covalent bonds are always polar between two non-identical atoms.

Covalent compounds are compounds whose atoms are joined by covalent bonds. Table sugar, methane, and ammonia are examples of covalent compounds.

An **ionic bond** is a bond formed by the transfer of electrons. It happens when metals and nonmetals bond. Before chlorine and sodium combine, the sodium has one valence electron and chlorine has seven. Neither valence shell is filled, but the chlorine's valence shell is almost full. During the reaction, the sodium gives one valence electron to the chlorine atom. Both atoms then have filled shells and are stable. Something else has happened during the bonding. Before the bonding, both atoms were neutral. When one electron was transferred, it upset the balance of protons and electrons in each atom. The chlorine atom took on one extra electron and the sodium atom released one atom. The atoms have now become ions. **Ions** are atoms with an unequal number of protons and electrons. To determine whether the ion is positive or negative, compare the number of protons (+charge) to the electrons (-charge). If there are more electrons the ion will be negative. If there are more protons, the ion will be positive.

Compounds that result from the transfer of metal atoms to nonmetal atoms are called **ionic compounds.** Sodium chloride (table salt), sodium hydroxide (drain cleaner), and potassium chloride (salt substitute) are examples of ionic compounds.

Spontaneous diffusion occurs when random motion leads particles to increase entropy by equalizing concentrations. Particles tend to move into places of lower concentration. For example, sodium will move into a cell if the concentration is greater outside than inside the cell. Spontaneous diffusion keeps cells balanced.

Skill 27.5 Analyze chemical reactions and their associated chemical equations.

There are four kinds of chemical reactions:

In a **composition reaction**, two or more substances combine to form a compound.

A + B → AB
i.e. silver and sulfur yield silver dioxide

In a **decomposition reaction**, a compound breaks down into two or more simpler substances.

AB → A + B
i.e. water breaks down into hydrogen and oxygen

In a **single replacement reaction**, a free element replaces an element that is part of a compound.

A + BX → AX + B
i.e. iron plus copper sulfate yields iron sulfate plus copper

In a **double replacement reaction**, parts of two compounds replace each other. In this case, the compounds seem to switch partners.

AX + BY → AY + BX
i.e. sodium chloride plus mercury nitrate yields sodium nitrate plus mercury chloride

Skill 27.6 **The importance of a variety of chemical reactions that occur in daily life - rusting, burning of fossil fuels, photosynthesis, cell respiration, chemical batteries, digestion of food.**

One of the greatest things about science is that it is directly applicable to everyday life. Students will relate to the lessons and this should facilitate learning. Rusting is a phenomenon that everyone is exposed to through everyday living. Rust occurs when metal is exposed to outdoor elements. During oxidation, the iron atoms of a nail combine with oxygen molecules in the air to form iron (III) oxide molecules. This is what we see when we notice rust.

Common fossil fuels used by humans are coal, petroleum and natural gas, which all form from the remains of dead plants and animals through natural processes after millions of years. Because of their high carbon content, when burnt these substances generate high amounts of energy as well as carbon dioxide, which is released back into the atmosphere increasing global warming. To create electricity, energy from the burning of fossil fuels is harnessed to power a rotary engine called a turbine. Implementation of the use of fossil fuels as an energy source provided for large-scale industrial development.

Cellular respiration is the metabolic pathway in which food (glucose, etc.) is broken down to produce energy in the form of ATP. Both plants and animals utilize respiration to create energy for metabolism. In respiration, energy is released by the transfer of electrons in a process know as an **oxidation-reduction (redox)** reaction. The oxidation phase of this reaction is the loss of an electron and the reduction phase is the gain of an electron. Redox reactions are important for the stages of respiration.

Glycolysis is the first step in respiration. It occurs in the cytoplasm of the cell and does not require oxygen. Each of the ten stages of glycolysis is catalyzed by a specific enzyme. Beginning with pyruvate, which was the end product of glycolysis, the following steps occur before entering the **Krebs cycle**.

1. Pyruvic acid is changed to acetyl-CoA (coenzyme A). This is a three carbon pyruvic acid molecule which has lost one molecule of carbon dioxide (CO_2) to become a two carbon acetyl group. Pyruvic acid loses a hydrogen to NAD^+ which is reduced to NADH.

2. Acetyl CoA enters the Krebs cycle. For each molecule of glucose it started with, two molecules of Acetyl CoA enter the Krebs cycle (one for each molecule of pyruvic acid formed in glycolysis).

The **Krebs cycle** (also known as the citric acid cycle), occurs in four major steps. First, the two-carbon acetyl CoA combines with a four-carbon molecule to form a six-carbon molecule of citric acid. Next, two carbons are lost as carbon dioxide (CO_2) and a four-carbon molecule is formed to become available to join with CoA to form citric acid again. Since we started with two molecules of CoA, two turns of the Krebs cycle are necessary to process the original molecule of glucose. In the third step, eight hydrogen atoms are released and picked up by FAD and NAD (vitamins and electron carriers). Lastly, for each molecule of CoA (remember there were two to start with) you get:

       3 molecules of NADH x 2 cycles
       1 molecule of $FADH_2$ x 2 cycles
       1 molecule of ATP x 2 cycles

Therefore, this completes the breakdown of glucose. At this point, a total of four molecules of ATP have been made; two from glycolysis and one from each of the two turns of the Krebs cycle. Six molecules of carbon dioxide have been released; two prior to entering the Krebs cycle, and two for each of the two turns of the Krebs cycle. Twelve carrier molecules have been made; ten NADH and two $FADH_2$. These carrier molecules will carry electrons to the electron transport chain. ATP is made by substrate level phosphorylation in the Krebs cycle. Notice that the Krebs cycle in itself does not produce much ATP, but functions mostly in the transfer of electrons to be used in the electron transport chain where the most ATP is made.

In the **Electron Transport Chain,** NADH transfers electrons from glycolysis and the Kreb's cycle to the first molecule in the chain of molecules embedded in the inner membrane of the mitochondrion. Most of the molecules in the electron transport chain are proteins. Nonprotein molecules are also part of the chain and are essential for the catalytic functions of certain enzymes. The electron transport chain does not make ATP directly. Instead, it breaks up a large free energy drop into a more manageable amount. The chain uses electrons to pump H^+ across the mitochondrion membrane. The H^+ gradient is used to form ATP synthesis in a process called **chemiosmosis** (oxidative phosphorylation). ATP synthetase and energy generated by the movement of hydrogen ions coming off of NADH and $FADH_2$ builds ATP from ADP on the inner membrane of the mitochondria. Each NADH yields three molecules of ATP (10 x 3) and each $FADH_2$ yields two molecules of ATP (2 x 2). Thus, the electron transport chain and oxidative phosphorylation produces 34 ATP.

So, the net gain from the whole process of respiration is 36 molecules of ATP:

Glycolysis - 4 ATP made, 2 ATP spent = net gain of 2 ATP
Acetyl CoA- 2 ATP used
Krebs cycle - 1 ATP made for each turn of the cycle = net gain of 2 ATP
Electron transport chain - 34 ATP gained

Below is a diagram of the relationship between cellular respiration and photosynthesis.

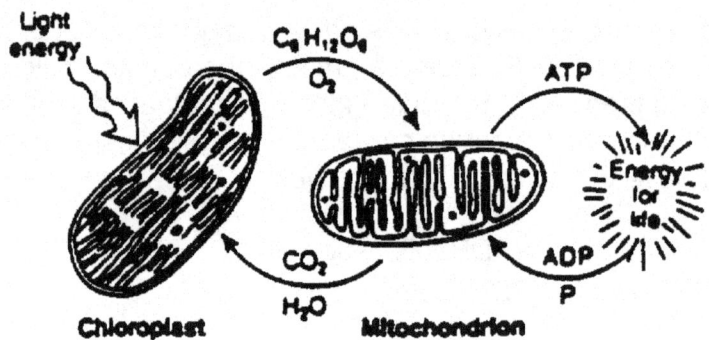

Photosynthesis is an anabolic process that stores energy in the form of a three carbon sugar. We will use glucose as an example for this section. Photosynthesis is done only by organisms that contain chloroplasts (plants, some bacteria, some protists). The **chloroplast** is the site of photosynthesis. It is similar to the mitochondria due to the increased surface area of the thylakoid membrane. It also contains a fluid called stroma between the stacks of thylakoids. The thylakoid membrane contains pigments (chlorophyll) that are capable of capturing light energy.

Photosynthesis reverses the electron flow. Water is split by the chloroplast into hydrogen and oxygen. The oxygen is given off as a waste product as carbon dioxide is reduced to sugar (glucose). This requires the input of energy, which comes from the sun.

Photosynthesis occurs in two stages: the light reactions and the Calvin cycle (dark reactions). The conversion of solar energy to chemical energy occurs in the light reactions. Electrons are transferred by the absorption of light by chlorophyll and cause the water to split, releasing oxygen as a waste product. The chemical energy that is created in the light reaction is in the form of NADPH. ATP is also produced by a process called photophosphorylation. These forms of energy are produced in the thylakoids and are used in the Calvin cycle to produce sugar.

The second stage of photosynthesis is the **Calvin cycle**. Carbon dioxide in the air is incorporated into organic molecules already in the chloroplast. The NADPH produced in the light reaction is used as reducing power for the reduction of the carbon to carbohydrate. ATP from the light reaction is also needed to convert carbon dioxide to carbohydrate (sugar). The process of photosynthesis is made possible by the presence of the sun. The formula for photosynthesis is:

$$CO_2 + H_2O + \text{energy (from sunlight)} \rightarrow C_6H_{12}O_6 + O_2$$

Chemical digestion of food in humans occurs as a series of exothermic reactions. Mechanically speaking, the teeth and saliva begin digestion by breaking food down into smaller pieces and lubricating it so it can be swallowed. The lips, cheeks, and tongue form a bolus or ball of food. It is carried down the pharynx by the process of peristalsis (wave-like contractions) and enters the stomach through the sphincter, which closes to keep food from going back up. In the stomach, pepsinogen and hydrochloric acid form pepsin, the enzyme that hydrolyzes proteins. The food is broken down further by this chemical action and is churned into acid chyme. The pyloric sphincter muscle opens to allow the food to enter the small intestine. Most nutrient absorption occurs in the small intestine. Its large surface area, accomplished by its length and protrusions called villi and microvilli, allow for a great absorptive surface into the bloodstream. Chyme is neutralized after coming from the acidic stomach to allow the enzymes found there to function. Accessory organs function in the production of necessary enzymes and bile. The pancreas makes many enzymes to break down food in the small intestine. The liver makes bile, which breaks down and emulsifies fatty acids. Any food left after the trip through the small intestine enters the large intestine. The large intestine functions to reabsorb water and produce vitamin K. The feces, or remaining waste, are passed out through the anus.

Skill 27.7   Applications of chemical properties of matter in physical, life, and earth/space science and technology - materials science, biochemistry, transportation, medicine, telecommunications.

Materials Science is based on the physics and chemistry of the solid state and embraces all aspects of engineering materials, including metals and their alloys, ceramic materials such as glasses, bricks, and porcelain insulators, polymers such as plastics, and rubbers together with semi-conducting and composite material. Materials Science extends from the extraction of the materials from their mineral sources and their refining and fabrication into finished products. It examines their chemical, crystal, molecular, and electronic structure because structure influences not only a material's magnetic and electronic characteristics but also its mechanical properties such as strength. It studies the degradation of materials in service by wear, corrosion, and oxidation and is concerned with developing methods of combating these; it considers the proper selection of materials for particular applications and the development of new materials for today's sophisticated technology.

Biochemistry is the study of chemical and physiochemical properties of living organisms. The thorough study and understanding of the structure and function of cellular components is based on the knowledge of the chemical properties of matter. An example is the study of cell metabolism. The understanding of processing nutrients is dependent on the understanding of chemical properties of the cell.

Advances have also been made in the field of **transportation** because of the understanding of chemical properties of matter. Scientists have been able to improve the quality of gasoline used in automobiles making it more environmentally friendly. The current research is focusing on the use of biofuels, renewable biological material, primarily plant matter or products derived from plant matter, as a fuel source.

In the **medical** field it was the understanding of the properties of bone and internal tissues of the body that allowed researchers to create magnetic resonance imaging (MRI). Scientists first created the nuclear magnetic machine to determine the structure of chemicals by measuring the vibration of atoms exposed to magnetic fields. They soon realized that this machine, connected to a computer, created the MRI and would allow them to take pictures of bone and internal tissue without the use of radioactivity. Created from an understanding of chemical properties of matter, this invention has also resulted in great strides in the diagnoses of disease and abnormalities.

Telecommunications has become dependent on the study of chemical properties of matter in improving and designing new modes of communication. The chemical properties of matter such as density and percent composition play a large role in the use of fiber optics to transport digital data. The modified chemical vapor deposition process that is used to create the form for optical fibers was created and continues to be refined because of an understanding of the chemical properties of matter.

COMPETENCY 028   UNDERSTANDING ENERGY AND INTERACTIONS BETWEEN MATTER AND ENERGY

Skill 28.1   Work, power, and potential and kinetic energy.

Work and energy:

Work is done on an object when an applied force moves through a distance.

Power is the work done divided by the amount of time that it took to do it. (Power = Work / time)

Technically, **energy is the ability to do work or supply heat.** Work is the transfer of energy to move an object a certain distance. It is the motion against an opposing force. Lifting a chair into the air is work; the opposing force is gravity. Pushing a chair across the floor is work; the opposing force is friction.

Heat, on the other hand, is not a form of energy but a method of transferring energy.

This energy, according to the First Law of Thermodynamics, is conserved. That means energy is neither created nor destroyed in ordinary physical and chemical processes (non-nuclear). Energy is merely changed from one form to another. Energy in all of its forms must be conserved. In any system, $\Delta E = q + w$ (E = energy, q = heat and w = work).

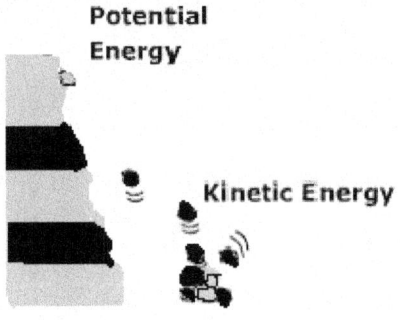

Energy exists in two basic forms, potential and kinetic. Kinetic energy is the energy of a moving object. Potential energy is the energy stored in matter due to position relative to other objects.

In any object, solid, liquid or gas, the atoms and molecules that make up the object are constantly moving (vibrational, translation and rotational motion) and colliding with each other. They are not stationary.

Due to this motion, the object's particles have varying amounts of kinetic energy. A fast moving atom can push a slower moving atom during a collision, so it has energy. All moving objects have energy and that energy depends on the object's mass and velocity. Kinetic energy is calculated: K.E. = ½ mv^2.

The temperature exhibited by an object is proportional to the average kinetic energy of the particles in the substance. Increase the temperature of a substance and its particles move faster so their average kinetic energies increase as well. But temperature is NOT energy, it is not conserved.

The energy an object has due to its position or arrangement of its parts is called potential energy. Potential energy due to position is equal to the mass of the object times the gravitational pull on the object times the height of the object, or:

PE = mgh

Where PE = potential energy; m = mass of object; g = gravity; and h = height.

Heat is energy that is transferred between objects caused by differences in their temperatures. Heat passes spontaneously from an object of higher temperature to one of lower temperature. This transfer continues until both objects reach the same temperature. Both kinetic energy and potential energy can be transformed into heat energy. When you step on the brakes in your car, the kinetic energy of the car is changed to heat energy by friction between the brake and the wheels. Other transformations can occur from kinetic to potential as well. Since most of the energy in our world is in a form that is not easily used, man and nature has developed some clever ways of changing one form of energy into another form that may be more useful.

Skill 28.2   Heat energy and the difference between heat and temperature.

Heat and temperature are different physical quantities. **Heat** is a measure of energy. **Temperature** is the measure of how hot (or cold) a body is with respect to a standard object.

Two concepts are important in the discussion of temperature changes. Objects are in thermal contact if they can affect each other's temperatures. Set a hot cup of coffee on a desk top. The two objects are in thermal contact with each other and will begin affecting each other's temperatures. The coffee will become cooler and the desktop warmer. Eventually, they will have the same temperature. When this happens, they are in **thermal equilibrium.**

We can not rely on our sense of touch to determine temperature because the heat from a hand may be conducted more efficiently by certain objects, making them feel colder. **Thermometers** are used to measure temperature. A small amount of mercury in a capillary tube will expand when heated. The thermometer and the object whose temperature it is measuring are put in contact long enough for them to reach thermal equilibrium. Then the temperature can be read from the thermometer scale.

Three temperature scales are used:

Celsius: The freezing point of water is set at 0 and the steam (boiling) point is 100. The interval between the two is divided into 100 equal parts called degrees Celsius.

Fahrenheit: The freezing point of water is 32 degrees and the boiling point is 212. The interval between is divided into 180 equal parts called degrees Fahrenheit.

Temperature readings can be converted from one to the other as follows.

Fahrenheit to Celsius   **Celsius to Fahrenheit**

$C = 5/9 (F - 32)$ $F = (9/5) C + 32$

Kelvin Scale has degrees the same size as the Celsius scale, but the zero point is moved to the triple point of water. Water inside a closed vessel is in thermal equilibrium in all three states (ice, water, and vapor) at 273.15 degrees Kelvin. This temperature is equivalent to .01 degrees Celsius. Because the degrees are the same in the two scales, temperature changes are the same in Celsius and Kelvin.

Temperature readings can be converted from Celsius to Kelvin:

Celsius to Kelvin   **Kelvin to Celsius**

$K = C + 273.15$ $C = K - 273.15$

Heat is a measure of energy. If two objects that have different temperatures come into contact with each other, heat flows from the hotter object to the cooler one.

Heat Capacity of an object is the amount of heat energy that it takes to raise the temperature of the object by one degree.

Heat capacity (C) per unit mass (m) is called **specific heat** (c):

$$c = \frac{C}{m} = \frac{Q/\Delta}{m}$$

Specific heats for many materials have been calculated and can be found in tables.

There are a number of ways that heat is measured. In each case, the measurement is dependent upon raising the temperature of a specific amount of water by a specific amount. These conversions of heat energy and work are called the **mechanical equivalent of heat**.

The **calorie** is the amount of energy that it takes to raise one gram of water one degree Celsius.

The **kilocalorie** is the amount of energy that it takes to raise one kilogram of water by one degree Celsius. Food calories are kilocalories.

In the International System of Units **(SI),** the calorie is equal to 4.184 **joules**.

A British thermal unit **(BTU)** = 252 calories = 1.054 kJ

Skill 28.3 **The principles of electricity and magnetism and their applications - electric circuits, motors, audio speakers, nerve impulses, lightning.**

An **electric circuit** is a path along which electrons flow. A simple circuit can be created with a dry cell, wire, a bell, or a light bulb. When all are connected, the electrons flow from the negative terminal, through the wire to the device and back to the positive terminal of the dry cell. If there are no breaks in the circuit, the device will work. The circuit is closed. Any break in the flow will create an open circuit and cause the device to shut off.

The device (bell, bulb) is an example of a **load**. A load is a device that uses energy. Suppose that you also add a buzzer so that the bell rings when you press the buzzer button. The buzzer is acting as a **switch**. A switch is a device that opens or closes a circuit. Pressing the buzzer makes the connection complete and the bell rings. When the buzzer is not engaged, the circuit is open and the bell is silent.

A **series circuit** is one where the electrons have only one path along which they can move. When one load in a series circuit goes out, the circuit is open. An example of this is a set of Christmas tree lights that is missing a bulb. None of the bulbs will work.

A **parallel circuit** is one where the electrons have more than one path to move along. If a load goes out in a parallel circuit, the other load will still work because the electrons can still find a way to continue moving along the path.

When an electron goes through a load, it does work and therefore loses some of its energy. The measure of how much energy is lost is called the **potential difference**. The potential difference between two points is the work needed to move a charge from one point to another.

Potential difference is measured in a unit called the volt. **Voltage** is potential difference. The higher the voltage, the more energy the electrons have. This energy is measured by a device called a voltmeter. To use a voltmeter, place it in a circuit parallel with the load you are measuring.

Current is the number of electrons per second that flow past a point in a circuit. Current is measured with a device called an ammeter. To use an ammeter, put it in series with the load you are measuring.

As electrons flow through a wire, they lose potential energy. Some is changed into heat energy because of resistance. **Resistance** is the ability of the material to oppose the flow of electrons through it. All substances have some resistance, even if they are a good conductor such as copper. This resistance is measured in units called **ohms**. A thin wire will have more resistance than a thick one because it will have less room for electrons to travel. In a thicker wire, there will be more possible paths for the electrons to flow. Resistance also depends upon the length of the wire. The longer the wire, the more resistance it will have. Potential difference, resistance, and current form a relationship know as **Ohm's Law**. Current **(I)** is measured in amperes and is equal to potential difference **(V)** divided by resistance **(R)**.

$$I = V / R$$

If you have a wire with resistance of 5 ohms and a potential difference of 75 volts, you can calculate the current by

$$I = 75 \text{ volts} / 5 \text{ ohms}$$
$$I = 15 \text{ amperes}$$

A current of 10 or more amperes will cause a wire to get hot. 22 amperes is about the maximum for a house circuit. Anything above 25 amperes can start a fire.

Electrostatics is the study of stationary electric charges. A plastic rod that is rubbed with fur or a glass rod that is rubbed with silk will become electrically charged and will attract small pieces of paper. The charge on the plastic rod rubbed with fur is negative and the charge on glass rod rubbed with silk is positive.

Electrically charged objects share these characteristics:

1. Like charges repel one another.
2. Opposite charges attract each other.
3. Charge is conserved. A neutral object has no net change. If the plastic rod and fur are initially neutral, when the rod becomes charged by the fur a negative charge is transferred from the fur to the rod. The net negative charge on the rod is equal to the net positive charge on the fur.

Materials through which electric charges can easily flow are called **conductors**. Metals which are good conductors include silicon and boron. On the other hand, an **insulator** is a material through which electric charges do not move easily, if at all. An example of an insulator would be non-metal elements of the periodic table. A simple device used to indicate the existence of a positive or negative charge is called an **electroscope**. An electroscope is made up of a conducting knob and attached to it are very lightweight conducting leaves usually made of foil (gold or aluminum). When a charged object touches the knob, the leaves push away from each other because like charges repel. It is not possible to tell whether if the charge is positive or negative.

Charging by induction:

Touch the knob with a finger while a charged rod is nearby. The electrons will be repulsed and flow out of the electroscope through the hand. If the hand is removed while the charged rod remains close, the electroscope will retain the charge.

When an object is rubbed with a charged rod, the object will take on the same charge as the rod. However, charging by induction gives the object the opposite charge as that of the charged rod.

Grounding charge:

Charge can be removed from an object by connecting it to the earth through a conductor. The removal of static electricity by conduction is called **grounding**.

Magnets have a north pole and a south pole. Like poles repel and opposing poles attract. A **magnetic field** is the space around a magnet where its force will affect objects. The closer you are to a magnet, the stronger the force. As you move away, the force becomes weaker.

Some materials act as magnets and some do not. This is because magnetism is a result of electrons in motion. The most important motion in this case is the spinning of the individual electrons. Electrons spin in pairs in opposite directions in most atoms. Each spinning electron has the magnetic field that it creates canceled out by the electron that is spinning in the opposite direction.

In an atom of iron, there are four unpaired electrons. The magnetic fields of these are not canceled out. Their fields add up to make a tiny magnet. There fields exert forces on each other setting up small areas in the iron called **magnetic domains** where atomic magnetic fields line up in the same direction.

You can make a magnet out of an iron nail by stroking the nail in the same direction repeatedly with a magnet. This causes poles in the atomic magnets in the nail to be attracted to the magnet. The tiny magnetic fields in the nail line up in the direction of the magnet. The magnet causes the domains pointing in its direction to grow in the nail. Eventually, one large domain results and the nail becomes a magnet.

A bar magnet has a north pole and a south pole. If you break the magnet in half, each piece will have a north and south pole.

The earth has a magnetic field. In a compass, a tiny, lightweight magnet is suspended and will line its south pole up with the North Pole magnet of the earth.

A magnet can be made out of a coil of wire by connecting the ends of the coil to a battery. When the current goes through the wire, the wire acts in the same way that a magnet does, it is called an **electromagnet**. The poles of the electromagnet will depend upon which way the electric current runs. An electromagnet can be made more powerful in three ways:

1. Make more coils.
2. Put an iron core (nail) inside the coils.
3. Use more battery power.

An **electric meter**, such as the one found on the side of a house, contains an aluminum disk that sits directly in a magnetic field created by electricity flowing through a conductor. The more the electricity flows (current), the stronger the magnetic field is. The stronger the magnetic field, the faster the disk turns. The disk is connected to a series of gears that turn a dial. Meter readers record the number from that dial.

In a **motor**, electricity is used to create magnetic fields that oppose each other and cause the rotor to move. The wiring loops attached to the rotating shaft have a magnetic field opposing the magnetic field caused by the wiring in the housing of the motor that cannot move. The repelling action of the opposing magnetic fields turns the rotor.

Nerve action depends on depolarization and an imbalance of electrical charges across the neuron. A polarized nerve has a positive charge outside the neuron. A depolarized nerve has a negative charge outside the neuron. Neurotransmitters turn off the sodium pump which results in depolarization of the membrane. This wave of depolarization (as it moves from neuron to neuron) carries an **electrical impulse**.

Lightning is a natural phenomena. With bolt temperatures hotter than the surface of the sun and shockwaves beaming out in all directions, its strength can cause death. Lightning is created in electrically charged storm systems. The upper portion of the cloud is positive and the lower portion is negative. When there is a charge separation in a cloud, there is also an electric field that is associated with that separation. Like the cloud, this field is negative in the lower region and positive in the upper region. The strength of the electric field is directly related to the amount of charge built up in the cloud. As atomic collisions occur, the charges at the top and bottom of the cloud increase, and the electric field becomes more intense. The repulsion of electrons causes the earth's surface to acquire a strong positive charge. The strong electric field created on Earth becomes the conductive path for the negative cloud bottom to contact the positive earth surface. The importance of this electric field delineation, is that it results in the separation of electrons. The electrons are now free to move much more easily than they could before the separation- the air is considered ionized and allows for electrical flow through the air, sometimes projecting into the atmosphere, and at other times landing on Earth's surface.

Skill 28.4 Properties of light - reflection, refraction, dispersion- to describe the function of optical systems and phenomena.

Shadows illustrate one of the basic properties of light. Light travels in a straight line. If you put your hand between a light source and a wall, you will interrupt the light and produce a shadow.

When light hits a surface, it is **reflected.** The angle of the incoming light (angle of incidence) is the same as the angle of the reflected light (angle of reflection). It is this reflected light that allows you to see objects. You see the objects when the reflected light reaches your eyes.

Different surfaces reflect light differently. Rough surfaces scatter light in many different directions. A smooth surface reflects the light in one direction. If it is smooth and shiny (like a mirror) you see your image in the surface.

When light enters a different medium, it bends. This bending, or change of speed, is called **refraction**.

Light can be **diffracted**, or bent around the edges of an object. Diffraction occurs when light goes through a narrow slit. As light passes through it, the light bends slightly around the edges of the slit. You can demonstrate this by pressing your thumb and forefinger together, making a very thin slit between them. Hold them about 8 cm from your eye and look at a distant source of light. The pattern you observe is caused by the diffraction of light.

Light and other electromagnetic radiation can be polarized because the waves are transverse. The distinguishing characteristic of transverse waves is that they are perpendicular to the direction of the motion of the wave. Polarized light has vibrations confined to a single plane that is perpendicular to the direction of motion. Light is able to be polarized by passing it through special filters that block all vibrations except those in a single plane. By blocking out all but one place of vibration, polarized sunglasses cut down on glare.

Light can travel through thin fibers of glass or plastic without escaping the sides. Light on the inside of these fibers is reflected so that it stays inside the fiber until it reaches the other end. Such fiber optics are being used to carry telephone messages. Sound waves are converted to electric signals which are coded into a series of light pulses which move through the optical fiber until they reach the other end. At that time, they are converted back into sound.

The image that you see in a bathroom **mirror** is a virtual image because it only seems to be where it is. However, a curved mirror can produce a real image. A real image is produced when light passes through the point where the image appears. A real image can be projected onto a screen.

Cameras use a convex lens to produce an image on the film. A **convex lens** is thicker in the middle than at the edges. The image size depends upon the focal length (distance from the focus to the lens). The longer the focal length, the larger the image. A **converging lens** produces a real image whenever the object is far enough from the lens so that the rays of light from the object can hit the lens and be focused into a real image on the other side of the lens.

Eyeglasses can help correct deficiencies of sight by changing where the image seen is focused on the retina of the eye. If a person is nearsighted, the lens of his eye focuses images in front of the retina. In this case, the corrective lens placed in the eyeglasses will be concave so that the image will reach the retina. In the case of farsightedness, the lens of the eye focuses the image behind the retina. The correction will call for a convex lens to be fitted into the glass frames so that the image is brought forward into sharper focus.

The most common type of **microscope** is the optical microscope. This is an instrument containing one or more lenses that produce an enlarged image of an object placed in the focal plane of the lens(es). Microscopes can largely be separated into two classes: optical theory microscopes and scanning microscopes. Optical theory microscopes are microscopes which function through lenses to magnify the image generated by the passage of a wave through the sample, this is what one thinks of in the high school classroom. The waves used are either electromagnetic in optical microscopes or electron beams in electron microscopes. Common types are the Compound Light, Stereo, and the electron microscope. Optical microscopes use refractive lenses, typically of glass and occasionally of plastic, to focus light into the eye or another light detector. Typical magnification of a light microscope is up to 1500x. Electron microscopes, which use beams of electrons instead of light, are designed for very high magnification and resolution. The most common of these would be the scanning electron microscopes used in spectroscopy studies.

A **rainbow** is an optical phenomenon. The rainbow's appearance is caused by dispersion of sunlight as it is refracted by raindrops. Hence, rainbows are commonly seen after a rainfall, or near fountains and waterfalls. A rainbow does not actually exist at a specific location in the sky, but rather is an optical phenomenon whose apparent position depends on the observer's location. All raindrops refract and reflect the sunlight in the same way, but only the light from some raindrops will reach the observer's eye. These raindrops create the perceived rainbow (as experienced by that observer).

Skill 28.5 Understanding the properties, production, and transmission of sound.

Sound waves are produced by a vibrating body. The vibrating object moves forward and compresses the air in front of it, then reverses direction so that the pressure on the air is lessened and expansion of the air molecules occurs. One compression and expansion creates one longitudinal wave. Sound can be transmitted through any gas, liquid, or solid. However, it cannot be transmitted through a vacuum, because there are no particles present to vibrate and bump into their adjacent particles to transmit the wave.

The vibrating air molecules move back and forth parallel to the direction of motion of the wave as they pass the energy from adjacent air molecules (closer to the source) to air molecules farther away from the source.

The **pitch** of a sound depends on the **frequency** that the ear receives. High-pitched sound waves have high frequencies. High notes are produced by an object that is vibrating at a greater number of times per second than one that produces a low note.

The **intensity** of a sound is the amount of energy that crosses a unit of area in a given unit of time. The loudness of the sound is subjective and depends upon the effect on the human ear. Two tones of the same intensity but different pitches may appear to have different loudness. The intensity level of sound is measured in decibels. Normal conversation is about 60 decibels. A power saw is about 110 decibels.

The **amplitude** of a sound wave determines its loudness. Loud sound waves have large amplitudes. The larger the sound wave, the more energy is needed to create the wave.

An oscilloscope is useful in studying waves because it gives a picture of the wave that shows the crest and trough of the wave. **Interference** is the interaction of two or more waves that meet. If the waves interfere constructively, the crest of each one meets the crests of the others. They combine into a crest with greater amplitude. As a result, you hear a louder sound. If the waves interfere destructively, then the crest of one meets the trough of another. They produce a wave with lower amplitude that produces a softer sound.

If you have two tuning forks that produce different pitches, then one will produce sounds of a slightly higher frequency. When you strike the two forks simultaneously, you may hear beats. **Beats** are a series of loud and soft sounds. This is because when the waves meet, the crests combine at some points and produce loud sounds. At other points, they nearly cancel each other out and produce soft sounds.

Skill 28.6   Properties and characteristics of waves - wavelength, frequency, interference- to describe a variety of waves.

The electromagnetic spectrum is measured in frequency (f) in hertz and wavelength (λ) in meters. The frequency times the wavelength of every **electromagnetic wave** equals the speed of light (3.0×10^9 meters/second).

Roughly, the range of wavelengths of the electromagnetic spectrum is:

	f	λ
Radio waves	$10^5 - 10^{-1}$ hertz	$10^3 - 10^9$ meters
Microwaves	$10^{-1} - 10^{-3}$ hertz	$10^9 - 10^{11}$ meters
Infrared radiation	$10^{-3} - 10^{-6}$ hertz	$10^{11.2} - 10^{14.3}$ meters
Visible light	$10^{-6.2} - 10^{-6.9}$ hertz	$10^{14.3} - 10^{15}$ meters
Ultraviolet radiation	$10^{-7} - 10^{-9}$ hertz	$10^{15} - 10^{17.2}$ meters
X-Rays	$10^{-9} - 10^{-11}$ hertz	$10^{17.2} - 10^{19}$ meters
Gamma Rays	$10^{-11} - 10^{-15}$ hertz	$10^{19} - 10^{23.25}$ meters

Radio waves are used for transmitting data. Common examples are television, cell phones, and wireless computer networks. Microwaves are used to heat food and deliver Wi-Fi service. Infrared waves are utilized in night vision goggles. Visible light we are all familiar with as the human eye is most sensitive to this wavelength range. UV light causes sunburns and would be even more harmful if most of it were not captured in the Earth's ozone layer. X-rays aid us in the medical field and gamma rays are most useful in the field of astronomy.

Seismic Waves are body waves that travel through the solid Earth. Primary Waves (P Waves) are the fastest traveling and the first to arrive, hence "primary." They are conducted as pressure differential and travel through any material. Secondary Waves (S Waves) are slower than P waves, "second" to arrive hence "secondary." They are conducted as a shearing motion and will not travel through liquids or gas because these materials lack shear strength.
There are three factors that cause different materials to conduct seismic waves differently.

Density: Seismic energy passing through more dense materials must displace more mass than seismic waves passing through a less dense material. All other factors being equal, higher densities slow down the propagation of seismic waves. Rocks in the deeper earth are denser because the weight of the overlying materials compresses the atoms closer together.

Incompressibility: Seismic waves are propagated by a rebounding action similar to the basketball: as p-wave energy hits a material, it causes rapid compression, and then the material rebounds (springs back) and passes the energy along. The faster the material rebounds, the faster seismic waves can travel through it. All materials have some compressive strength, but gasses are more compressible than liquids and liquids are more compressible than solids. Thus P waves travel through gasses very slowly (called sound waves) and they travel through solids rapidly.

Rigidity: rigidity or shear strength is the material's resistance to shearing force-- bending. Rigidity is measured in dynes/cm2. You can tell that water has zero rigidity: try turning ajar of water with a fish in it--the jar turns, the water does not, and the fish is still facing the same direction. This is because the sides of the jar slide across the water without affecting it-water has no shear strength. Put gelatin into the jar and you have a material with shear strength and the fish will turn with the jar. Also consider the act of diving into the water: you put your hands in front to break the water, and, if you do it right, you slide right into the water without pain, again because water has no shear strength, but if you do it wrong and face an area of your body toward the water you confront the incompressibility of the water. The change in rigidity will effect the conduction of the seismic waves.

Water Waves

Everything from earthquakes to ship wakes creates waves; however, the most common cause is wind. As wind passes over the water's surface, friction forces it to ripple. The strength of the wind, the distance the wind blows (fetch) and the length of the gust (duration) determine how big the ripples will become. Waves are divided into several parts. The crest is the highest point on a wave, while the trough or valley between two waves, is the lowest point. Wavelength is the horizontal distance, either between the crests or troughs of two consecutive waves. Wave height is a vertical distance between a wave's crest and the next trough. Wave period measures the size of the wave in time. A wave period can be measured by picking a stationary point and counting the seconds it takes for two consecutive crests or troughs to pass it.

In deep water, a wave is a forward motion of energy, not water. In fact, the water does not even move forward with a wave. If we followed a single drop of water during a passing wave, we would see it move in a vertical circle, returning to a point near its original position at the wave's end. These vertical circles are more obvious at the surface. As depth increases, their effects slowly decrease until completely disappearing about half a wavelength below the surface.

Longitudinal waves, such as **sound waves**, are characterized by the particle motion being parallel to the wave motion. The particles carry the sound to your ear.

COMPETENCY 029 **UNDERSTANDING ENERGY TRANSFORMATIONS AND THE CONSERVATION OF MATTER AND ENERGY.**

Skill 29.1 **Describes the processes that generate energy in the sun and other stars.**

All stars derive their energy through the thermonuclear fusion of light elements into heavy elements. The minimum temperature required for the fusion of hydrogen is 5 million degrees. Elements with more protons in their nuclei require higher temperatures. For instance, to fuse Carbon requires a temperature of about 1 billion degrees.

A star that is composed of mostly hydrogen is a young star. As a star gets older its hydrogen is consumed and tremendous energy and light is released through fusion. This is a three-step process: (1) two hydrogen nuclei (protons) fuse to form a heavy hydrogen called deuterium and release an electron and 4.04 MeV energy, (2) the deuterium fuses with another hydrogen nucleus (proton) to form a helium-3 and release a neutron and 3.28 MeV energy, and (3) and the helium-3 fuses with another helium-3 to form a helium-4 and release two hydrogens and 10.28 MeV energy.

In stars with central temperatures greater than 600-700 million degrees, carbon fusion is thought to take over the dominant role rather than hydrogen fusion. Carbon fusions can produce magnesium, sodium, neon, or helium. Some of the reactions release energy and alpha particles or protons.

Skill 29.2 **The law of conservation of matter to analyze a variety of situations - the water cycle, food chains, decomposition, balancing chemical equations.**

The law of conservation states that matter can be neither created nor destroyed. In layman's terms, matter is recycled indefinitely. Let's look at the water cycle. In the water cycle, water is recycled through the processes of evaporation and precipitation. Precipitation is part of a continuous process in which water at the Earth's surface evaporates, condenses into clouds, and returns to Earth. The water present now is the water that has been here since our atmosphere formed.

Food chains are linked in nature. Autotrophs produce their own energy, and are consumed by heterotrophs. Herbivores are eaten by carnivores. In most cases, each species has at least one indigenous predator who keeps the size of the population within limits. That predator has its own predator. Where a predator doesn't exist, man has become the hunter, nature has controlled populations through natural catastrophies and succession, or the population becomes too large and individuals eventually out compete one another for resources, while weaker members die.

Decomposition is nature's own compost pile. Over time, all organic matter decomposes, returning vital elements to the soil, increasing its richness for the next generation of growth. There are even specific bacteria and fungi whose aid in this process ensures that nutrients are not lost.

It is important to keep true to the law of conservation when balancing equations. This means that there must be the same number of atoms on both sides of the equation. Remember that the subscript numbers indicate the number of atoms in the elements. If there is no subscript, assume there is only one atom. The quantity of molecules are indicated by the number in front of an element or compound. If no number appears, assume that it is one molecule. Many chemical reactions give off energy. Like matter, energy can change form but it can be neither created nor destroyed during a chemical reaction.

Skill 29.3   Sources of electrical energy and processes of energy transformation for human uses - fossil fuels, solar panels, hydroelectric plants.

Burning of fossil fuels causes a great deal of air pollution as sulfur oxide, unburned hydrocarbons, and carbon monoxide are released into the air. Natural gas burns much cleaner and has the advantage of being able to be pumped through pipes to where it can be used.

Solar energy is radiation from the sun. Solar energy must be stored for use when the sun is not shining. Storage methods include heating water or rocks or converting the sun's rays into electricity using a photoelectric cell. Photoelectric cells are very expensive to make. Enormous amounts of water or rocks must be heated and insulated in storage to make use of the solar energy.

Hydroelectricity is produced by moving water. Building the dam and changing the flow of the stream can be harmful to the environment, can kill fish, and frequently destroys plant life. Hydroelectricity is dependent on the cycle of rain and snow, so heavier rainfall and snowfall in the mountains makes hydroelectricity production more reasonable.

Wood chips and other low-grade wood wastes are one type of biomass fuel. Other common biomass fuel sources are agricultural crop residues and farm animal wastes. Biomass includes gasoline made from corn or soybeans. Biomass is a renewable fuel that can be continuously produced. Since biomass is locally grown and harvested, there are no transportation costs and local jobs are preserved or created. Compared to fossil fuels, biomass fuels are historically lower-priced. Using biomass in place of fossil fuel reduces the atmospheric buildup of greenhouse gases, which cause climate change, and can also reduce the levels of gases that cause acid rain. Particulate emissions are relatively low.

Wind is captured by wind turbines and used to generate electricity. Wind power generation is clean; it doesn't cause air, soil or water pollution. However, wind farms must be located on large tracts of land or along coastlines to capture the greatest wind movement. Devoting those areas to wind power generation sometimes conflicts with other priorities, such as agriculture, urban development, or waterfront views from homes in prime locations.

Heat within the earth is called geothermal energy. Geothermal power plants produce little if any pollution or environmental hazards and can operate continuously for many years.

A nuclear reactor is a device for controlling and using the energy from a nuclear chain reaction. Strict safety regulations must be followed to protect people working in the plant and those in surrounding communities from radiation and the very hot water that is produced. Nuclear energy has some advantages over fossil fuels like coal and oil. Nuclear reactors do not add pollutants to the air. A greater amount of energy can be produced from smaller amounts of nuclear fuels. However, they require enormous amounts of water for cooling and they produce radioactive wastes that must be stored for thousands of years before they lose their radioactivity.

Skill 29.4 Exothermic and endothermic chemical reactions and their applications.

Interacting objects in the universe constantly exchange and transform energy. Total energy remains the same, but the form of the energy readily changes. Energy often changes from kinetic (motion) to potential (stored) or potential to kinetic. In reality, available energy, energy that is easily utilized, is rarely conserved in energy transformations. Heat energy is an example of relatively "useless" energy often generated during energy transformations. Exothermic reactions release heat and endothermic reactions require heat energy to proceed. For example, the human body is notoriously inefficient in converting chemical energy from food into mechanical energy. The digestion of food is exothermic and produces substantial heat energy.

Skill 29.5 The transfer of energy in a variety of situations - the production of heat, light, sound, and magnetic effects by electrical energy; the process of photosynthesis; weather processes; food webs; food/energy pyramids.

Thermodynamics is the study of energy and energy transfer. The first law of thermodynamics states the energy of the universe is constant. Thus, interactions involving energy deal with the transfer and transformation of energy, not the creation or destruction of energy.

Electricity is an important source of energy. Ovens and electric heaters convert electrical energy into heat energy. Electrical energy energizes the filament of a light bulb to produce light. Finally, the movement of electrical charges creates magnetic fields. Charges moving in a magnetic field experience a force, which is a transfer of energy.

The process of photosynthesis converts light energy from the sun into chemical energy (sugar). Cellular respiration later converts the sugar into ATP, a major energy source of all living organisms. Plants and certain types of bacteria carry out photosynthesis. The actions of the green pigment chlorophyll allow the conversion of unusable light energy into usable chemical energy.

Energy transfer plays an important role in weather processes. The three main types of heat transfer to the atmosphere are radiation, conduction, and convection. Radiation is the transfer of heat by electromagnetic waves. Sun light is an example of radiation. Conduction is the transfer of energy from one substance to another, or within a substance. Convection is the transfer of heat energy in a fluid. Air in the atmosphere acts as a fluid for the transfer of heat energy. Convection, resulting indirectly from the energy generated by sun light, is responsible for many weather phenomena including wind and clouds.

Energy transfer is also a key concept in the creation of food webs and food pyramids. Food webs and pyramids show the feeding relationships between organisms in an ecosystem. The primary producers of an ecosystem produce organic compounds from an energy source and inorganic materials. Primary consumers obtain energy by feeding on producers. Finally, secondary consumers obtain energy by feeding on primary consumers.

Skill 29.6 The law of conservation of energy to analyze a variety of physical phenomena- specific heat, nuclear reactions, efficiency of simple machines, collisions.

The principle of conservation states that certain measurable properties of an isolated system remain constant despite changes in the system. Two important principles of conservation are the conservation of mass and charge.

The principle of conservation of mass states that the total mass of a system is constant. Examples of conservation in mass in nature include the burning of wood, rusting of iron, and phase changes of matter. When wood burns, the total mass of the products, such as soot, ash, and gases, equals the mass of the wood and the oxygen that reacts with it. When iron reacts with oxygen, rust forms. The total mass of the iron-rust complex does not change. Finally, when matter changes phase, mass remains constant. Thus, when a glacier melts due to atmospheric warming, the mass of liquid water formed is equal to the mass of the glacier.

The principle of conservation of charge states that the total electrical charge of a closed system is constant. Thus, in chemical reactions and interactions of charged objects, the total charge does not change. Chemical reactions and the interaction of charged molecules are essential and common processes in living organisms and systems.

The kinetic theory states that matter consists of molecules, possessing kinetic energies, in continual random motion. The state of matter (solid, liquid, or gas) depends on the speed of the molecules and the amount of kinetic energy the molecules possess. The molecules of solid matter merely vibrate allowing strong intermolecular forces to hold the molecules in place. The molecules of liquid matter move freely and quickly throughout the body and the molecules of gaseous matter move randomly and at high speeds.

Matter changes state when energy is added or taken away. The addition of energy, usually in the form of heat, increases the speed and kinetic energy of the component molecules. Faster moving molecules more readily overcome the intermolecular attractions that maintain the form of solids and liquids. In conclusion, as the speed of molecules increases, matter changes state from solid to liquid to gas (melting and evaporation).

As matter loses heat energy to the environment, the speed of the component molecules decrease. Intermolecular forces have greater impact on slower moving molecules. Thus, as the speed of molecules decrease, matter changes from gas to liquid to solid (condensation and freezing).

Skill 29.7 Applications of energy transformations and the conservation of matter and energy in life and earth/space science.

The principle of conservation states that certain measurable properties of an isolated system remain constant despite changes in the system. Two important principles of conservation are the conservation of mass and charge.

The principle of conservation of mass states that the total mass of a system is constant. Examples of conservation in mass in nature include the burning of wood, rusting of iron, and phase changes of matter. When wood burns, the total mass of the products, such as soot, ash, and gases, equals the mass of the wood and the oxygen that reacts with it. When iron reacts with oxygen, rust forms. The total mass of the iron-rust complex does not change. Finally, when matter changes phase, mass remains constant. Thus, when a glacier melts due to atmospheric warming, the mass of liquid water formed is equal to the mass of the glacier.

The principle of conservation of charge states that the total electrical charge of a closed system is constant. Thus, in chemical reactions and interactions of charged objects, the total charge does not change. Chemical reactions and the interaction of charged molecules are essential and common processes in living organisms and systems.

DOMAIN III.   THE LIFE SCIENCES

COMPETENCY 030   UNDERSTANDING THE STRUCTURE AND FUNCTION OF LIVING THINGS.

Skill 30.1   Characteristics of organisms from the major taxonomic groups.

The organization of living systems builds by levels from small to increasingly more large and complex. All aspects, whether it be a cell or an ecosystem, have the same requirements to sustain life. Life is organized from simple to complex in the following way:

> **Organelles** make up **cells** which make up **tissues** which make up **organs**. Groups of organs make up **organ systems**. Organ systems work together to provide life for the **organism.**

Several characteristics have been described to identify living versus non-living substances.

1. **Living things are made of cells**; they grow, are capable of reproduction and respond to stimuli.

2. **Living things must adapt to environmental changes or perish**.

3. **Living things carry on metabolic processes**. They use and make energy.

Carolus Linnaeus is termed the father of taxonomy. **Taxonomy** is the science of classification. Linnaeus based his system on morphology (study of structure). Later on, evolutionary relationships (phylogeny) were also used to sort and group species. The modern classification system uses binomial nomenclature. This consists of a two word name for every species. The genus is the first part of the name and the species is the second part. Notice, in the levels explained below, that Homo sapiens is the scientific name for humans. Starting with the kingdom, the groups get smaller and more alike as one moves down the levels in the classification of humans:

Kingdom: Animalia, **Phylum:** Chordata, **Subphylum:** Vertebrata, **Class:** Mammalia, **Order:** Primate, **Family:** Hominidae, **Genus:** Homo, **Species:** sapiens

Species are defined by the ability to successfully reproduce with members of their own kind.

Classification of the major groups of animals.

Annelida - the segmented worms. The Annelida have specialized tissue. The circulatory system is more advanced in these worms and is a closed system with blood vessels. The nephridia are their excretory organs. They are hermaphrodidic and each worm fertilizes the other upon mating. They support themselves with a hydrostatic skeleton and have circular and longitudinal muscles for movement.

Mollusca - clams, octopus, soft bodied animals. These animals have a muscular foot for movement. They breathe through gills and most are able to make a shell for protection from predators. They have an open circulatory system, with sinuses bathing the body regions.

Arthropoda - insects, crustaceans and spiders; this is the largest group of the animal kingdom. Phylum arthropoda accounts for about 85% of all the animal species. Animals in the phylum arthropoda possess an exoskeleton made of chitin. They must molt to grow. Insects, for example, go through four stages of development. They begin as an egg, hatch into a larva, form a pupa, then emerge as an adult. Arthropods breathe through gills, trachae or book lungs. Movement varies, with members being able to swim, fly, and crawl. There is a division of labor among the appendages (legs, antennae, etc). This is an extremely successful phylum, with members occupying diverse habitats.

Echinodermata - sea urchins and starfish; these animals have spiny skin. Their habitat is marine. They have tube feet for locomotion and feeding.

Chordata - all animals with a notocord or a backbone. The classes in this phylum include Agnatha (jawless fish), Chondrichthyes (cartilage fish), Osteichthyes (bony fish), Amphibia (frogs and toads; gills which are replaced by lungs during development), Reptilia (snakes, lizards; the first to lay eggs with a protective covering), Aves (birds; warm-blooded with wings consisting of a particular shape and composition designed for flight), and Mammalia (warm blooded animals with body hair that bear their young alive, and possess mammary glands for milk production).

Kingdom Monera - bacteria and blue-green algae, prokaryotic, having no true nucleus, unicellular.

Kingdom Protista - eukaryotic, unicellular, some are photosynthetic, some are consumers.

Kingdom Fungi - eukaryotic, multicellular, absorptive consumers, contain a chitin cell wall.

Bacteria are classified according to their morphology (shape). **Bacilli** are rod shaped, **cocci** are round, and **spirillia** are spiral shaped. The **gram stain** is a staining procedure used to identify bacteria. Gram positive bacteria pick up the stain and turn purple. Gram negative bacteria do not pick up the stain and are pink in color. Microbiologists use methods of locomotion, reproduction, and how the organism obtains its food to classify protista.

Methods of locomotion - Flagellates have a flagellum, ciliates have cilia, and ameboids move through use of pseudopodia.

Methods of reproduction - binary fission is simply dividing in half and is asexual. All new organisms are exact clones of the parent. Sexual modes provide more diversity. Bacteria can reproduce sexually through conjugation, where genetic material is exchanged.

Methods of obtaining nutrition - photosynthetic organisms or producers, convert sunlight to chemical energy, consumers or heterotrophs eat other living things. Saprophytes are consumers that live off dead or decaying material.

Skill 30.2   Analyze how structure complements function in cells.

Parts of Eukaryotic Cells

1. Nucleus - The brain of the cell. The nucleus contains:

> **chromosomes**- DNA, RNA and proteins tightly coiled to conserve space while providing a large surface area.
> **chromatin** - loose structure of chromosomes. Chromosomes are called chromatin when the cell is not dividing.
> **nucleoli** - where ribosomes are made. These are seen as dark spots in the nucleus.
> **nuclear membrane** - contains pores which let RNA out of the nucleus. The nuclear membrane is continuous with the endoplasmic reticulum which allows the membrane to expand or shrink if needed.

2. Ribosomes - the site of protein synthesis. Ribosomes may be free floating in the cytoplasm or attached to the endoplasmic reticulum. There may be up to a half a million ribosomes in a cell, depending on how much protein is made by the cell.

3. **Endoplasmic Reticulum** - These are folded and provide a large surface area. They are the "roadway" of the cell and allow for transport of materials. The lumen of the endoplasmic reticulum helps to keep materials out of the cytoplasm and headed in the right direction. The endoplasmic reticulum is capable of building new membrane material. There are two types:

>   **Smooth Endoplasmic Reticulum** - contain no ribosomes on their surface.

>   **Rough Endoplasmic Reticulum** - contain ribosomes on their surface. This form of ER is abundant in cells that make many proteins, like in the pancreas, which produces many digestive enzymes.

4. **Golgi Complex or Golgi Apparatus** - This structure is stacked to increase surface area. The Golgi Complex functions to sort, modify and package molecules that are made in other parts of the cell. These molecules are either sent out of the cell or to other organelles within the cell.

5. **Lysosomes** - found mainly in animal cells. These contain digestive enzymes that break down food, substances not needed, viruses, damaged cell components, and eventually the cell itself. It is believed that lysosomes are responsible for the aging process.

6. **Mitochondria** - large organelles that make ATP to supply energy to the cell. Muscle cells have many mitochondria because they use a great deal of energy. The folds inside the mitochondria are called cristae. They provide a large surface where the reactions of cellular respiration occur. Mitochondria have their own DNA and are capable of reproducing themselves if a greater demand is made for additional energy. Mitochondria are found only in animal cells.

7. **Plastids** - found in photosynthetic organisms only. They are similar to the mitochondria due to their double membrane structure. They also have their own DNA and can reproduce if increased capture of sunlight becomes necessary. There are several types of plastids:

>   **Chloroplasts** - green, function in photosynthesis. They are capable of trapping sunlight.
>   **Chromoplasts** - make and store yellow and orange pigments; they provide color to leaves, flowers and fruits.
>   **Amyloplasts** - store starch and are used as a food reserve. They are abundant in roots like potatoes.

8. **Cell Wall** - found in plant cells only, it is composed of cellulose and fibers. It is thick enough for support and protection, yet porous enough to allow water and dissolved substances to enter. Cell walls are cemented to each other.

9. Vacuoles - hold stored food and pigments. Vacuoles are very large in plants. This allows them to fill with water in order to provide turgor pressure. Lack of turgor pressure causes a plant to wilt.

10. Cytoskeleton - composed of protein filaments attached to the plasma membrane and organelles. They provide a framework for the cell and aid in cell movement. They constantly change shape and move about. Three types of fibers make up the cytoskeleton:

> **Microtubules** - largest of the three; makes up cilia and flagella for locomotion. Flagella grow from a basal body. Some examples are sperm cells, and tracheal cilia. Centrioles are also composed of microtubules. They form the spindle fibers that pull the cell apart into two cells during cell division. Centrioles are not found in the cells of higher plants.
>
> **Intermediate Filaments** - they are smaller than microtubules but larger than microfilaments. They help the cell to keep its shape.
>
> **Microfilaments** - smallest of the three, they are made of actin and small amounts of myosin (like in muscle cells). They function in cell movement such as cytoplasmic streaming, endocytosis, and ameboid movement. This structure pinches the two cells apart after cell division, forming two cells.

Skill 30.3   Analyze how structure complements function in tissues, organs, organ systems, and organisms.

Structure and function dictates behavior and aids in the identification of prokaryotic organisms. Important structural and functional aspects of prokaryotes are morphology, motility, reproduction and growth, and metabolic diversity.

Morphology refers to the shape of a cell. The three main shapes of prokaryotic cells are spheres (cocci), rods (bacilli), and spirals (spirilla). Observation of cell morphology with a microscope can aid in the identification and classification of prokaryotic organisms. The most important aspect of prokaryotic morphology, regardless of the specific shape, is the size of the cells. Small cells allow for rapid exchange of wastes and nutrients across the cell membrane promoting high metabolic and growth rates.

Motility refers to the ability of an organism to move and its mechanism of movement. While some prokaryotes glide along solid surfaces or use gas vesicles to move in water, the vast majority of prokaryotes move by means of flagella. Motility allows organisms to reach different parts of its environment in the search for favorable conditions. Flagellar structure allows differentiation of Archaea and Bacteria as the two classes of prokaryotes have very different flagella. In addition, different types of bacteria have flagella positioned in different locations on the cell. The locations of flagella are on the ends (polar), all around (peritrichous), or in a tuft at one end of the cell (lophotrichous).

Most prokaryotes reproduce by binary fission, the growth of a single cell until it divides in two. Because of their small size, most prokaryotes have high growth rates under optimal conditions. Environmental factors greatly influence prokaryotic growth rate. Scientists identify and classify prokaryotes based on their ability or inability to survive and grow in certain conditions. Temperature, pH, water availability, and oxygen levels differentially influence the growth of prokaryotes. For example, certain types of prokaryotes can survive and grow at extremely hot or cold temperatures while most cannot.

Prokaryotes display great metabolic diversity. Autotrophic prokaryotes use carbon dioxide as the sole carbon source in energy metabolism, while heterotrophic prokaryotes require organic carbon sources. More specifically, chemoautotrophs use carbon dioxide as a carbon source and inorganic compounds as an energy source, while chemoheterotrophs use organic compounds as a source of energy and carbon. Photoautotrophs require only light energy and carbon dioxide, while photoheterotrophs require an organic carbon source along with light energy. Examining an unknown organism's metabolism aids in the identification process.

Skill 30.4   Identify human body systems and describes their functions.

Skeletal System - The skeletal system functions in support. Vertebrates have an endoskeleton, with muscles attached to bones. Skeletal proportions are controlled by area to volume relationships. Body size and shape are limited due to the forces of gravity. Surface area is increased to improve efficiency in all organ systems.

> The **axial skeleton** consists of the bones of the skull and vertebrae. The **appendicular skeleton** consists of the bones of the legs, arms, tail, and shoulder girdle. Bone is a connective tissue. Parts of the bone include compact bone which gives strength, spongy bone which contains red marrow to make blood cells, yellow marrow in the center of long bones to store fat cells, and the periosteum which is the protective covering on the outside of the bone.

A **joint** is defined as a place where two bones meet. Joints enable movement. **Ligaments** attach bone to bone. **Tendons** attach bones to muscles.

Muscular System - Functions in movement. There are three types of muscle tissue. Skeletal muscle is voluntary. These muscles are attached to bones. Smooth muscle is involuntary. It is found in organs and enables functions such as digestion and respiration. Cardiac muscle is a specialized type of smooth muscle and is found in the heart. Muscles can only contract; therefore they work in antagonistic pairs to allow back and forward movement. Muscle fibers are made of groups of myofibrils which are made of groups of sarcomeres. Actin and myosin are proteins which make up the sarcomere.

> **Physiology of muscle contraction** - A nerve impulse strikes a muscle fiber. This causes calcium ions to flood the sarcomere. Calcium ions allow ATP to expend energy. The myosin fibers creep along the actin, causing the muscle to contract. Once the nerve impulse has passed, calcium is pumped out and the contraction ends.

Nervous System - The neuron is the basic unit of the nervous system. It consists of an axon, which carries impulses away from the cell body, the dendrite, which carries impulses toward the cell body, and the cell body, which contains the nucleus. Synapses are spaces between neurons. Chemicals called neurotransmitters are found close to the synapse. The myelin sheath, composed of Schwann cells, covers the neurons and provides insulation.

> **Physiology of the nerve impulse** - Nerve action depends on depolarization and an imbalance of electrical charges across the neuron. A polarized nerve has a positive charge outside the neuron. A depolarized nerve has a negative charge outside the neuron. Neurotransmitters turn off the sodium pump which results in depolarization of the membrane. This wave of depolarization (as it moves from neuron to neuron) carries an electrical impulse. This is actually a wave of opening and closing gates that allows for the flow of ions across the synapse. Nerves have an action potential. There is a threshold of the level of chemicals that must be met or exceeded in order for muscles to respond. This is called the "all or none" response.

> The **reflex arc** is the simplest nerve response. The brain is bypassed. When a stimulus (like touching a hot stove) occurs, sensors in the hand send the message directly to the spinal cord. This stimulates motor neurons that contract the muscles to move the hand.

> **Voluntary nerve responses** involve the brain. Receptor cells send the message to sensory neurons that lead to association neurons. The message is taken to the brain. Motor neurons are stimulated and the message is transmitted to effector cells that cause the end effect.

Organization of the Nervous System - The somatic nervous system is controlled consciously. It consists of the central nervous system (brain and spinal cord) and the peripheral nervous system (nerves that extend from the spinal cord to the muscles). The autonomic nervous system is unconsciously controlled by the hypothalamus of the brain. Smooth muscles, the heart and digestion are some processes controlled by the autonomic nervous system. The sympathetic nervous system works opposite of the parasympathetic nervous system. For example, if the sympathetic nervous system stimulates an action, the parasympathetic nervous system would end that action.

> **Neurotransmitters** - these are chemicals released by exocytosis. Some neurotransmitters stimulate, while others inhibit, action.
>
> **Acetylcholine** - the most common neurotransmitter; it controls muscle contraction and heartbeat. The enzyme acetylcholinesterase breaks it down to end the transmission.
>
> **Epinephrine** - responsible for the "fight or flight" reaction. It causes an increase in heart rate and blood flow to prepare the body for action. It is also called adrenaline.
>
> **Endorphins and enkephalins** - these are natural pain killers and are released during serious injury and childbirth.

Digestive System - The function of the digestive system is to break food down and absorb it into the blood stream where it can be delivered to all cells of the body for use in cellular respiration. The teeth and saliva begin digestion by breaking food down into smaller pieces and lubricating it so it can be swallowed. The lips, cheeks, and tongue form a bolus (ball) of food. It is carried down the pharynx by the process of peristalsis (wave like contractions) and enters the stomach through the cardiac sphincter that closes to keep food from going back up. In the stomach, pepsinogen and hydrochloric acid form pepsin, the enzyme that breaks down proteins. The food is broken down further by this chemical action and is turned into chyme. The pyloric sphincter muscle opens to allow the food to enter the small intestine. Most nutrient absorption occurs in the small intestine. Its large surface area, accomplished by its length and protrusions called villi and microvilli allow for a great absorptive surface. Upon arrival into the small intestine, chyme is neutralized to allow the enzymes found there to function. Any food left after the trip through the small intestine enters the large intestine. The large intestine functions to reabsorb water and produce vitamin K. The feces, or remaining waste, are passed out through the anus.

> **Accessory organs** - although not part of the digestive tract, these organs function in the production of necessary enzymes and bile. The pancreas makes many enzymes to break down food in the small intestine. The liver makes bile which breaks down and emulsifies fatty acids.

Respiratory System - This system functions in the gas exchange of oxygen (needed) and carbon dioxide (waste). It delivers oxygen to the bloodstream and picks up carbon dioxide for release out of the body. Air enters the mouth and nose, where it is warmed, moistened and filtered of dust and particles. Cilia in the trachea trap unwanted material in mucus, which can be expelled. The trachea splits into two bronchial tubes and the bronchial tubes divide into smaller and smaller bronchioles in the lungs. The internal surface of the lung is composed of alveoli, which are thin walled air sacs. These allow for a large surface area for gas exchange. The alveoli are lined with capillaries. Oxygen diffuses into the bloodstream and carbon dioxide diffuses out to be exhaled. The oxygenated blood is carried to the heart and delivered to all parts of the body.

The thoracic cavity holds the lungs. A muscle, the diaphragm, below the lungs is an adaptation that makes inhalation possible. As the volume of the thoracic cavity increases, the diaphragm muscle flattens out and inhalation occurs. When the diaphragm relaxes, exhalation occurs.

Circulatory System

The function of the circulatory system is to carry oxygenated blood and nutrients to all cells of the body and return carbon dioxide waste to be expelled from the lungs. Be familiar with the parts of the heart and the path blood takes from the heart to the lungs, through the body and back to the heart. Unoxygenated blood enters the heart through the inferior and superior vena cava. The first chamber it encounters is the right atrium. It goes through the tricuspid valve to the right ventricle, on to the pulmonary arteries, and then to the lungs where it is oxygenated. It returns to the heart through the pulmonary vein into the left atrium. It travels through the bicuspid valve to the left ventricle where it is pumped to all parts of the body through the aorta.

Sinoatrial node (SA node) - the pacemaker of the heart. Located on the right atrium, it is responsible for contraction of the right and left atrium.

Atrioventricular node (AV node) - located on the left ventricle, it is responsible for contraction of the ventricles.

Blood vessels include:

arteries - lead away from the heart. All arteries carry oxygenated blood except the pulmonary artery going to the lungs. Arteries are under high pressure.

arterioles - arteries branch off to form these smaller passages.

capillaries - arterioles branch off to form tiny capillaries that reach every cell. Blood moves slowest here due to the small size; only one red blood cell may pass at a time to allow for diffusion of gases into and out of cells. Nutrients are also absorbed by the cells from the capillaries.

venules - capillaries combine to form larger venules. The vessels are now carrying waste products from the cells.

veins - venules combine to form larger veins, leading back to the heart. Veins and venules have thinner walls than arteries because they are not under as much pressure. Veins contain valves to prevent the backward flow of blood due to gravity.

Components of the blood include:

plasma – 60% of the blood is plasma. It contains salts called electrolytes, nutrients, and waste. It is the liquid part of blood.

erythrocytes - also called red blood cells; they contain hemoglobin which carries oxygen molecules.

leukocytes - also called white blood cells. White blood cells are larger than red cells. They are phagocytic and can engulf invaders. White blood cells are not confined to the blood vessels and can enter the interstitial fluid between cells.

platelets - assist in blood clotting. Platelets are made in the bone marrow.

Blood clotting - the neurotransmitter that initiates blood vessel constriction following an injury is called serotonin. A material called prothrombin is converted to thrombin with the help of thromboplastin. The thrombin is then used to convert fibrinogen to fibrin, which traps red blood cells to form a scab and stop blood flow.

Lymphatic System (Immune System)

Nonspecific defense mechanisms – They do not target specific pathogens, but are a whole body response. Results of nonspecific mechanisms are seen as symptoms of an infection. These mechanisms include the skin, mucous membranes and cells of the blood and lymph (ie: white blood cells, macrophages). Fever is a result of an increase of white blood cells. Pyrogens are released by white blood cells, which set the body's thermostat to a higher temperature. This inhibits the growth of microorganisms. It also increases metabolism to increase phagocytosis and body repair.

Specific defense mechanisms - They recognize foreign material and respond by destroying the invader. These mechanisms are specific in purpose and diverse in type. They are able to recognize individual pathogens. They are able to differentiate between foreign material and self. Memory of the invaders provides immunity upon further exposure.

> **antigen** - any foreign particle that invades the body.
> **antibody** - manufactured by the body, they recognize and latch onto antigens, hopefully destroying them.
> **immunity** - this is the body's ability to recognize and destroy an antigen before it causes harm. Active immunity develops after recovery from an infectious disease (chicken pox) or after a vaccination (mumps, measles, rubella). Passive immunity may be passed from one individual to another. It is not permanent. A good example is the immunities passed from mother to nursing child.

Excretory System

The function of the excretory system is to rid the body of nitrogenous wastes in the form of urea. The functional unit of excretion is the nephrons, which make up the kidneys. Antidiuretic hormone (ADH), which is made in the hypothalamus and stored in the pituitary, is released when differences in osmotic balance occur. This will cause more water to be reabsorbed. As the blood becomes more dilute, ADH release ceases.

The Bowman's capsule contains the glomerulus, a tightly packed group of capillaries. The glomerulus is under high pressure. Waste and fluids leak out due to pressure. Filtration is not selective in this area. Selective secretion by active and passive transport occur in the proximal convoluted tubule. Unwanted molecules are secreted into the filtrate. Selective secretion also occurs in the loop of Henle. Salt is actively pumped out of the tube and much water is lost due to the hyperosmosity of the inner part (medulla) of the kidney. As the fluid enters the distal convoluted tubule, more water is reabsorbed. Urine forms in the collecting duct which leads to the ureter then to the bladder where it is stored. Urine is passed from the bladder through the urethra. The amount of water reabsorbed back into the body is dependent upon how much water or fluids an individual has consumed. Urine can be very dilute or very concentrated if dehydration is present.

Endocrine System

The function of the endocrine system is to manufacture proteins called hormones. Hormones are released into the bloodstream and are carried to a target tissue where they stimulate an action. Hormones may build up over time to cause their effect, as in puberty or the menstrual cycle.

Hormone activation - Hormones are specific and fit receptors on the target tissue cell surface. The receptor activates an enzyme which converts ATP to cyclic AMP. Cyclic AMP (cAMP) is a second messenger from the cell membrane to the nucleus. The genes found in the nucleus turn on or off to cause a specific response.

There are two classes of hormones. **Steroid hormones** come from cholesterol. Steroid hormones cause sexual characteristics and mating behavior. Hormones include estrogen and progesterone in females and testosterone in males. **Peptide hormones** are made in the pituitary, adrenal glands (kidneys), and the pancreas. They include the following:

Follicle stimulating hormone (FSH) - production of sperm or egg cells

Luteinizing hormone (LH) - functions in ovulation

Luteotropic hormone (LTH) - assists in production of progesterone

Growth hormone (GH) - stimulates growth

Antidiuretic hormone (ADH) - assists in retention of water

Oxytocin - stimulates labor contractions at birth and let-down of milk

Melatonin - regulates circadian rhythms and seasonal changes

Epinephrine (adrenaline) - causes fight or flight reaction of the nervous system

Thyroxin - increases metabolic rate

Calcitonin - removes calcium from the blood

Insulin - decreases glucose level in blood

Glucagon - increases glucose level in blood

Hormones work on a feedback system. The increase or decrease in one hormone may cause the increase or decrease in another. Release of hormones causes a specific response.

Reproductive System

Sexual reproduction greatly increases diversity due to the many combinations possible through meiosis and fertilization. Gametogenesis is the production of the sperm and egg cells. Spermatogenesis begins at puberty in the male. One spermatozoa produces four sperm. The sperm mature in the seminiferous tubules located in the testes. Oogenesis, the production of egg cells is usually complete by the birth of a female. Egg cells are not released until menstruation begins at puberty. Meiosis forms one ovum with all the cytoplasm and three polar bodies which are reabsorbed by the body. The ovum is stored in the ovaries and released each month from puberty to menopause.

Path of the sperm - sperm are stored in the seminiferous tubules in the testes where they mature. Mature sperm are found in the epididymis located on top of the testes. After ejaculation, the sperm travels up the vas deferens where they mix with semen made in the prostate and seminal vesicles and travel out the urethra.

Path of the egg - eggs are stored in the ovaries. Ovulation releases the egg into the fallopian tubes which are ciliated to move the egg along. Fertilization normally occurs in the fallopian tube. If pregnancy does not occur, the egg passes through the uterus and is expelled through the vagina during menstruation. Levels of progesterone and estrogen stimulate menstruation. In the event of pregnancy, hormonal levels are affected by the implantation of a fertilized egg, so menstruation does not occur.

Pregnancy - if fertilization occurs, the zygote implants in about two to three days in the uterus. Implantation promotes secretion of human chorionic gonadotropin (HCG). This is what is detected in pregnancy tests. The HCG keeps the level of progesterone elevated to maintain the uterine lining in order to feed the developing embryo until the umbilical cord forms. Labor is initiated by oxytocin, which causes labor contractions and dilation of the cervix. Prolactin and oxytocin cause the production of milk.

Skill 30.5   Describe how organisms obtain and use energy and matter.

All organisms can be classes by the manner in which they obtain energy: chemoautotrophs, photoautotrophs, and heterotrophs.

Chemoautotrophs- These organisms are able to obtain energy via the oxidation of inorganic molecules (i.e., hydrogen gas and hydrogen sufide) or methane. This process is known as chemosynthesis. Most chemoautotrophs are bacteria or archaea that thrive in oxygen-poor environments, such as deep sea vents.

Photoautotrophs- Instead of obtaining energy from simple inorganic compounds like the chemoautotrophs, organisms of this type receive energy from sunlight. They employ the process of photosynthesis to create sugar from light, carbon dioxide and water. Higher order plants and algae as well as some bacteria and protists are photoautotrophs.

Heterotrophs- Any organism that requires organic molecules for as its source of energy is a heterotroph. These organisms are consumers in the food chain and must obtain nutrition from autotrophs or other heterotrophs. All animals are heterotrophs, as are some fungi and bacteria.

Photosynthesis is the process by which plants make carbohydrates from the energy of the sun, carbon dioxide, and water. Oxygen is a waste product. Photosynthesis occurs in the chloroplast where the pigment chlorophyll traps sun energy. It is divided into two major steps:

> **Light Reactions** - Sunlight is trapped, water is split, and oxygen is given off. ATP is made and hydrogens reduce NADP to $NADPH_2$. The light reactions occur in light. The products of the light reactions enter into the dark reactions (Calvin cycle).

> **Dark Reactions** - Carbon dioxide enters during the dark reactions which can occur with or without the presence of light. The energy transferred from $NADPH_2$ and ATP allow for the fixation of carbon into glucose.

Respiration - during times of decreased light, plants break down the products of photosynthesis through cellular respiration. Glucose, with the help of oxygen, breaks down and produces carbon dioxide and water as waste. Approximately fifty percent of the products of photosynthesis are used by the plant for energy.

Transpiration - water travels up the xylem of the plant through the process of transpiration. Water sticks to itself (cohesion) and to the walls of the xylem (adhesion). As it evaporates through the stomata of the leaves, the water is pulled up the column from the roots. Environmental factors such as heat and wind increase the rate of transpiration. High humidity will decrease the rate of transpiration.

Animal respiration takes in oxygen and gives off waste gases. For instance a fish uses its gills to extract oxygen from the water. Bubbles are evidence that waste gasses are expelled. Respiration without oxygen is called anaerobic respiration. Anaerobic respiration in animal cells is also called lactic acid fermentation. The end product is lactic acid.

Animal reproduction can be asexual or sexual. Geese lay eggs. Animals such as bear cubs, deer, and rabbits are born alive. Some animals reproduce frequently while others do not. Some animals only produce one baby yet others produce many (clutch size).

Animal digestion – some animals only eat meat (carnivores) while others only eat plants (herbivores). Many animals do both (omnivores). Nature has created animals with structural adaptations so they may obtain food through sharp teeth or long facial structures. Digestion's purpose is to break down carbohydrates, fats, and proteins. Many organs are needed to digest food. The process begins with the mouth. Certain animals, such as birds, have beaks to puncture wood or allow for large fish to be consumed. The tooth structure of a beaver is designed to cut down trees. Tigers are known for their sharp teeth used to rip hides from their prey. Enzymes are catalysts that help speed up chemical reactions by lowering effective activation energy. Enzyme rate is affected by temperature, pH, and the amount of substrate. Saliva is an enzyme that changes starches into sugars.

Animal circulation – The blood temperature of all mammals stays constant regardless of the outside temperature. This is called warm-blooded, while cold-blooded animals' (amphibians) circulation will vary with the temperature.

Skill 30.6 **Apply chemical properties to describe the structure and function of the basic chemical components - proteins, carbohydrates, lipids, nucleic acids - of living things.**

A compound consists of two or more elements. There are four major chemical compounds found in the cells and bodies of living things. These include carbohydrates, lipids, proteins and nucleic acids.

Monomers are the simplest unit of structure. **Monomers** can be combined to form **polymers**, or long chains, making a large variety of molecules possible. Monomers combine through the process of condensation reaction (also called dehydration synthesis). In this process, one molecule of water is removed between each of the adjoining molecules. In order to break the molecules apart in a polymer, water molecules are added between monomers, thus breaking the bonds between them. This is called hydrolysis.

Carbohydrates contain a ratio of two hydrogen atoms for each carbon and oxygen $(CH_2O)_n$. Carbohydrates include sugars and starches. They function in the release of energy. **Monosaccharides** are the simplest sugars and include glucose, fructose, and galactose. They are major nutrients for cells. In cellular respiration, the cells extract the energy in glucose molecules. **Disaccharides** are made by joining two monosaccharides by condensation to form a glycosidic linkage (covalent bond between two monosaccharides). Maltose is formed from the combination of two glucose molecules, lactose is formed from joining glucose and galactose, and sucrose is formed from the combination of glucose and fructose. **Polysaccharides** consist of many monomers joined. They are storage material hydrolyzed as needed to provide sugar for cells or building material for structures protecting the cell. Examples of polysaccharides include starch, glycogen, cellulose and chitin.

> **Starch** - major energy storage molecule in plants. It is a polymer consisting of glucose monomers.
> **Glycogen** - major energy storage molecule in animals. It is made up of many glucose molecules.
> **Cellulose** - found in plant cell walls, its function is structural. Many animals lack the enzymes necessary to hydrolyze cellulose, so it simply adds bulk (fiber) to the diet.
> **Chitin** - found in the exoskeleton of arthropods and fungi. Chitin contains an amino sugar (glycoprotein).

Lipids are composed of glycerol (an alcohol) and three fatty acids. Lipids are **hydrophobic** (water fearing) and will not mix with water. There are three important families of lipids, fats, phospholipids and steroids.

Fats consist of glycerol (alcohol) and three fatty acids. Fatty acids are long carbon skeletons. The nonpolar carbon-hydrogen bonds in the tails of fatty acids are why they are hydrophobic. Fats are solids at room temperature and come from animal sources (butter, lard).

Phospholipids are a vital component in cell membranes. In a phospholipid, one or two fatty acids are replaced by a phosphate group linked to a nitrogen group. They consist of a **polar** (charged) head that is hydrophilic or water loving and a **nonpolar** (uncharged) tail which is hydrophobic or water fearing. This allows the membrane to orient itself with the polar heads facing the interstitial fluid found outside the cell and the internal fluid of the cell.

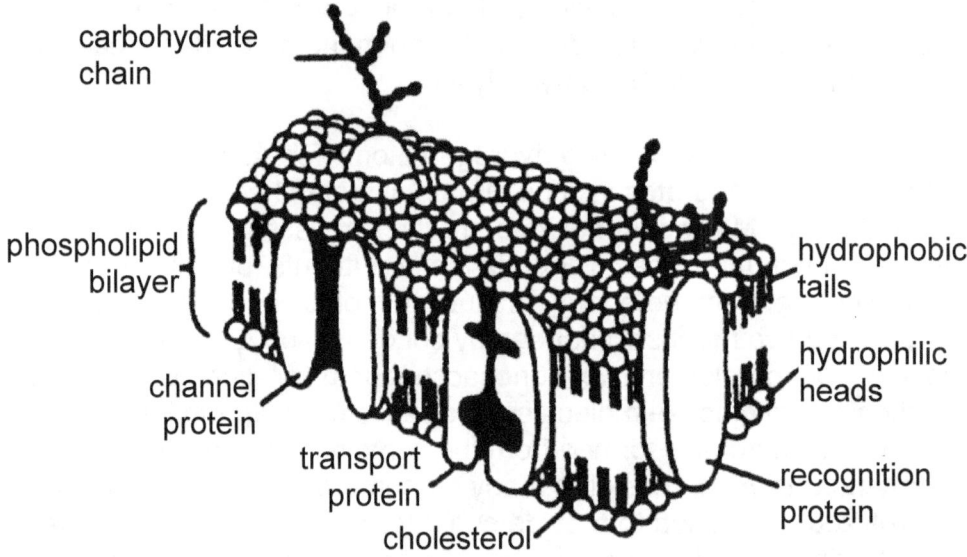

Steroids are insoluble and are composed of a carbon skeleton consisting of four inter-connected rings. An important steroid is cholesterol, which is the precursor from which other steroids are synthesized. Hormones, including cortisone, testosterone, estrogen, and progesterone, are steroids. Their insolubility keeps them from dissolving in body fluids.

Proteins compose about fifty percent of the dry weight of animals and bacteria. Proteins function in structure and aid in support (connective tissue, hair, feathers, quills), storage of amino acids (albumin in eggs, casein in milk), transport of substances (hemoglobin), hormonal to coordinate body activities (insulin), membrane receptor proteins, contraction (muscles, cilia, flagella), body defense (antibodies), and as enzymes to speed up chemical reactions.

All proteins are made of twenty **amino acids**. An amino acid contains an amino group and an acid group. The radical group varies and defines the amino acid. Amino acids form through condensation reactions with the removal of water. The bond that is formed between two amino acids is called a peptide bond. Polymers of amino acids are called polypeptide chains. An analogy can be drawn between the twenty amino acids and the alphabet. Millions of words can be formed using an alphabet of only twenty-six letters. This diversity is also possible using only twenty amino acids. This results in the formation of many different proteins, whose structure defines the function.

There are four levels of protein structure: primary, secondary, tertiary, and quaternary.

Primary structure is the protein's unique sequence of amino acids. A slight change in primary structure can affect a protein's conformation and its ability to function. **Secondary structure** is the coils and folds of polypeptide chains. The coils and folds are the result of hydrogen bonds along the polypeptide backbone. The secondary structure is either in the form of an alpha helix or a pleated sheet. The alpha helix is a coil held together by hydrogen bonds. A pleated sheet is the polypeptide chain folding back and forth. The hydrogen bonds between parallel regions hold it together. **Tertiary structure** is formed by bonding between the side chains of the amino acids. Disulfide bridges are created when two sulfhydryl groups on the amino acids bond together to form a strong covalent bond. **Quaternary structure** is the overall structure of the protein from the aggregation of two or more polypeptide chains. An example of this is hemoglobin. Hemoglobin consists of two kinds of polypeptide chains.

Nucleic acids consist of DNA (deoxyribonucleic acid) and RNA (ribonucleic acid). Nucleic acids contain the instructions for the amino acid sequence of proteins and the instructions for replicating. The monomer of nucleic acids is called a nucleotide. A nucleotide consists of a 5 carbon sugar, (deoxyribose in DNA, ribose in RNA), a phosphate group, and a nitrogenous base. The base sequence codes for the instructions. There are five bases: adenine, thymine, cytosine, guanine, and uracil. Uracil is found only in RNA and replaces the thymine. A summary of nucleic acid structure can be seen in the table below:

	SUGAR	PHOSPHATE	BASES
DNA	deoxy-ribose	present	adenine, thymine, cytosine, guanine
RNA	ribose	present	adenine, uracil, cytosine, guanine

Due to the molecular structure, adenine will always pair with thymine in DNA or uracil in RNA. Cytosine always pairs with guanine in both DNA and RNA. This allows for the symmetry of the DNA molecule seen below.

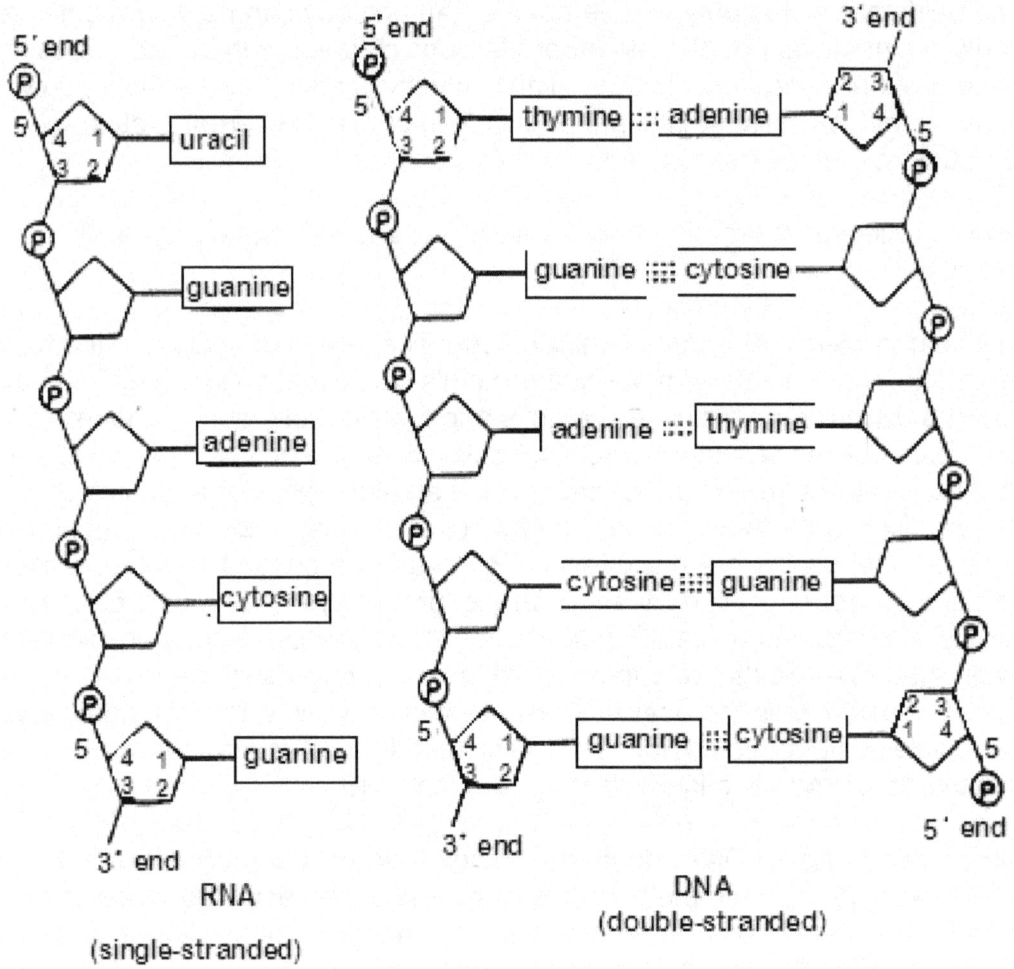

RNA
(single-stranded)

DNA
(double-stranded)

Adenine and thymine (or uracil) are linked by two covalent bonds and cytosine and guanine are linked by three covalent bonds. The guanine and cytosine bonds are harder to break apart than thymine (uracil) and adenine because of the greater number of these bonds. The DNA molecule is called a double helix due to its twisted ladder shape.

COMPETENCY 031 UNDERSTANDING REPRODUCTION AND THE MECHANISMS OF HEREDITY

Skill 31.1 Compare and contrast sexual and asexual reproduction

The obvious advantage of asexual reproduction is that it does not require a partner. This is a huge advantage for organisms, such as the hydra, which do not move around. Not having to move around to reproduce also allows organisms to conserve energy. Asexual reproduction also tends to be faster. There are disadvantages, as in the case of regeneration, in plants if the plant is not in good condition or in the case of spore-producing plants, if the surrounding conditions are not suitable for the spores to grow. As asexual reproduction produces only exact copies of the parent organism, it does not allow for genetic variation, which means that mutations, or weaker qualities, will always be passed on. This can also be detrimental to a species well-adapted to a particular environment when the conditions of that environment change suddenly. On the whole, asexual reproduction is more reliable because it requires fewer steps and less can go wrong.

Sexual reproduction shares genetic information between gametes, thereby producing variety in the species. This can result in a better species with an improved chance of survival. There is the disadvantage that sexual reproduction requires a partner, which in turn with many organisms requires courtship, finding a mate, and mating. Another disadvantage is that sexually reproductive organisms require special mechanisms.

Skill 31.2 The organization of heredity material - DNA, genes, chromosomes.

DNA and DNA REPLICATION

The modern definition of a gene is a unit of genetic information. DNA makes up genes which in turn make up the chromosomes. DNA is wound tightly around proteins in order to conserve space. The DNA/protein combination makes up the chromosome. DNA controls the synthesis of proteins, thereby controlling the total cell activity. DNA is capable of making copies of itself.

Review of DNA structure:

1. Made of nucleotides; a five carbon sugar, phosphate group and nitrogen base (either adenine, guanine, cytosine or thymine).

2. Consists of a sugar/phosphate backbone that is covalently bonded. The bases are joined down the center of the molecule and are attached by hydrogen bonds that are easily broken during replication.

3. The amount of adenine equals the amount of thymine and the amount of cytosine equals the amount of guanine.

4. The shape is that of a twisted ladder called a double helix. The sugar/phosphates make up the sides of the ladder and the base pairs make up the rungs of the ladder.

DNA Replication

Enzymes control each step of the replication of DNA. The molecule untwists. The hydrogen bonds between the bases break and serve as a pattern for replication. Free nucleotides found inside the nucleus join on to form a new strand. Two new pieces of DNA are formed which are identical. This is a very accurate process. There is only one mistake for every billion nucleotides added. This is because there are enzymes (polymerases) present that proofread the molecule. In eukaryotes, replication occurs in many places along the DNA at once. The molecule may open up at many places like a broken zipper. In prokaryotic circular plasmids, replication begins at a point on the plasmid and goes in both directions until it meets itself.

Base pairing rules are important in determining a new strand of DNA sequence. For example say our original strand of DNA had the sequence as follows:

1. A T C G G C A A T A G C   This may be called our sense strand as it contains a sequence that makes sense or codes for something.
The complementary strand (or other side of the ladder) would follow base pairing rules (A bonds with T and C bonds with G) and would read:

2. T A G C C G T T A T C G   When the molecule opens up and nucleotides join on, the base pairing rules create two new identical strands of DNA

1. A T C G G C A A T A G C   and   A T C G G C A A T A G C
   T A G C C G T T A T C G   2.T A G C C G T T A T C G

Gregor Mendel is recognized as the father of genetics. His work in the late 1800's is the basis of our knowledge of genetics. Although unaware of the presence of DNA or genes, Mendel realized there were factors (now known as genes) that were transferred from parents to their offspring. Mendel worked with pea plants and fertilized the plants himself, keeping track of subsequent generations which led to the Mendelian laws of genetics. Mendel found that two "factors" governed each trait, one from each parent. Traits or characteristics came in several forms, known as alleles. For example, the trait of flower color had white alleles and purple alleles. Mendel formed three laws:

Law of dominance - in a pair of alleles, one trait may cover up the allele of the other trait. Example: brown eyes are dominant to blue eyes.

Law of segregation - only one of the two possible alleles from each parent is passed on to the offspring from each parent. (During meiosis, the haploid number insures that half the sex cells get one allele, half get the other).

Law of independent assortment - alleles sort independently of each other. (Many combinations are possible depending on which sperm ends up with which egg. Compare this to the many combinations of hands possible when dealing a deck of cards).

monohybrid cross - a cross using only one trait.

dihybrid cross - a cross using two traits. More combinations are possible.

Punnet squares - these are used to show the possible ways that genes combine and indicate probability of the occurrence of a certain genotype or phenotype. One parent's genes are put at the top of the box and the other parent at the side of the box. Genes combine on the square just like numbers that are added in addition tables we learned in elementary school.

Example: Monohybrid Cross - four possible gene combinations

Example: Dihybrid Cross - sixteen possible gene combinations

Skill 31.3 **Describes how an inherited trait can be determined by one or many genes and how more than one trait can be influenced by a single gene**

Please see skill 12.4 for a review of the various ways genes work together to build the complete phenotype of an individual organism.

Skill 31.4 Distinguish between dominant and recessive traits and predict the probable outcomes of genetic combinations.

SOME DEFINITIONS TO KNOW -

Dominant - the stronger of the two traits. If a dominant gene is present, it will be expressed. Shown by a capital letter.

Recessive - the weaker of the two traits. In order for the recessive gene to be expressed, there must be two recessive genes present. Shown by a lower case letter.

Homozygous - (purebred) having two of the same genes present; an organism may be homozygous dominant with two dominant genes or homozygous recessive with two recessive genes.

Heterozygous - (hybrid) having one dominant gene and one recessive gene. The dominant gene will be expressed due to the Law of Dominance.

Genotype - the genes the organism has. Genes are represented with letters. AA, Bb, and tt are examples of genotypes.

Phenotype - how the trait is expressed in an organism. Blue eyes, brown hair, and red flowers are examples of phenotypes.

Incomplete dominance - neither gene masks the other; a new phenotype is formed. For example, red flowers and white flowers may have equal strength. A heterozygote (Rr) would have pink flowers. If a problem occurs with a third phenotype, incomplete dominance is occurring.

Codominance - genes may form new phenotypes. The ABO blood grouping is an example of co-dominance. A and B are of equal strength and O is recessive. Therefore, type A blood may have the genotypes of AA or AO, type B blood may have the genotypes of BB or BO, type AB blood has the genotype A and B, and type O blood has two recessive O genes.

Linkage - genes that are found on the same chromosome usually appear together unless crossing over has occurred in meiosis. (Example - blue eyes and blonde hair)

Lethal alleles - these are usually recessive due to the early death of the offspring. If a 2:1 ratio of alleles is found in offspring, a lethal gene combination is usually the reason. Some examples of lethal alleles include sickle cell anemia, tay-sachs and cystic fibrosis. Usually the coding for an important protein is affected.

Inborn errors of metabolism - these occur when the protein affected is an enzyme. Examples include PKU (phenylketonuria) and albanism.

Polygenic characters - many alleles code for a phenotype. There may be as many as twenty genes that code for skin color. This is why there is such a variety of skin tones. Another example is height. A couple of medium height may have very tall offspring.

Sex linked traits - the Y chromosome found only in males (XY) carries very little genetic information, whereas the X chromosome found in females (XX) carries very important information. Since men have no second X chromosome to cover up a recessive gene, the recessive trait is expressed more often in men. Women need the recessive gene on both X chromosomes to show the trait. Examples of sex linked traits include hemophilia and color-blindness.

Sex influenced traits - traits are influenced by the sex hormones. Male pattern baldness is an example of a sex influenced trait. Testosterone influences the expression of the gene. Mostly men loose their hair due to this trait.

Skill 31.5   Evaluate the influence of environmental and genetic factors on the traits of an organism.

Environmental factors can influence the structure and expression of genes. For instance, viruses can insert their DNA into the host's genome changing the composition of the host DNA. In addition, mutagenic agents found in the environment cause mutations in DNA and carcinogenic agents promote cancer, often by causing DNA mutations.

Many viruses can insert their DNA into the host genome causing mutations. Many times viral insertion of DNA does not harm the host DNA because of the location of the insertion. Some insertions, however, can have grave consequences for the host. Oncogenes are genes that increase the malignancy of tumor cells. Some viruses carry oncogenes that, when inserted into the host genome, become active and promote cancerous growth. In addition, insertion of other viral DNA into the host genome can stimulate expression of host proto-oncogenes, genes that normally promote cell division. For example, insertion of a strong viral promoter in front of a host proto-oncogene may stimulate expression of the gene and lead to uncontrolled cell growth (i.e. cancer).

In addition to viruses, physical and chemical agents found in the environment can damage gene structure. Mutagenic agents cause mutations in DNA. Examples of mutagenic agents are x-rays, uv light, and ethidium bromide. Carcinogenic agents are any substances that promote cancer. Carcinogens are often, but not always, mutagens. Examples of agents carcinogenic to humans are asbestos, uv light, x-rays, and benzene.

Skill 31.6 Current applications of genetic research- *cloning*, reproduction, health, industry, agriculture.

Research in molecular genetics is highly technical and very useful to society. Molecular genetics is generously funded from various sources and the scientific community focuses on this area of research because of it's use to humanity.

A major goal of molecular genetics is to correlate the sequence of a gene with its function. The primary objective is to obtain the sequence of a gene.

Scientists use a number of techniques to isolate genes. The foremost technique is complementary DNA (cDNA) cloning, which is very widely used because of its reliability. The aim of many cloning experiments is to obtain a sequence of DNA that directs the production of a specific protein. The principle involved in this cDNA cloning is that an mRNA population isolated from a specific developmental stage contains mRNAs specific for any protein expressed during that stage. Thus, if the mRNA can be isolated, the gene can be obtained. It is not possible to clone mRNA directly, but by copying the isolated DNA, a copy of the mRNA can be obtained and subsequently cloned. In this method, a single strand of the DNA is copied. The second strand is generated by the DNA polymerase and the result is a double stranded DNA.

The objective of cloning experiments is to isolate many genes from a wide variety of living organisms.

These isolated genes can be used in genetic engineering in a variety of ways.

1. Cloning a clinically important genes for diagnostic purposes. The gene which encodes for one type of hemophilia has been used for this purpose. This field of research is proving to be of invaluable help to people suffering from various diseases like cancer, diabetes etc. and research in this area hold lot of promise for future generations.

2. Isolated genes could be used to identify similar genes from other organisms. Thus, it can serve as a heterologous probe.

3. The nucleic acid sequence of a gene can be derived. If a partial or complete sequence of the protein that a gene encodes is available, the gene can be confirmed in this manner. If the protein product is not known then the sequence of the gene is compared with those of the known genes to try to derive a function for that gene. Knowing the function of a gene is very important in clinical diagnostic purposes.

4. Isolating a gene that causes disease in a particular crop, will help the farmers to deal with that disease.

COMPETENCY 32.0 UNDERSTANDING ADAPTATION OF ORGANISMS AND THE THEORY OF EVOLUTION.

Skill 32.1 Similarities and differences among various types of organisms and methods of classifying organisms.

Scientists believe that there are over ten million species of living things. Of these, 1.5 million have been named and classified. Systems of classification show similarities and also assist scientists with an internationally accepted system of organization.

Taxonomy is the science of classification. **Carolus Linnaeus** is known as the father of taxonomy. Linnaeus based his system on morphology (study of structure). Later on, evolutionary relationships (phylogeny) were also used to sort and group species. The modern classification system uses binomial nomenclature. This consists of a two word name for every species. The genus is the first part of the name and the species is the second part. Notice in the levels explained below that *Homo sapiens* is the scientific name for humans. Starting with the kingdom, the groups get smaller and more alike as one moves down the levels in the classification of humans:

Kingdom: Animalia - Phylum: Chordata - Subphylum: Vertebrata - Class: Mammalia - Order: Primate - Family: Hominidae - Genus: Homo - Species: sapiens

The typical graphic representation of a classification is called a **phylogenetic tree**, which represents a hypothesis of the relationships between organisms based on branching of lineages through time within a group. Every time you see a phylogenetic tree, you should be aware that it is making statements on the degree of similarity between organisms, or the particular pattern in which the various lineages diverged (phylogenetic history).

Cladistics is the study of phylogenetic relationships of organisms by analysis of shared, derived character states. Cladograms are constructed to show evolutionary pathways. Character states are polarized in cladistic analysis to be plesiomorphous (ancestral features), symplesiomorphous (shared ancestral features), apomorphous (derived characteristics), and synapomorphous (shared, derived features).

Prokaryotes are separated from eukaryotes in the current five kingdom system. The prokaryotes belong to the kingdom monera while the eukaryotes belong to either kingdom protista, plantae, fungi, or animalia. Recent comparisons of nucleic acids and proteins between different groups of organisms have led to problems concerning the five kingdom system. Based on these comparisons, alternative kingdom systems have emerged. Six and eight kingdoms as well as a three domain system have been proposed as a more accurate classification system. It is important to note that classification systems evolve as more information regarding characteristics and evolutionary histories of organisms arise.

Species are defined by the ability to successfully reproduce with members of their own kind.

Several different morphological criteria are used to classify organisms:

1. **Ancestral characters** - characteristics that are unchanged after evolution (eg.: 5 digits on the hand of an ape).

2. **Derived characters** - characteristics that have evolved more recently (eg.: the absence of a tail on an ape).

3. **Conservative characters** - traits that change slowly.

4. **Homologous characters** - characteristics with the same genetic basis but used for a different function. (eg.: wing of a bat, arm of a human. The bone structure is the same, but the limbs are used for different purposes).

5. **Analogous characters** – structures that differ, but used for similar purposes (eg.- the wing of a bird and the wing of a butterfly).

6. **Convergent evolution** - development of similar adaptations by organisms that are unrelated.

Biological characteristics are also used to classify organisms. Protein comparison, DNA comparison, and DNA analysis of fossilized structures are powerful comparative methods used to measure evolutionary relationships between species. Taxonomists consider the organism's life history, biochemical (DNA) makeup, behavior, and how the organisms are distributed geographically. The fossil record is also used to show evolutionary relationships.

Skill 32.2 Traits in a population that enhance survival and reproductive success.

Anatomical structures and physiological processes that evolve over geological time to increase the overall reproductive success of an organism in its environment are known as biological adaptations. Such evolutionary changes occur through natural selection, the process by which individual organisms with favorable traits survive to reproduce more frequently than those with unfavorable traits. The heritable components of such favorable traits are often passed down to offspring during reproduction, increasing the frequency of the favorable trait in a population over many generations.

Adaptations increase long-term reproductive success by making an organism better suited for survival under particular environmental conditions and pressures. These biological changes can increase an organism's ability to obtain air, water, food and nutrients, to cope with environmental variables and defend themselves. The term adaptation may apply to changes in biological processes that, for example, enable on organism to produce venom or to regulate body temperature, and also to structural adaptations, such as an organisms' skin color and shape. Adaptations can occur in behavioral traits and survival mechanisms as well.

One well-known structural change that demonstrates the concept of adaptation is the development of the primate and human opposable thumb, the first digit of the hand that can be moved around to touch other digits and to grasp objects. The history of the opposable thumb is one of complexly linked structural and behavioral adaptations in response to environmental stressors.

Early apes first appearing in the Tertiary Period were mostly tree dwelling organisms that foraged for food and avoided predators high above the ground. The apes' need to quickly and effectively navigate among branches led to the eventual development of the opposable thumb through the process of natural selection, as apes with more separated thumbs demonstrated higher survival and reproductive rates. This structural adaptation made the ape better suited for its environment by increasing dexterity while climbing trees, moving through the canopy, gathering food and gripping tools such as sticks and branches.

Following the development of the opposable thumb in primates, populations of early human ancestors began to appear in a savannah environment with fewer trees and more open spaces. The need to cross such expanses and utilize tools led to the development of bipedalism in certain primates and hominids. Bipedalism was both a structural adaptation in the physical changes that occurred in the skull, spine and other parts of the body to accommodate upright walking, as well as a behavioral adaptation that led primates and hominids to walk on only two feet. Freeing of the hands for tool use led, in turn, to other adaptations, and evolutionists attribute the gradual increase in brain size and expansion of motor skills in hominids largely to appearance of the opposable thumb. Thus, the developments of many of the most important adaptations of primates and humans demonstrate closely connected evolutionary histories.

Skill 32.3   How populations and species change through time.

Darwin defined the theory of Natural Selection in the mid-1800's. Darwin theorized that nature selects the traits that are advantageous to the organism through the study of finches on the Galapagos Islands. He further theorized that those that do not possess the desirable trait die and do not pass on their genes. However, those more fit to survive reproduce, thus increasing the prevalence of that gene in the population. According to Darwin's theory, there are four principles used to define natural selection:

1. The individuals in a certain species vary from generation to generation.
2. Some of the variations are determined by the genetic makeup of the species.
3. More individuals are produced than will survive.
4. Some genes allow for better survival of an animal.

Causes of evolution - Certain factors increase the chances of variability in a population, thus leading to evolution. Items that increase variability include mutations, sexual reproduction, immigration, and large population size. Items that decrease variation would be natural selection, emigration, small population size, and random mating.

Sexual selection – Combinations of genes determine the makeup of the gene pool. Animals that use mating behaviors may be successful or unsuccessful. An animal that lacks attractive plumage or has a weak mating call will not attract the female, thereby eventually limiting the prevalence of that gene in the gene pool. Mechanical isolation, where sex organs do not fit the female, has an obvious disadvantage.

Skill 32.4 Apply knowledge of the mechanisms and processes of biological evolution- variation, mutation, environmental factors, natural selection.

Different evolutionary patterns affect the diversity of organisms in different ways. Genetic drift, convergent evolution, and three patterns of selection lead to a decrease in diversity. On the other hand, some aspects of punctuated equilibrium and one pattern of selection, promote increased diversity.

Genetic drift along with natural selection is one of the main mechanisms of evolution. Genetic drift is the chance deviation in the frequency of alleles (traits) resulting from the randomness of zygote formation and selection. Because only a small percentage of all possible zygotes become mature adults, parents do not necessarily pass all of their alleles on to their offspring. Genetic drift is particularly important in small populations because chance deviations in allelic frequency can quickly alter the genotypic make-up of the population. In extreme cases, certain alleles may completely disappear from the gene pool. Genetic drift is particularly influential when environmental events and conditions produce small, isolated populations. The loss of traits associated with genetic drift in small populations can decrease genetic diversity.

Convergent evolution is the process of organisms developing similar characteristics while evolving in different locations or ecosystems. Such organisms do not descend from a common ancestor, but develop similar characteristics because they react and adapt to environmental pressures in the same way. Thus, convergent evolution reduces biological diversity by making distinct species more like each other.

Punctuated equilibrium is the model of evolution that states that organismal forms diverge, and species form, rapidly over relatively short periods. Between the times of rapid speciation, the characteristics of species are relatively stable. Punctuationalists use fossil records to support their claim. Punctuated equilibrium affects the diversity of organisms in two ways. First, during the period of rapid change and speciation, the diversity of organisms increases dramatically. Second, during the intervening periods, the diversity of organisms remains nearly unchanged.

Patterns of selection are the effects of selection on phenotypes. The four main patterns of selection are stabilizing, disruptive, directional, and balancing. Stabilizing selection is the selection against extreme values of a trait and selection for the average or intermediate values. Conversely, disruptive selection favors individuals at both extremes of the distribution of a characteristic or trait while directional selection progressively favors one extreme of a characteristic distribution. Finally, balancing selection maintains multiple alleles in a population, often by favoring heterozygote individuals. Balancing selection is the only pattern of selection that increases genetic diversity. Stabilizing, disruptive, and directional selection all decrease diversity by eliminating unfavorable characteristics from the population.

Skill 32.5 Evidence that supports the theory of evolution *for* life on Earth.

The wide range of evidence of evolution provides information on the natural processes by which the variety of life on earth developed.

1. **Palaeontology**: Palaeontology is the study of past life based on fossil records and their relation to different geologic time periods.
When organisms die, they often decompose quickly or are consumed by scavengers, leaving no evidence of their existence. However, occasionally some organisms are preserved. The remains or traces of the organisms from a past geological age when embedded in rocks by natural processes are called fossils. They are very important to the understanding of the evolutionary history of life on earth as they provide evidence of evolution and detailed information on the ancestry of organisms.

Petrification is the process by which a dead animal gets fossilized. For this to occur, a dead organism must be buried quickly to avoid weathering and decomposition. When the organism is buried, the organic matter decays. The mineral salts from the mud (in which the organism is buried) will infiltrate into the bones and gradually fill up the pores. The bones will harden and be preserved as fossils. If dead organisms are covered by wind-blown sand and the sand is subsequently turned into mud by heavy rain or floods, the same process of mineral infiltration may occur. Besides petrification, the organisms may be well preserved in ice, in hardened resin of coniferous trees (amber), in tar or in anaerobic acidic peat. Fossilization can sometimes be a trace or an impression of a form – e.g., leaves and footprints.

Fossils are also obtained from the horizontal layers of sedimentary rocks (these are formed by silt or mud on top of each other) called strata; each layer consists fossils. The oldest layer of sedimentary rock is the one the deepest layer and the fossils found in this layer are the oldest. This is how the paleontologists determine the relative ages of discovered fossils.

Some organisms only appear in particular layers of sedimentary rock indicating that thy lived only during a specific time period and became extinct. A succession of animals and plants can also be seen in fossil records, which supports the theory that organisms tend to progressively increase in complexity.

According to fossil records, some modern species of plants and animals are found to be almost identical to the species that lived in ancient geological ages. They are existing species of ancient lineage that have remained unchanged morphologically and maybe physiologically as well. Hence they are called "living fossils". Some examples of living fossils are tuatara, nautilus, horseshoe crab, gingko and metasequoia.

2. Anatomy: Comparative anatomical studies reveal that some structural features are basically similar – e.g., flowers generally have sepals, petals, stigma, style and ovaries but the size, color, number of petals, sepals etc., may differ from species to species.

The degree of resemblance between two organisms indicates how closely they are related in evolution.

4. Groups with little in common are supposed to have diverged from a common ancestor much earlier in geological history than groups which have more in common.
5. To decide how closely two organisms are, anatomists look for structures which may serve different purposes in the adult, but are basically similar (homologous).
6. In cases where similar structures serve different functions in adults, it is important to trace their origin and embryonic development.

When a group of organisms share a specialized homologous structure, that is used to perform a variety of functions in order to adapt to different environmental conditions, this is called adaptive radiation. The gradual spreading of organisms with adaptive radiation is known as divergent evolution. Examples of divergent evolution are – pentadactyl limb and insect mouthparts

Under similar environmental conditions, fundamentally different structures in different groups of organisms may undergo modifications to serve similar functions. This is called convergent evolution. Structures, which have no close phylogenetic links but show adaptation to perform the same functions, are called analogous structures. Examples are – wings of bats, birds and insects; jointed legs of insects and vertebrates; eyes of vertebrates and cephalopods.

Organs that are smaller and simpler in structure than corresponding parts in the ancestral species are called vestigial organs. These are usually degenerated or underdeveloped and were functional in ancestral species but now have become non functional. Examples of vestigial structures include hind limbs of whales; vestigial leaves of some xerophytes; vestigial wings of flightless birds like ostriches, etc.

3. Geographical distribution:

Continental distribution: All organisms are adapted to their environment to a greater or lesser extent. It is generally assumed that the same type of species would be found in a similar habitat in a similar geographic area.

An example would be that Africa has short tailed (old world) monkeys, elephants, lions and giraffes. South America has long-tailed monkeys, pumas, jaguars and llamas.

Evidence for migration and isolation: The fossil record shows that evolution of camels started in North America, from which they migrated across the Bering strait into Asia and Africa and through the Isthmus of Panama into south America.

Continental drift: Fossils of the ancient amphibians, arthropods and ferns are found in South America, Africa, India, Australia and Antarctica which can be dated to the Paleozoic Era, at which time they were all in a single landmass called Gondwana.

Oceanic Island distribution: Most small isolated islands only have native species. Plant life in Hawaii could have arrived as airborne spores or as seeds in the droppings of birds. A few large mammals present in remote islands were brought by human settlers.

4. Evidence from comparative embryology: Comparative embryology shows how embryos start off looking the same. As they develop, their similarities slowly decrease until they take the form of their particular class.
Example: Adult vertebrates are diverse, yet their embryos are quite similar at very early stages. Fishlike structures still form in early embryos of reptiles, birds and mammals. In fish embryos, a two-chambered heart, some veins, and parts of arteries develop and persist in adult fishes. The same structures form early human embryos but do not persist as in adults.

5. Physiology and Biochemistry:
Evolution of widely distributed proteins and molecules: All organisms make use of DNA and/or RNA. ATP is the metabolic currency. The Genetic code is the same for almost every organism. For example a piece of RNA in a bacterium cell codes for the same protein as in a human cell.

Comparison of the DNA sequence allows organisms to be grouped by sequence similarity, and the resulting phylogenetic trees are typically consistent with traditional taxonomy. These analyses are often used to strengthen or correct taxonomic classifications. DNA sequence comparison is considered strong enough evidence to be used to correct erroneous assumptions in the phylogenetic tree in cases where other evidence is missing. The sequence of the 168S rRNA gene, a vital gene encoding a part of the ribosome was used to find the broad phylogenetic relationships between all life.

Proteomic evidence also supports the universal ancestry of life. Vital proteins such as ribosomes, DNA polymerase, and RNA polymerase are found in the most primitive bacteria and in the most complex mammals.

Since metabolic processes do not leave fossils, research into the evolution of the basic cellular processes is done largely by comparison of existing organisms.

COMPETENCY 033 UNDERSTANDING REGULATORY MECHANISMS AND BEHAVIOR

Skill 33.1 How organisms respond to internal and external stimuli.

Behavior animal behavior is responsible for courtship leading to mating, communication between species, territoriality, aggression between animals, and dominance within a group. Behaviors may also include body posture, mating calls, display of feathers/fur, coloration or bearing of teeth and claws.

Innate behavior behaviors that are inborn or instinctual. An environmental stimulus such as the length of day or temperature results in a behavior. Hibernation among some animals is an example of innate behavior.

Learned behavior behavior that is modified due to past experience is called learned behavior.

Response to stimuli is one of the key characteristics of any living thing. Any detectable change in the internal or external environment (the stimulus) may trigger a response in an organism. Just like physical characteristics, organisms" responses to stimuli are adaptations that allow them to better survive. While these responses may be more noticeable in animals that can move quickly, all organisms are actually capable of responding to changes.

Single cell organisms

These organisms are able to respond to basic stimuli such as the presence of light, heat, or food. Changes in the environment are typically sensed via **cell surface receptors**. These organisms may respond to such stimuli by making **changes in internal biochemical pathways, initiating reproduction or phagocytosis**. Those capable of **simple motility**, using flagella for instance, may respond by moving toward food or away from heat.

Plants

Plants typically do not possess sensory organs and so **individual cells recognize stimuli** through a variety of pathways. When **several cells respond to stimuli together**, a response becomes apparent. Logically then, the responses of plants occur on a rather **longer timescale** than those of animals. Plants are capable of **responding to a few basic stimuli including light, water and gravity**. Some common examples of plants responses to stimuli include the way plants turn and grow toward the sun, the sprouting of seeds when exposed to warmth and moisture, and the growth of roots in the direction of gravity.

Animals

Lower members of the animal kingdom have responses similar to those seen in single cell organisms. However, higher animals have developed complex systems to detect and respond to stimuli. The **nervous system, sensory organs (eyes, ears, skin, etc), and muscle tissue all allow animals to sense and quickly respond to changes in their environment.**

As in other organisms, many responses to stimuli in animals are **involuntary**. For example, pupils dilate in response to the reduction of light. Such reactions are typically called **reflexes**. However, many animals are also capable of **voluntary response**. In many animal species, voluntary reactions are **instinctive**. For instance, a zebra's response to a lion is a *voluntary* one, but, *instinctively*, it will flee quickly as soon as the lion's presence is sensed. Complex responses, which may or may not be instinctive, are typically termed **behavior**. An example is the annual migration of birds when seasons change. Even more **complex social behavior** is seen in animals that live in large groups.

Skill 33.2 Knowledge of structures and physiological processes that maintain stable internal conditions.

All of the body's systems contribute to homeostasis through a complex set of interactions. The following is a discussion of how different systems to contribute to maintaining homeostasis.

Skeletal system – Because calcium is an important component of bones, the skeletal system serves as a calcium reserve for the body. Proper levels of calcium are important for many bodily functions. When levels are too low, the body mobilizes calcium from bones to use for other purposes. For example, proper muscle function requires a certain concentration of calcium ions in the blood and body tissue. Thus, the skeletal system acts as a reservoir for maintaining calcium homeostasis by absorbing or releasing calcium as needed.

Muscular system – The muscular system contributes to homeostasis in two ways. First, muscle contraction produces heat as a by-product. This heat helps maintain the body's ial temperature. An example of this function is involuntary shivering (i.e. rapid muscle contraction) that occurs when the body temperature drops. Second, the muscular system (in coordination with the skeletal system) allows organisms to move to environments that are more favorable from a homeostatic perspective.

Circulatory system – The circulatory system plays a vital role in homeostasis. The circulatory system delivers nutrients and removes waste from all the body's tissue by pumping blood through blood vessels. Nutrient delivery and waste removal maintains a favorable environment for all the body's cells. In addition, the circulatory system acts in coordination with the integumentary system in the maintenance of body temperature. When the body temperature rises, blood vessels near the skin dilate allowing greater blood flow and greater release of heat from the body. In contrast, when body temperature drops, blood vessels near the skin constrict limiting the amount of heat lost from the skin.

Immune system – The entire function of the immune system is homeostatic in nature. The immune system protects the body's internal environment from invading microorganisms, viruses, and cancerous cells. These actions maintain a favorable environment for cellular function.

Skill 33.3 Feedback mechanisms that allow organisms to maintain stable internal conditions.

Feedback loops in human systems serve to regulate bodily functions in relation to environmental conditions. Positive feedback loops enhance the body's response to external stimuli and promote processes that involve rapid deviation from the initial state. For example, positive feedback loops function in stress response and the regulation of growth and development. Negative feedback loops help maintain stability in spite of environmental changes and function in homeostasis. For example, negative feedback loops function in the regulation of blood glucose levels and the maintenance of body temperature.

Feedback loops regulate the secretion of classical vertebrate hormones in humans. The pituitary gland and hypothalamus respond to varying levels of hormones by increasing or decreasing production and secretion. High levels of a hormone cause down-regulation of the production and secretion pathways, while low levels of a hormone cause up-regulation of the production and secretion pathways.

"Fight or flight" refers to the human body's response to stress or danger. Briefly, as a response to an environmental stressor, the hypothalamus releases a hormone that acts on the pituitary gland, triggering the release of another hormone, adrenocorticotropin (ACTH), into the bloodstream. ACTH then signals the adrenal glands to release the hormones cortisol, epinephrine, and norepinephrine. These three hormones act to prepare the body to respond to a threat by increasing blood pressure and heart rate, speeding reaction time, diverting blood to the muscles, and releasing glucose for use by the muscles and brain. The stress-response hormones also down-regulate growth, development, and other non-essential functions. Finally, cortisol completes the "fight or flight" feedback loop by acting on the hypothalamus to stop hormonal production after the threat has passed.

Skill 33.4 How evolutionary history affects behavior.

Like so many factors affecting behavior, evolutionary history also does affect behavior. Ecology influences the evolutionary behavior. There are two factors that have profound influence on the behavior of animals – one is the phylogenetic characters and the other is the adaptive significance.

The factors that do prevent certain groups of animals from doing certain things, for example, flying, are called phylogenetic constraints. Mammals can't fly, where as birds can. The fact that birds fly and mammals can't fly is no coincidence. The evolutionary history of these characters has helped the birds to fully develop their ability to fly. At the same time, if the birds are not threatened by predators, there is every chance that these same birds may gradually lose their ability to fly. This is evident in New Zealand, where there were no mammals, until the Europeans settled there. There used to be a higher proportion of flightless birds residing there.

Finally, all behavior is subject to natural selection as with the other traits of an animal. This emphasizes the fact that animals that are well adapted sire many offspring compared to those not so organized in their behavioral adaptation.

Evolutionary Stable strategy (ESS) is another driving force in the evolution of behavior. There are two factors that influence animal behavior – one is the optimal behavior, giving maximum benefit to the animal, and the other is the behavior adapted by the population of which it is a member.

It is important to bear in mind that evolution is not only driven by the physical environment of the animal, but also the interaction between other individuals.

COMPETENCY 034 UNDERSTANDING THE RELATIONSHIP BETWEEN ORGANISMS AND ENVIRONMENT

Skill 34.1   Abiotic and biotic components of an ecosystem.

Biotic factors are living things in an ecosystem; plants, animals, bacteria, fungi, etc. If one population in a community increases, it affects the ability of another population to succeed by limiting the available amount of food, water, shelter and space.

Abiotic factors are non-living aspects of an ecosystem; soil quality, rainfall, and temperature. Changes in climate and soil can cause effects at the beginning of the food chain, thus limiting or accelerating the growth of populations.

Skill 34.2   The interrelationships among producers, consumers, and decomposers in an ecosystem.

Ecosystems are successful primarily because of the interrelationships and recycling that occurs between its three main groups: the producers, the consumers, and the decomposers. The autotrophs produce food, the heterotrophs consume them, and the decomposers clean up any waste left behind, thus keeping the planet tidy and returning essential nutrients to the soil.

Definitions of feeding relationships:

Parasitism - two species occupy a similar place; the parasite benefits from the relationship, the host is harmed.

Commensalism - two species occupy a similar place; neither species is harmed or benefits from the relationship.

Mutualism (symbiosis)- two species occupy a similar place; both species benefit from the relationship.

Competition - two species occupy the same habitat or eat the same food are said to be in competition with each other.

Predation - animals eat other animals are called predators. The animals they feed on are called the prey. Population growth depends upon competition for food, water, shelter, and space. The amount of predators determines the amount of prey, which in turn affects the number of predators.

Skill 34.3 Identify factors that influence the size and growth of population in an Ecosystem.

Populations are ultimately limited by their productivity levels. **Carrying Capacity** is the total amount of life a habitat can support. Once the habitat runs out of food, water, shelter, or space, the carrying capacity decreases, and then stabilizes. Growth of a population depends upon the health of the organisms, who depend upon food (among other things), for survival.

A limiting factor is the component of a biological process that determines how quickly or slowly the process proceeds. Photosynthesis is the main biological process determining the rate of ecosystem productivity, the rate at which an ecosystem creates biomass. Thus, in evaluating the productivity of an ecosystem, potentially limiting factors are light intensity, gas concentrations, and mineral availability. The Law of the Minimum states that the required factor in a given process that is most scarce controls the rate of the process.

One potentially limiting factor of ecosystem productivity is light intensity because photosynthesis requires light energy. Light intensity can limit productivity in two ways. First, too little light limits the rate of photosynthesis because the required energy is not available. Second, too much light can damage the photosynthetic system of plants and microorganisms thus slowing the rate of photosynthesis. Decreased photosynthesis equals decreased productivity.

Another potentially limiting factor of ecosystem productivity is gas concentrations. Photosynthesis requires carbon dioxide. Thus, increased concentrations of carbon dioxide often results in increased productivity. While carbon dioxide is often not the ultimate limiting factor of productivity, increased concentration can indirectly increase rates of photosynthesis in several ways. First, increased carbon dioxide concentration often increases the rate of nitrogen fixation (available nitrogen is another limiting factor of productivity). Second, increased carbon dioxide concentration can decrease the pH of rain, improving the water source of photosynthetic organisms.

Finally, mineral availability also limits ecosystem productivity. Plants require adequate amounts of nitrogen and phosphorus to build many cellular structures. The availability of the inorganic minerals phosphorus and nitrogen often is the main limiting factor of plant biomass production. In other words, in a natural environment phosphorus and nitrogen availability most often limits ecosystem productivity, rather than carbon dioxide concentration or light intensity.

Skill 34.4 Analyze adaptive characteristics that result in a population's or species' unique niche in an ecosystem.

The term 'Niche' describes the relational position of a species or population in an ecosystem. Niche includes how a population responds to the abundance of its resources and enemies (e.g., by growing when resources are abundant and predators, parasites and pathogens are scarce).

Niche also indicates the life history of an organism, habitat and place in the food chain.
According to the competitive exclusion principle, no two species can occupy the same niche in the same environment for a long time.

The full range of environmental conditions (biological and physical) under which an organism can exist describes its fundamental niche. Because of the pressure from superior competitors, superior are driven to occupy a niche much narrower than their previous niche. This is known as the 'realized niche.'

Examples of niche:

1. Oak trees:
* live in forests
* absorb sunlight by photosynthesis
* provide shelter for many animals
* act as support for creeping plants
* serve as a source of food for animals
* cover their ground with dead leaves in the autumn

If the oak trees were cut down or destroyed by fire or storms they would no longer be doing their job and this would have a disastrous effect on all the other organisms living in the same habitat.

2. Hedgehogs:
* eat a variety of insects and other invertebrates which live underneath the dead leaves and twigs in the garden
* the spines are a superb environment for fleas and ticks
* put the nitrogen back into the soil when they urinate
* eat slugs and protect plants from them

If there were no hedgehogs around, the population of slugs would increase and the nutrients in the dead leaves and twigs would not be recycled.

Skill 34.5   Describe and analyze energy flow through various types of Ecosystems.

Trophic levels are based on the feeding relationships that determine energy flow and chemical cycling.

Autotrophs are the primary producers of the ecosystem. **Producers** mainly consist of plants. **Primary consumers** are the next trophic level. The primary consumers are the herbivores that eat plants or algae. **Secondary consumers** are the carnivores that eat the primary consumers. **Tertiary consumers** eat the secondary consumer. These trophic levels may go higher depending on the ecosystem. **Decomposers** are consumers that feed off animal waste and dead organisms. This pathway of food transfer is known as the food chain.

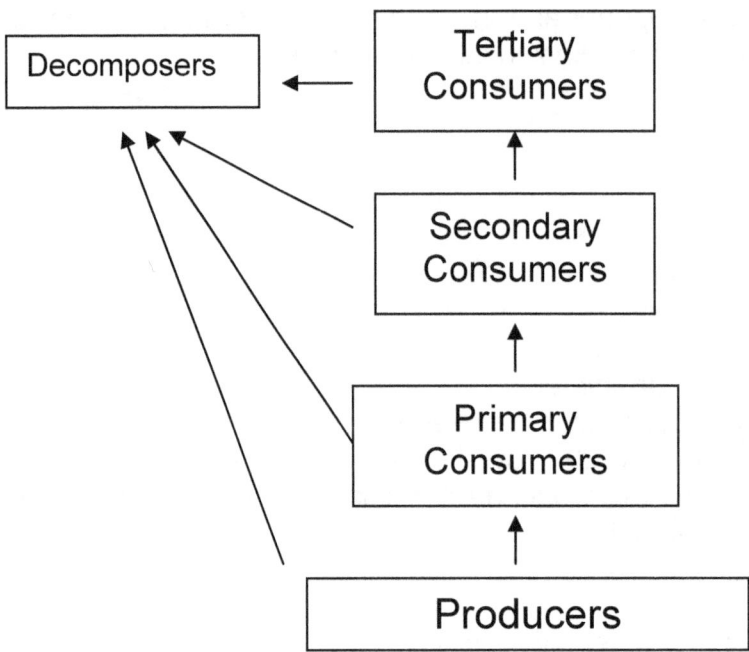

Most food chains are more elaborate, becoming food webs.

Skill 34.6 How populations and species modify and affect ecosystems.

Ecology is the study of organisms, where they live and their interactions with the environment. A **population** is a group of the same species in a specific area. A **community** is a group of populations residing in the same area.

The environment is ever changing because of natural events and the actions of humans, animals, plants, and other organisms. Even the slightest changes in environmental conditions can greatly influence the function and balance of communities, ecosystems, and ecoregions. For example, subtle changes in salinity and temperature of ocean waters over time can greatly influence the range and population of certain species of fish. In addition, a slight increase in average atmospheric temperature can promote tree growth in a forest, but a corresponding increase in the viability of pathogenic bacteria can decrease the overall growth and productivity of the forest.

Another important concept in ecological change is succession. Ecological succession is the transition in the composition of species in an ecosystem, often after an ecological disturbance in the community. Primary succession begins in an environment virtually void of life, such as a volcanic island. Secondary succession occurs when a natural event disrupts an ecosystem, leaving the soil intact. An example of secondary succession is the reestablishment of a forest after destruction by a forest fire.

Factors that drive the process of succession include interspecies competition, environmental conditions, inhibition, and facilitation. In a developing ecosystem, species compete for scarce resources. The species that compete most successfully dominate. Environmental conditions, as previously discussed, influence the viability of species. Finally, the activities of certain species can inhibit or facilitate the growth and development of other species. Inhibition results from exploitative competition or interference competition. In facilitation, a species or group of species lays the foundation for the establishment of other, more advanced species. For example, the presence of a certain bacterial population can change the pH of the soil, allowing for the growth of different types of plants and trees.

Communities that are ecologically similar in regards to temperature, rainfall and the species that live there are called **biomes**. Specific biomes include:

Marine - covers 75% of the earth. This biome is organized by the depth of the water. The intertidal zone is from the tide line to the edge of the water. The littoral zone is from the water's edge to the open sea. It includes coral reef habitats and is the most densely populated area of the marine biome. The open sea zone is divided into the epipelagic zone and the pelagic zone. The epipelagic zone receives more sunlight and has a larger number of species. The ocean floor is called the benthic zone and is populated with bottom feeders.

Tropical Rain Forest - temperature is constant (25 degrees C), rainfall exceeds 200 cm. per year. Located around the area of the equator, the rain forest has abundant, diverse species of plants and animals.

Savanna - temperatures range from 0-25 degrees C depending on the location. Rainfall is from 90 to 150 cm per year. Plants include shrubs and grasses. The savanna is a transitional biome between the rain forest and the desert.

Desert - temperatures range from 10-38 degrees C. Rainfall is under 25 cm per year. Plant species include xerophytes and succulents. Lizards, snakes and small mammals are common animals.

Temperate Deciduous Forest - temperature ranges from -24 to 38 degrees C. Rainfall is between 65 to 150 cm per year. Deciduous trees are common, as well as deer, bear and squirrels.

Taiga - temperatures range from -24 to 22 degrees C. Rainfall is between 35 to 40 cm per year. Taiga is located very north and very south of the equator, getting close to the poles. Plant life includes conifers and plants that can withstand harsh winters. Animals include weasels, mink, and moose.

Tundra - temperatures range from -28 to 15 degrees C. Rainfall is limited, ranging from 10 to 15 cm per year. The tundra is located even further north and south than the Taiga. Common plants include lichens and mosses. Animals include polar bears and musk ox.

Polar or Permafrost - temperature ranges from -40 to 0 degrees C. It rarely gets above freezing. Rainfall is below 10 cm per year. Most water is bound up as ice. Life is limited.

DOMAIN IV.   **EARTH AND SPACE SCIENCE**

COMPETENCY 035   **UNDERSTANDING THE STRUCTURE AND FUNCTION OF EARTH SYSTEMS**

Skill 35.1   **The structure of Earth and analyzes constructive and destructive processes that produce geological change.**

Orogeny is the term given to natural mountain building.

A mountain is terrain that has been raised high above the surrounding landscape by volcanic action, or some form of tectonic plate collisions. The plate collisions could be intercontinental or ocean floor collisions with a continental crust (subduction). The physical composition of mountains would include igneous, metamorphic, or sedimentary rocks; some may have rock layers that are tilted or distorted by plate collision forces.

There are many different types of mountains. The physical attributes of a mountain range depends upon the angle at which plate movement thrust layers of rock to the surface. Many mountains (Adirondacks, Southern Rockies) were formed along high angle faults.

Folded mountains (Alps, Himalayas) are produced by the folding of rock layers during their formation. The Himalayas are the highest mountains in the world and contain Mount Everest which rises almost 9 km above sea level. The Himalayas were formed when India collided with Asia. The movement which created this collision is still in process at the rate of a few centimeters per year.

Fault-block mountains (Utah, Arizona, and New Mexico) are created when plate movement produces tension forces instead of compression forces. The area under tension produces normal faults and rock along these faults is displaced upward. Dome mountains are formed as magma tries to push up through the crust but fails to break the surface. Dome mountains resemble a huge blister on the earth's surface. Upwarped mountains (Black Hills of S.D.) are created in association with a broad arching of the crust. They can also be formed by rock thrust upward along high angle faults.

Volcanism is the term given to the movement of magma through the crust and its emergence as lava onto the earth's surface. Volcanic mountains are built up by successive deposits of volcanic materials. An active volcano is one that is presently erupting or building to an eruption. A dormant volcano is one that is between eruptions but still shows signs of internal activity that might lead to an eruption in the future. An extinct volcano is said to be no longer capable of erupting. Most of the world's active volcanoes are found along the rim of the Pacific Ocean, which is also a major earthquake zone. This curving belt of active faults and volcanoes is often called the Ring of Fire.

The world's best known volcanic mountains include: Mount Etna in Italy and Mount Kilimanjaro in Africa. The Hawaiian Islands are actually the tops of a chain of volcanic mountains that rise from the ocean floor.

There are three types of volcanic mountains: shield volcanoes, cinder cones and composite volcanoes.

Shield Volcanoes are associated with quiet eruptions. Lava emerges from the vent or opening in the crater and flows freely out over the earth's surface until it cools and hardens into a layer of igneous rock. A repeated lava flow builds this type of volcano into the largest volcanic mountain. Mauna Loa found in Hawaii, is the largest volcano on earth.

Cinder Cone Volcanoes are associated with explosive eruptions as lava is hurled high into the air in a spray of droplets of various sizes. These droplets cool and harden into cinders and particles of ash before falling to the ground. The ash and cinder pile up around the vent to form a steep, cone-shaped hill called the cinder cone. Cinder cone volcanoes are relatively small but may form quite rapidly.

Composite Volcanoes are described as being built by both lava flows and layers of ash and cinders. Mount Fuji in Japan, Mount St. Helens in Washington, USA and Mount Vesuvius in Italy are all famous composite volcanoes.

Mechanisms of producing mountains

Mountains are produced by different types of mountain-building processes. Most major mountain ranges are formed by the processes of folding and faulting.

Folded Mountains are produced by the folding of rock layers. Crustal movements may press horizontal layers of sedimentary rock together from the sides, squeezing them into wavelike folds. Up-folded sections of rock are called anticlines; down-folded sections of rock are called synclines. The Appalachian Mountains are an example of folded mountains with long ridges and valleys in a series of anticlines and synclines formed by folded rock layers.

Faults are fractures in the earth's crust which have been created by either tension or compression forces transmitted through the crust. These forces are produced by the movement of separate blocks of crust.

Faultings are categorized on the basis of the relative movement between the blocks on both sides of the fault plane. The movement can be horizontal, vertical or oblique.

A dip-slip fault occurs when the movement of the plates is vertical and opposite. The displacement is in the direction of the inclination, or dip, of the fault. Dip-slip faults are classified as normal faults when the rock above the fault plane moves down relative to the rock below.

Reverse faults are created when the rock above the fault plane moves up relative to the rock below. Reverse faults having a very low angle to the horizontal are also referred to as thrust faults.

Faults in which the dominant displacement is horizontal movement along the trend or strike (length) of the fault are called **strike-slip faults**. When a large strike-slip fault is associated with plate boundaries it is called a **transform fault**. The San Andreas Fault in California is a well-known transform fault.

Faults that have both vertical and horizontal movement are called **oblique-slip faults**.

When lava cools, igneous rock is formed. This formation can occur either above ground or below ground.

Intrusive rock includes any igneous rock that was formed below the earth's surface. Batholiths are the largest structures of intrusive type rock and are composed of near granite materials; they are the core of the Sierra Nevada Mountains.

Extrusive rock includes any igneous rock that was formed at the earth's surface.

Dikes are old lava tubes formed when magma entered a vertical fracture and hardened. Sometimes magma squeezes between two rock layers and hardens into a thin horizontal sheet called a **sill**. A **laccolith** is formed in much the same way as a sill, but the magma that creates a laccolith is very thick and does not flow easily. It pools and forces the overlying strata creating an obvious surface dome.

A **caldera** is normally formed by the collapse of the top of a volcano. This collapse can be caused by a massive explosion that destroys the cone and empties most if not all of the magma chamber below the volcano. The cone collapses into the empty magma chamber forming a caldera.

An inactive volcano may have magma solidified in its pipe. This structure, called a volcanic neck, is resistant to erosion and today may be the only visible evidence of the past presence of an active volcano.
When lava cools, igneous rock is formed. This formation can occur either above ground or below ground.

Glaciation

A continental glacier covered a large part of North America during the most recent ice age. Evidence of this glacial coverage remains as abrasive grooves, large boulders from northern environments dropped in southerly locations, glacial troughs created by the rounding out of steep valleys by glacial scouring, and the remains of glacial sources called cirques that were created by frost wedging the rock at the bottom of the glacier. Remains of plants and animals found in warm climate have been discovered in the moraines and out wash plains help to support the theory of periods of warmth during the past ice ages.

The Ice Age began about 2 -3 million years ago. This age saw the advancement and retreat of glacial ice over millions of years. Theories relating to the origin of glacial activity include Plate Tectonics, where it can be demonstrated that some continental masses, now in temperate climates, were at one time blanketed by ice and snow. Another theory involves changes in the earth's orbit around the sun, changes in the angle of the earth's axis, and the wobbling of the earth's axis. Support for the validity of this theory has come from deep ocean research that indicates a correlation between climatic sensitive micro-organisms and the changes in the earth's orbital status.

About 12,000 years ago, a vast sheet of ice covered a large part of the northern United States. This huge, frozen mass had moved southward from the northern regions of Canada as several large bodies of slow-moving ice, or glaciers. A time period in which glaciers advance over a large portion of a continent is called an ice age. A glacier is a large mass of ice that moves or flows over the land in response to gravity. Glaciers form among high mountains and in other cold regions.

There are two main types of glaciers: valley glaciers and continental glaciers. Erosion by valley glaciers is characteristic of U-shaped erosion. They produce sharp peaked mountains such as the Matterhorn in Switzerland. Erosion by continental glaciers often rides over mountains in their paths leaving smoothed, rounded mountains and ridges.

Skill 35.2   The form and function of surface and subsurface water.

World weather patterns are greatly influenced by ocean surface currents in the upper layer of the ocean. These currents continuously move along the ocean surface in specific directions. Ocean currents that flow deep below the surface are called sub-surface currents. These currents are influenced by such factors as the location of landmasses in the current's path and the earth's rotation.

Surface currents are caused by winds and are classified by temperature. Cold currents originate in the Polar Regions and flow through surrounding water that is measurably warmer. Those currents with a higher temperature than the surrounding water are called warm currents and can be found near the equator. These currents follow swirling routes around the ocean basins and the equator.

The Gulf Stream and the California Current are the two main surface currents that flow along the coastlines of the United States. The Gulf Stream is a warm current in the Atlantic Ocean that carries warm water from the equator to the northern parts of the Atlantic Ocean. Benjamin Franklin studied and named the Gulf Stream. The California Current is a cold current that originates in the Arctic regions and flows southward along the west coast of the United States.

Differences in water density also create ocean currents. Water found near the bottom of oceans is the coldest and the densest. Water tends to flow from a denser area to a less dense area. Currents that flow because of a difference in the density of the ocean water are called density currents. Water with a higher salinity is denser than water with a lower salinity. Water that has salinity different from the surrounding water may form a density current.

The movement of ocean water is caused by the wind, the sun's heat energy, the earth's rotation, the moon's gravitational pull on earth, and by underwater earthquakes. Most ocean waves are caused by the impact of winds. Wind blowing over the surface of the ocean transfers energy (friction) to the water and causes waves to form. Waves are also formed by seismic activity on the ocean floor. A wave formed by an earthquake is called a seismic sea wave. These powerful waves can be very destructive, with wave heights increasing to 30 m or more near the shore. The crest of a wave is its highest point. The trough of a wave is its lowest point. The distance from wave top to wave top is the wavelength. The wave period is the time between the passing of two successive waves.

Two percent of all the available water is fixed and held in ice or the bodies of organisms. Available water includes surface water (lakes, ocean, and rivers) and ground water (aquifers, wells). 96% of all available water is from ground water. Water is recycled through the processes of evaporation and precipitation. The water present now is the water that has been here since our atmosphere formed.

Water flows and is collected in a predictable manner. In most situations it runs across land and into small streams that feed larger bodies of water. All of the land that acts like a funnel for water flowing into a single larger body of water is known as a watershed or drainage basin. The watershed includes the streams and rivers that bear the water and the surfaces across which the water runs. Thus, the pollution load and general state of all the land within a watershed has an effect on the health and cleanliness of the body of water to which it drains. Large land features, such as mountains, separate watersheds from one another.

However, some portion of water from one watershed may enter the groundwater and ultimately flow towards another, adjacent watershed.

Not all water flows to the streams, rivers, and lakes that comprise the above ground water supply. Some water remains in the soil as ground water. Additionally, underground rivers are found in areas of karst topography, though these are relatively rare. It is more common for water to collect in underground aquifers. Aquifers are layers of permeable rock or loose material (gravel, sand, or silt) that hold water. Aquifers may be either confined or unconfined. Confined aquifers are deep in the ground and below the water table. Unconfined aquifers border on the water table. The water table is the level at which ground water exists and is always equal to atmospheric pressure. To visualize the entire ground water system, we can imagine a hole dug in wet sand at the beach and a small pool of water within the hole. The wet sand corresponds to the aquifer, the hole to a well or lake, and the level of water in the hole to the water table.

In some cases, people have created reservoirs, artificial storage areas that make large amounts of water readily available. Reservoirs are most often created by damming rivers. A dam is built from cement, soil, or rock and the river fills the newly created reservoir. A reservoir may be created by building a dam either across a valley or around the entire perimeter of an artificial lake (a bunded dam). The former technique is more common and relies on natural features to form a watertight reservoir. However, such a feature must exist to allow this type of construction. A fully bunded dam does not require such a natural feature but does necessitate more construction since a waterproof structure must be built all the way around the reservoir. This structure is typically made from clay and/or cement. Since no river feeds such reservoirs, mechanical pumps are used to fill them from nearby water sources. Occasionally, watertight roofs are added to these reservoirs so they can be used to hold treated water. These are known as service reservoirs.

Skill 35.3 The composition and structure of the atmosphere and its properties.

Dry air is composed of three basic components; dry gas, water vapor, and solid particles (dust from soil, etc.).

The most abundant dry gases in the atmosphere are:

  (N2) Nitrogen   78.09 %   makes up about 4/5 of gases in atmosphere
  (O2) Oxygen   20.95 %
  (AR) Argon   0.93 %
  (CO2) Carbon Dioxide   0.03 %

The atmosphere is divided into four main layers based on temperature. These layers are labeled Troposphere, Stratosphere, Mesosphere, Thermosphere.

Troposphere - this layer is the closest to the earth's surface and all weather phenomena occurs here as it is the layer with the most water vapor and dust. Air temperature decreases with increasing altitude. The average thickness of the Troposphere is 7 miles (11 km).

Stratosphere - this layer contains very little water, clouds within this layer are extremely rare. The Ozone layer is located in the upper portions of the stratosphere. Air temperature is fairly constant but does increase somewhat with height due to the absorption of solar energy and ultra violet rays from the ozone layer.

Mesosphere - air temperature again decreases with height in this layer. It is the coldest layer with temperatures in the range of -100^0 C at the top..

Thermosphere - extends upward into space. Oxygen molecules in this layer absorb energy from the sun, causing temperatures to increase with height. The lower part of the thermosphere is called the Ionosphere. Here charged particles or ions and free electrons can be found. When gases in the Ionosphere are excited by solar radiation, the gases give off light and glow in the sky. These glowing lights are called the Aurora Borealis in the Northern Hemisphere and Aurora Australis in Southern Hemisphere. The upper portion of the Thermosphere is called the Exosphere. Gas molecules are very far apart in this layer. Layers of Exosphere are also known as the Van Allen Belts and are held together by earth's magnetic field.

Skill 35.4 The interactions that occur among the biosphere, geosphere, hydrosphere, and atmosphere .

When we look at the various phenomena in geology and meteorology we can see manifestations of many scientific principles. It is not simply that certain examples exist in the natural world. Rather it is the reverse; scientists observed natural phenomena to formulate and refine the theories and laws of science.

The environment is ever changing because of natural events and the actions of humans, animals, plants, and other organisms. Even the slightest changes in environmental conditions can greatly influence the function and balance of communities, ecosystems, and ecoregions. For example, subtle changes in salinity and temperature of ocean waters over time can greatly influence the range and population of certain species of fish. In addition, a slight increase in average atmospheric temperature can promote tree growth in a forest, but a corresponding increase in the viability of pathogenic bacteria can decrease the overall growth and productivity of the forest.

Another important concept in ecological change is succession. Ecological succession is the transition in the composition of species in an ecosystem, often after an ecological disturbance in the community. Primary succession begins in an environment virtually void of life, such as a volcanic island. Secondary succession occurs when a natural event disrupts an ecosystem, leaving the soil intact. An example of secondary succession is the reestablishment of a forest after destruction by a forest fire.

There are many simple chemical and physical principles that underlie the various phenomena affecting Earth. For instance, basic principles of radioactive decay are exploited to allow carbon dating. Chemical laws also dictate the dissolution and precipitation that govern the formation of many rocks. Chemical principles are also important in various processes involving water: the dissolution of carbonic acid that causes acid rain; reactions between rock and water lead to karst topography; and solubility rules govern what compounds leach out of soil and into groundwater. The laws of thermodynamics are also extremely important in the natural world since they predict everything from how weather systems to move to how water flows to the ultimate heat death of the universe.

Additionally, there are many laws that pertain especially to earth and atmospheric science. A few are listed below:

Law of Superposition states that higher layers of sedimentary rock are younger than those beneath it. This is a logical statement, but it means that layers of sedimentary rock can be viewed as a biogeological timeline of sorts.

Principle of Uniformitarianism states that geological processes took place in the past in the same manner that they take place now. This is an important fact if we are to speculate on past geological events.

Buys-Ballot's Law states that wind travels counterclockwise around low-pressure zones in the Northern Hemisphere and clockwise in the Southern Hemisphere. This is a consequence of the original observation, that in the Northern Hemisphere, if you stand with your back to the wind, the low-pressure area will be on your left.

Skill 35.5 How human activity and natural processes, both gradual and catastrophic, can alter earth systems.

An important topic in science is the effect of natural disasters and events on society and the effect human activity has on inducing such events. Naturally occurring geological, weather, and environmental events can greatly affect the lives of humans. In addition, the activities of humans can induce such events that would not normally occur.

Nature-induced hazards include floods, landslides, avalanches, volcanic eruptions, wildfires, earthquakes, hurricanes, tornadoes, droughts, and disease. Such events often occur naturally, because of changing weather patterns or geological conditions. Property damage, resource destruction, and the loss of human life are the possible outcomes of natural hazards. Thus, natural hazards are often extremely costly on both an economic and personal level.

While many nature-induced hazards occur naturally, human activity can often stimulate such events. For example, destructive land use practices such as mining can induce landslides or avalanches if not properly planned and monitored. In addition, human activities can cause other hazards including global warming and waste contamination. Global warming is an increase in the Earth's average temperature resulting, at least in part, from the burning of fuels by humans. Global warming is hazardous because it disrupts the Earth's environmental balance and can negatively affect weather patterns. Ecological and weather pattern changes can promote the natural disasters listed above. Finally, improper hazardous waste disposal by humans can contaminate the environment. One important effect of hazardous waste contamination is the stimulation of disease in human populations. Thus, hazardous waste contamination negatively affects both the environment and the people that live in it.

Skill 35.6 Identify the sources of energy (e.g., solar, geothermal) in earth systems and describes mechanisms of energy transfer - convection, radiation.

Energy is transferred in Earth's atmosphere in three ways. Earth gets most of its energy from the sun in the form of waves. This transfer of energy by waves is termed **radiation**. The transfer of thermal energy through matter by actual contact of molecules is called **conduction**. For example, heated rocks and sandy beaches transfer heat to the surrounding air. The transfer of thermal energy due to air density differences is called **convection**. Convection currents circulate in a constant exchange of cold, dense air for less dense warm air.

Carbon Dioxide in the atmosphere absorbs energy from the sun. Carbon Dioxide also blocks the direct escape of energy from the Earth's surface. This process by which heat is trapped by gases, water vapor and other gases in the Earth's atmosphere is called the **Greenhouse Effect**.

Most of the Earth's water is found in the oceans and lakes. Through the **water cycle**, water evaporates into the atmosphere and condenses into clouds. Water then falls to the Earth in the form of precipitation, returning to the oceans and lakes and falling on land. Water on the land may return to the oceans and lakes as runoff or seep from the soil as groundwater.

COMPETENCY 036 UNDERSTANDING CYCLES IN EARTH SYSTEMS.

Skill 36.1 The rock cycle and how rocks, minerals, and soils are formed.

Three major subdivisions of rocks are sedimentary, metamorphic and igneous.

Lithification of sedimentary rocks

When fluid sediments are transformed into solid sedimentary rocks, the process is known as lithification. One very common process affecting sediments is compaction where the weights of overlying materials compress and compact the deeper sediments. The compaction process leads to cementation. Cementation is when sediments are converted to sedimentary rock.

Factors in crystallization of igneous rocks

Igneous rocks can be classified according to their texture, their composition, and the way they formed.

Molten rock is called magma. When molten rock pours out onto the surface of Earth, it is called lava.

As magma cools, the elements and compounds begin to form crystals. The slower the magma cools, the larger the crystals grow. Rocks with large crystals are said to have a coarse-grained texture. . Granite is an example of a coarse grained rock. Rocks that cool rapidly before any crystals can form have a glassy texture such as obsidian, also commonly known as volcanic glass.

Metamorphic rocks are formed by high temperatures and great pressures. The process by which the rocks undergo these changes is called metamorphism. The outcome of metamorphic changes include deformation by extreme heat and pressure, compaction, destruction of the original characteristics of the parent rock, bending and folding while in a plastic stage, and the emergence of completely new and different minerals due to chemical reactions with heated water and dissolved minerals.

Metamorphic rocks are classified into two groups, foliated (leaflike) rocks and unfoliated rocks. Foliated rocks consist of compressed, parallel bands of minerals, which give the rocks a striped appearance. Examples of such rocks include slate, schist, and gneiss. Unfoliated rocks are not banded and examples of such include quartzite, marble, and anthracite rocks.

Minerals are natural, non-living solids with a definite chemical composition and a crystalline structure. **Ores** are minerals or rock deposits that can be mined for a profit. **Rocks** are earth materials made of one or more minerals. A **Rock Facies** is a rock group that differs from comparable rocks (as in composition, age or fossil content).

Minerals must adhere to five criteria. They must be (1) non-living, (2) formed in nature, (3) solid in form, (4) their atoms form a crystalline pattern, (5) its chemical composition is fixed within narrow limits.

There are over 3000 minerals in Earth's crust. Minerals are classified by composition. The major groups of minerals are silicates, carbonates, oxides, sulfides, sulfates, and halides. The largest group of minerals is the silicates. Silicates are made of silicon, oxygen, and one or more other elements.

Soils are composed of particles of sand, clay, various minerals, tiny living organisms, and humus, plus the decayed remains of plants and animals. Soils are divided into three classes according to their texture. These classes are sandy soils, clay soils, and loamy soils.

Sandy soils are gritty, and their particles do not bind together firmly. They are porous which means that water passes through them rapidly. As a result sandy soils do not hold much water.

Clay soils are smooth and greasy, their particles bind together firmly. Clay soils are moist and usually do not allow water to pass through easily.

Loamy soils feel somewhat like velvet and their particles clump together. Loamy soils are made up of sand, clay, and silt. Loamy soils holds water but some water can pass through.

In addition to three main classes, soils are further grouped into three major types based upon their composition. These groups are pedalfers, pedocals, and laterites.

Pedalfers form in the humid, temperate climate of the eastern United States. Pedalfer soils contain large amounts of iron oxide and aluminum-rich clays, making the soil a brown to reddish brown color. This soil supports forest type vegetation.

Pedocals are found in the western United States where the climate is dry and temperate. These soils are rich in calcium carbonate. This type of soil supports grasslands and brush vegetation.

Laterites are found where the climate is wet and tropical. Large amounts of water flows through this soil. Laterites are red-orange soils rich in iron and aluminum oxides. There is little humus and this soil is not very fertile.

Skill 36.2 The water cycle and its relationship to earth systems.

Water that falls to Earth in the form of rain and snow is called **precipitation.** Precipitation is part of a continuous process in which water at the Earth's surface evaporates, condenses into clouds, and returns to Earth. This process is termed the **water cycle**. The water located below the surface is called groundwater.

The impacts of altitude upon climatic conditions are primarily related to temperature and precipitation. As altitude increases, climatic conditions become increasingly drier and colder. Solar radiation becomes more severe as altitude increases while the effects of convection forces are minimized. Climatic changes as a function of latitude follow a similar pattern (as a reference, latitude moves either north or south from the equator). The climate becomes colder and drier as the distance from the equator increases. Proximity to land or water masses produces climatic conditions based upon the available moisture. Dry and arid climates prevail where moisture is scarce; lush tropical climates can prevail where moisture is abundant. Climate, as described above, depends upon the specific combination of conditions making up an area's environment. Man impacts all environments by producing pollutants in earth, air, and water. It follows then, that man is a major player in world climatic conditions.

Skill 36.3 The nutrient (e.g., carbon, nitrogen) cycle and its relationship to earth systems.

Essential elements are recycled through an ecosystem. At times, the element needs to be "fixed" in a useable form. Cycles are dependent on plants, algae and bacteria to fix nutrients for use by animals.

> **Carbon cycle** - Ten percent of all available carbon in the air (from carbon dioxide gas) is fixed by photosynthesis. Plants fix carbon in the form of glucose, animals eat the plants and are able to obtain their source of carbon. When animals release carbon dioxide through respiration, the plants again have a source of carbon to fix.
>
> **Nitrogen cycle** - Eighty percent of the atmosphere is in the form of nitrogen gas. Nitrogen must be fixed and taken out of the gaseous form to be incorporated into an organism. Only a few genera of bacteria have the correct enzymes to break the triple bond between nitrogen atoms. These bacteria live within the roots of legumes (peas, beans, alfalfa) and add bacteria to the soil so it may be taken up by the plant. Nitrogen is necessary to make amino acids and the nitrogenous bases of DNA.

Phosphorus cycle - Phosphorus exists as a mineral and is not found in the atmosphere. Fungi and plant roots have structures called mycorrhizae that are able to fix insoluble phosphates into useable phosphorus. Urine and decayed matter returns phosphorus to the earth where it can be fixed in the plant. . Phosphorus is needed for the backbone of DNA and for the manufacture of ATP.

Skill 36.4   Applies knowledge of how human and natural processes affect earth systems.

See Skill 16.5

Skill 36.5   The dynamic interactions that occur among the various cycles in the biosphere, geosphere, hydrosphere, and atmosphere.

While the hydrosphere, lithosphere, and atmosphere can be described and considered separately, they are actually constantly interacting with one another. Energy and matter flows freely between these different spheres. For instance, in the water cycle, water beneath the Earth's surface and in rocks (in the lithosphere) is exchanged with vapor in the atmosphere and liquid water in lakes and the ocean (the hydrosphere). Similarly, significant events in one sphere almost always have effects in the other spheres. The recent increase in greenhouse gases provides an example of this ripple effect. Additional greenhouse gases produced by human activities were released into the atmosphere where they built up and caused widening holes in certain areas of the atmosphere and global warming. These increasing temperatures have had many effects on the hydrosphere: rising sea levels, increasing water temperature, and climate changes. These lead to even more changes in the lithosphere such as glacier retreat and alterations in the patterns of water-rock interaction (run-off, erosion, etc).

COMPETENCY 37.0 UNDERSTANDING THE ROLE OF ENERGY IN WEATHER AND CLIMATE

Skill 37.1 The elements of weather (e.g., humidity, wind speed, pressure, temperature) and how they are measured.

Humidity is the amount of water vapor in the air. Humidity may be measured as absolute humidity, relative humidity, and specific humidity. Relative humidity is the most frequently used measurement because it is used in weather forecasts. Relative humidity indicates the likelihood of precipitation, dew, or fog. **Relative humidity** is the actual amount of water vapor in a certain volume of air compared to the maximum amount of water vapor this air could hold at a given temperature. The air temperature at which water vapor begins to condense is called the dew point.

Wind speed is the speed of movement of air relative to a point on Earth. A wind occurs when air moves from one place to the next. High wind speeds can cause destruction, and strong winds are referred to as gales, hurricanes, and typhoons. An **anemometer** measures wind speed and a **wind vane** measures wind direction.

Cyclones are huge air masses of low pressure air. The wind in a low-pressure system blows counterclockwise. Air pressure in a low pressure system is lowest at the center and highest along its outer edges. Air masses of high pressure air are called **anticyclones**. The wind in a high-pressure system blows clockwise. Air pressure is greatest at the center of the air mass and lowest along its outer edge.

Temperature is measured with a thermometer. The scale may be in either degrees Celsius or degrees Fahrenheit. Fahrenheit is common in America, but the worldwide standard is Celsius. The conversions are: [°F] = [°C] (9/5 + 32) and conversely [°C] = ([°F] − 32)(5/9).

Skill 37.2 Compares and contrasts weather and climate

The weather in a region is called the climate of that region. Unlike the weather, which consists of hourly and daily changes in the atmosphere over a region, climate is the average of all weather conditions in a region over a period of time. Many factors are used to determine the climate of a region including temperature and precipitation. Climate varies from one place to another because of the unequal heating of the Earth's surface. This varied heating of the surface is the result of the unequal distribution of land masses, oceans, and polar ice caps.

Skill 37.3 Analyze weather charts and data to make weather predictions.

Every day every one of us is affected by weather regardless if it is the typical thunderstorm with the brief moist air coming down on us from a cumulonimbus cloud or a severe storm with pounding winds that can have wind factors that can cause either hurricanes or twisters called tornados. These are common terms that we can identify with including the term blizzards or ice storms.

The daily newscast relates terms such as dew point and relative humidity and barometric pressure. Suddenly all to common terms become clouded with terms more frequently used by a meteorologist or someone that forecasts weather. Dew point is the air temperature at which water vapor begins to condense. Relative humidity is the actual amount of water vapor in a certain volume of air compared to the maximum amount of water vapor that this air could hold at a given temperature.

Weather instruments that forecast weather include aneroid barometer and the mercury barometer that measures air pressure. The air exerts varying pressures on a metal diaphragm that will read air pressure. The mercury barometer operates when atmospheric pressure pushes on a pool of (mercury) in a glass tube. The higher the pressure the higher up the tube the mercury rises.

Relative humidity is measured by two kinds of weather instruments, the psychrometer and the hair gygrometer. Relative humidity simply indicates the amount of moisture in the air. Relative humidity is defined as a ration of existing amounts of water vapor and moisture in the air when compared to the maximum amount of moisture that the air can hold at the same given pressure and temperature. Relative humidity is stated as a percentage so for example the relative humidity can be 100%.

For example if you were to analyze relative humidity from data an example might be If a parcel of air is saturated, meaning it now holds all the moisture it can hold at a given temperature, the relative humidity is 100%.

Lesson Plans for teachers to analyze data and predict weather

http://www.srh.weather.gov/srh/jetstream/synoptic/ll_analyze.htm

Skill 37.4 How transfers of energy among earth systems affect weather and climate.

Air masses moving toward or away from the Earth's surface are called air currents. Air moving parallel to Earth's surface is called **wind**. Weather conditions are generated by winds and air currents carrying large amounts of heat and moisture from one part of the atmosphere to another. Wind speeds are measured by instruments called anemometers.

The wind belts in each hemisphere consist of convection cells that encircle Earth like belts. There are three major wind belts on Earth: (1) trade winds (2) prevailing westerlies, and (3) polar easterlies. Wind belt formation depends on the differences in air pressures that develop in the doldrums, the horse latitudes, and the Polar Regions. The Doldrums surround the equator. Within this belt heated air usually rises straight up into Earth's atmosphere. The Horse latitudes are regions of high barometric pressure with calm and light winds and the Polar Regions contain cold dense air that sinks to the Earth's surface.

Winds caused by local temperature changes include sea breezes, and land breezes.

Sea breezes are caused by the unequal heating of the land and an adjacent, large body of water. Land heats up faster than water. The movement of cool ocean air toward the land is called a sea breeze. Sea breezes usually begin blowing about mid-morning; ending about sunset.

A breeze that blows from the land to the ocean or a large lake is called a **land breeze.**

Monsoons are huge wind systems that cover large geographic areas and that reverse direction seasonally. The monsoons of India and Asia are examples of these seasonal winds. They alternate wet and dry seasons. As denser cooler air over the ocean moves inland, a steady seasonal wind called a summer or wet monsoon is produced.

Skill 37.5 Analyze how Earth's position, orientation, and surface features affect weather and climate.

Earth is the third planet away from the sun in our solar system. Earth's numerous types of motion and states of orientation greatly effect global conditions, such as seasons, tides and lunar phases. The Earth orbits the Sun with a period of 365 days. During this orbit, the average distance between the Earth and Sun is 93 million miles. The shape of the Earth's orbit around the Sun deviates from the shape of a circle only slightly. This deviation, known as the Earth's eccentricity, has a very small affect on the Earth's climate. The Earth is closest to the Sun at perihelion, occurring around January 2nd of each year, and farthest from the Sun at aphelion, occurring around July 2nd. Because the Earth is closest to the sun in January, the northern winter is slightly warmer than the southern winter.

Seasons

The rotation axis of the Earth is not perpendicular to the orbital (ecliptic) plane. The axis of the Earth is tilted 23.45° from the perpendicular. The tilt of the Earth's axis is known as the obliquity of the ecliptic, and is mainly responsible for the four seasons of the year by influencing the intensity of solar rays received by the Northern and Southern Hemispheres. The four seasons, spring, summer, fall and winter, are extended periods of characteristic average temperature, rainfall, storm frequency and vegetation growth or dormancy. The effect of the Earth's tilt on climate is best demonstrated at the solstices, the two days of the year when the Sun is farthest from the Earth's equatorial plane. At the Summer Solstice (June Solstice), the Earth's tilt on its axis causes the Northern Hemisphere to the lean toward the Sun, while the southern hemisphere leans away. Consequently, the Northern Hemisphere receives more intense rays from the Sun and experiences summer during this time, while the Southern Hemisphere experiences winter. At the Winter Solstice (December Solstice), it is the Southern Hemisphere that leans toward the sun and thus experiences summer. Spring and fall are produced by varying degrees of the same leaning toward or away from the Sun.

Tides

The orientation of and gravitational interaction between the Earth and the Moon are responsible for the ocean tides that occur on Earth. The term "tide" refers to the cyclic rise and fall of large bodies of water. Gravitational attraction is defined as the force of attraction between all bodies in the universe. At the location on Earth closest to the Moon, the gravitational attraction of the Moon draws seawater toward the Moon in the form of a tidal bulge. On the opposite side of the Earth, another tidal bulge forms in the direction away from the Moon because at this point, the Moon's gravitational pull is the weakest. "Spring tides" are especially strong tides that occur when the Earth, Sun and Moon are in line, allowing both the Sun and the Moon to exert gravitational force on the Earth and increase tidal bulge height. These tides occur during the full moon and the new moon. "Neap tides" are especially weak tides occurring when the gravitational forces of the Moon and the Sun are perpendicular to one another. These tides occur during quarter moons.

COMPETENCY 038 UNDERSTANDING THE CHARACTERISTICS OF THE SOLAR SYSTEM AND THE UNIVERSE

Skill 38.1 The properties and characteristics of celestial objects.

Astronomers use groups or patterns of stars called **constellations** as reference points to locate other stars in the sky. Familiar constellations include: Ursa Major (also known as the big bear) and Ursa Minor (known as the little bear). Within the Ursa Major, the smaller constellation, The Big Dipper is found. Within the Ursa Minor, the smaller constellation, The Little Dipper is found.

Different constellations appear as the earth continues its revolution around the sun with the seasonal changes.

Magnitude stars are 21 of the brightest stars that can be seen from earth. These are the first stars noticed at night. In the Northern Hemisphere there are 15 commonly observed first magnitude stars.

A vast collection of stars are defined as **galaxies**. Galaxies are classified as irregular, elliptical, and spiral. An irregular galaxy has no real structured appearance; most are in their early stages of life. An elliptical galaxy consists of smooth ellipses, containing little dust and gas, but composed of millions or trillion stars. Spiral galaxies are disk-shaped and have extending arms that rotate around its dense center. Earth's galaxy is found in the Milky Way and it is a spiral galaxy.

Terms related to deep space

A **pulsar** is defined as a variable radio source that emits signals in very short, regular bursts; believed to be a rotating neutron star.

A **quasar** is defined as an object that photographs like a star but has an extremely large redshift and a variable energy output; believed to be the active core of a very distant galaxy.

Black holes are defined as an object that has collapsed to such a degree that light can not escape from its surface; light is trapped by the intense gravitational field.

Skill 38.2 Apply knowledge of the earth-moon-sun system and the interactions among them - seasons, lunar phases, eclipses.

Lunar Phases

The Earth's orientation in respect to the solar system is also responsible for our perception of the phases of the moon. As the Earth orbits the Sun within a period of 365 days, the Moon orbits the Earth every 27 days. As the moon circles the Earth, its shape in the night sky appears to change. The changes in the appearance of the moon from Earth are known as "lunar phases." These phases vary cyclically according to the relative positions of the Moon, the Earth and the Sun. At all times, half of the Moon is facing the Sun and is thus illuminated by reflecting the Sun's light. As the Moon orbits the Earth and the Earth orbits the Sun, the half of the moon that faces the Sun changes. However, the Moon is in synchronous rotation around the Earth, meaning that nearly the same side of the moon faces the Earth at all times. This side is referred to as the near side of the moon. Lunar phases occur as the Earth and Moon orbit the Sun and the fractional illumination of the Moon's near side changes.

When the Sun and Moon are on opposite sides of the Earth, observers on Earth perceive a "full moon," meaning the moon appears circular because the entire illuminated half of the moon is visible. As the Moon orbits the Earth, the Moon "wanes" as the amount of the illuminated half of the Moon that is visible from Earth decreases. A gibbous moon is between a full moon and a half moon, or between a half moon and a full moon. When the Sun and the Moon are on the same side of Earth, the illuminated half of the moon is facing away from Earth, and the moon appears invisible. This lunar phase is known as the "new moon." The time between each full moon is approximately 29.53 days.

A list of all lunar phases includes:

- New Moon: the moon is invisible or the first signs of a crescent appear
- Waxing Crescent: the right crescent of the moon is visible
- First Quarter: the right quarter of the moon is visible
- Waxing Gibbous: only the left crescent is not illuminated
- Full Moon: the entire illuminated half of the moon is visible
- Waning Gibbous: only the right crescent of the moon is not illuminated
- Last Quarter: the left quarter of the moon is illuminated
- Waning Crescent: only the left crescent of the moon is illuminated

Viewing the moon from the Southern Hemisphere would cause these phases to occur in the opposite order.

Eclipses

Eclipses are defined as the passing of one object into the shadow of another object. A **Lunar eclipse** occurs when the moon travels through the shadow of the earth. A **Solar eclipse** occurs when the moon positions itself between the sun and earth.

Skill 38.3   Identify properties of the components of the solar system.

The **sun** is considered the nearest star to earth that produces solar energy. By the process of nuclear fusion, hydrogen gas is converted to helium gas. Energy flows out of the core to the surface, then radiation escapes into space.

Parts of the sun include: (1) **core:** the inner portion of the sun where fusion takes place, (2) **photosphere:** considered the surface of the sun which produces **sunspots** (cool, dark areas that can be seen on its surface), (3) **chromosphere:** hydrogen gas causes this portion to be red in color (also found here are solar flares (sudden brightness of the chromosphere) and solar prominences (gases that shoot outward from the chromosphere)), and (4) **corona**, the transparent area of sun visible only during a total eclipse.

Solar radiation is energy traveling from the sun that radiates into space. **Solar flares** produce excited protons and electrons that shoot outward from the chromosphere at great speeds reaching earth. These particles disturb radio reception and also affect the magnetic field on earth.

There are eight established planets in our solar system; Mercury, Venus, Earth, Mars, Jupiter, Saturn, Uranus, and Neptune. Pluto was an established planet in our solar system, but as of Summer 2006, it's status is being reconsidered. The planets are divided into two groups based on distance from the sun. The inner planets include: Mercury, Venus, Earth, and Mars. The outer planets include: Jupiter, Saturn, Uranus, and Neptune.

Planets

Mercury -- the closest planet to the sun. Its surface has craters and rocks. The atmosphere is composed of hydrogen, helium and sodium. Mercury was named after the Roman messenger god.

Venus -- has a slow rotation when compared to Earth. Venus and Uranus rotate in opposite directions from the other planets. This opposite rotation is called retrograde rotation. The surface of Venus is not visible due to the extensive cloud cover. The atmosphere is composed mostly of carbon dioxide. Sulfuric acid droplets in the dense cloud cover give Venus a yellow appearance. Venus has a greater greenhouse effect than observed on Earth. The dense clouds combined with carbon dioxide trap heat. Venus was named after the Roman goddess of love.

Earth -- considered a water planet with 70% of its surface covered by water. Gravity holds the masses of water in place. The different temperatures observed on earth allow for the different states (solid. Liquid, gas) of water to exist. The atmosphere is composed mainly of oxygen and nitrogen. Earth is the only planet that is known to support life.

Mars -- the surface of Mars contains numerous craters, active and extinct volcanoes, ridges, and valleys with extremely deep fractures. Iron oxide found in the dusty soil makes the surface seem rust colored and the skies seem pink in color. The atmosphere is composed of carbon dioxide, nitrogen, argon, oxygen and water vapor. Mars has polar regions with ice caps composed of water. Mars has two satellites. Mars was named after the Roman war god.

Jupiter --the largest planet in the solar system has 16 moons. The atmosphere is composed of hydrogen, helium, methane and ammonia. There are white colored bands of clouds indicating rising gas and dark colored bands of clouds indicating descending gases. The gas movement is caused by heat resulting from the energy of Jupiter's core. Jupiter has a Great Red Spot that is thought to be a hurricane type cloud. Jupiter has a strong magnetic field.

Saturn -- the second largest planet in the solar system has rings of ice, rock, and dust particles circling it. Saturn's atmosphere is composed of hydrogen, helium, methane, ammonia, and has 20 plus satellites. Saturn was named after the Roman god of agriculture.

Uranus -- the second largest planet in the solar system with retrograde revolution. Uranus is a gaseous planet. It has 10 dark rings and 15 satellites. Its atmosphere is composed of hydrogen, helium, and methane. Uranus was named after the Greek god of the heavens.

Neptune -- another gaseous planet with an atmosphere consisting of hydrogen, helium, and methane. Neptune has 3 rings and 2 satellites. Neptune was named after the Roman sea god because its atmosphere is the same color as the seas.

Pluto – once considered the smallest planet in the solar system; it's status as a planet is being reconsidered . Pluto's atmosphere probably contains methane, ammonia, and frozen water. Pluto has 1 satellite. It revolves around the sun every 250 years. Pluto was named after the Roman god of the underworld.

Comets, asteroids, and meteors.
Astronomers believe that rocky fragments may have been the remains of the birth of the solar system that never formed into a planet. **Asteroids** are found in the region between Mars and Jupiter.

Comets are masses of frozen gases, cosmic dust, and small rocky particles. Astronomers think that most comets originate in a dense comet cloud beyond Pluto. Comet consists of a nucleus, a coma, and a tail. A comet's tail always points away from the sun. The most famous comet, **Halley's Comet,** is named after the person whom first discovered it in 240 B.C. It returns to the skies near earth every 75 to 76 years.

Meteoroids are composed of particles of rock and metal of various sizes. When a meteoroid travels through the earth's atmosphere, friction causes its surface to heat up and it begins to burn. The burning meteoroid falling through the earth's atmosphere is called a **meteor** (also known as a "shooting star").

Meteorites are meteors that strike the earth's surface. A physical example of a meteorite's impact on the earth's surface can be seen in Arizona. The Barringer Crater is a huge meteor crater. There are many other meteor craters throughout the world.

Skill 38.4　Characteristics of stars and galaxies and their distribution in the universe.

A star is a ball of hot, glowing gas that is hot enough and dense enough to trigger nuclear reactions, which fuel the star. In comparing the mass, light production, and size of the Sun to other stars, astronomers find that the Sun is a perfectly ordinary star. It behaves exactly the way they would expect a star of its size to behave. The main difference between the Sun and other stars is that the Sun is much closer to Earth.

Most stars have masses similar to that of the Sun. The majority of stars' masses are between 0.3 to 3.0 times the mass of the Sun. Theoretical calculations indicate that in order to trigger nuclear reactions and to create its own energy—that is, to become a star—a body must have a mass greater than 7 percent of the mass of the Sun. Astronomical bodies that are less massive than this become planets or objects called brown dwarfs. The largest accurately determined stellar mass is of a star called V382 Cygni and is 27 times that of the Sun.

The range of brightness among stars is much larger than the range of mass. Astronomers measure the brightness of a star by measuring its magnitude and luminosity. Magnitude allows astronomers to rank how bright, comparatively, different stars appear to humans. Because of the way our eyes detect light, a lamp ten times more luminous than a second lamp will appear less than ten times brighter to human eyes. This discrepancy affects the magnitude scale, as does the tradition of giving brighter stars lower magnitudes. The lower a star's magnitude, the brighter it is. Stars with negative magnitudes are the brightest of all.

Magnitude is given in terms of absolute and apparent values. Absolute magnitude is a measurement of how bright a star would appear if viewed from a set distance away. Astronomers also measure a star's brightness in terms of its luminosity. A star's absolute luminosity or intrinsic brightness is the total amount of energy radiated by the star per second. Luminosity is often expressed in units of watts.

Skill 38.5   Scientific theories on the origin of the universe.

Two main hypotheses of the origin of the solar system are (1) **the tidal hypothesis** and (2) **the condensation hypothesis**.

The tidal hypothesis proposes that the solar system began with a near collision of the sun and a large star. Some astronomers believe that as these two stars passed each other, the great gravitational pull of the large star extracted hot gases out of the sun. The mass from the hot gases started to orbit the sun, which began to cool then condensing into the nine planets. (Few astronomers support this example).

The condensation hypothesis proposes that the solar system began with rotating clouds of dust and gas. Condensation occurred in the center forming the sun and the smaller parts of the cloud formed the nine planets. (This example is widely accepted by many astronomers).

Two main theories to explain the origins of the universe include (1) **The Big Bang Theory** and (2) **The Steady-State Theory.**

The Big Bang Theory has been widely accepted by many astronomers. It states that the universe originated from a magnificent explosion spreading mass, matter and energy into space. The galaxies formed from this material as it cooled during the next half-billion years.

The Steady-State Theory is the least accepted theory. It states that the universe is a continuously being renewed. Galaxies move outward and new galaxies replace the older galaxies. Astronomers have not found any evidence to prove this theory.

The future of the universe is hypothesized with the Oscillating Universe Hypothesis. It states that the universe will oscillate or expand and contract. Galaxies will move away from one another and will in time slow down and stop. Then a gradual moving toward each other will again activate the explosion or The Big Bang theory.

COMPETENCY 039 UNDERSTANDING THE HISTORY OF THE EARTH SYSTEM

Skill 39.1 The scope of the geologic time scale and its relationship to geologic processes

The biological history of the earth is partitioned into four major Eras which are further divided into major periods. The latter periods are refined into groupings called Epochs.

Earth's history extends over more than four billion years and is reckoned in terms of a scale. Paleontologists who study the history of the Earth have divided this huge period of time into four large time units called eons. Eons are divided into smaller units of time called eras. An era refers to a time interval in which particular plants and animals were dominant, or present in great abundance. The end of an era is most often characterized by (1) a general uplifting of the crust, (2) the extinction of the dominant plants or animals, and (3) the appearance of new life-forms.

Each era is divided into several smaller divisions of time called periods. Some periods are divided into smaller time units called epochs.

Methods of geologic dating

Estimates of the Earth's age have been made possible with the discovery of **radioactivity** and the invention of instruments that can measure the amount of radioactivity in rocks. The use of radioactivity to make accurate determinations of Earth's age is called Absolute Dating. This process depends upon comparing the amount of radioactive material in a rock with the amount that has decayed into another element. Studying the radiation given off by atoms of radioactive elements is the most accurate method of measuring the Earth's age. These atoms are unstable and are continuously breaking down or undergoing decay. The radioactive element that decays is called the parent element. The new element that results from the radioactive decay of the parent element is called the daughter element.

The time required for one half of a given amount of a radioactive element to decay is called the half-life of that element or compound.

Geologists commonly use Carbon Dating to calculate the age of a fossil substance.

Infer the history of an area using geologic evidence.

The determination of the age of rocks by cataloging their composition has been outmoded since the middle 1800s. Today a sequential history can be determined by the fossil content (principle of fossil succession) of a rock system as well as its superposition within a range of systems. This classification process was termed stratigraphy and permitted the construction of a Geologic Column in which rock systems are arranged in their correct chronological order.

Principles of catastrophism and uniformitarianism

Uniformitarianism - is a fundamental concept in modern geology. It simply states that the physical, chemical, and biological laws that operated in the geologic past operate in the same way today. The forces and processes that we observe presently shaping our planet have been at work for a very long time. This idea is commonly stated as "the present is the key to the past."

Catastrophism - the concept that the earth was shaped by catastrophic events of a short term nature.

Skill 39.2   Theories about the earth's origin and geologic history.

The dominant scientific theory about the origin of the Universe, and consequently the Earth, is the Big Bang Theory. According to this theory, an atom exploded about 10 to 20 billion years ago throwing matter in all directions. Although this theory has never been proven, and probably never will be, it is supported by the fact that distant galaxies in every direction are moving away from us at great speeds.

Earth, itself, is believed to have been created 4.5 billion years ago as a solidified cloud of gases and dust left over from the creation of the sun. As millions of years passed, radioactive decay released energy that melted some of Earth's components. Over time, the heavier components sank to the center of the Earth and accumulated into the core. As the Earth cooled, a crust formed with natural depressions. Water rising from the interior of the Earth filled these depressions and formed the oceans. Slowly, the Earth acquired the appearance it has today.

The **Heterotroph Hypothesis** supposes that life on Earth evolved from **heterotrophs**, the first cells. According to this hypothesis, life began on Earth about 3.5 billion years ago. Scientists have shown that the basic molecules of life formed from lightning, ultraviolet light, and radioactivity. Over time, these molecules became more complex and developed metabolic processes, thereby becoming heterotrophs. Heterotrophs could not produce their own food and fed off organic materials. However, they released carbon dioxide which allowed for the evolution of **autotrophs**, which could produce their own food through photosynthesis. The autotrophs and heterotrophs became the dominant life forms and evolved into the diverse forms of life we see today.

Proponents of **creationism** believe that the species we currently have were created as recounted in the Bible in the book of Genesis.. This retelling asserts that God created all life about 6,000 years ago in one mass creation event. However, scientific evidence casts doubt on creationism.

Evolution

The most significant evidence to support the history of evolution is fossils, which have been used to construct a fossil record. Fossils give clues as to the structure of organisms and the times at which they existed. However, there are limitations to the study of fossils, which leave huge gaps in the fossil record.

Scientists also try to relate two organisms by comparing their internal and external structures. This is called **comparative anatomy**. Comparative anatomy categorizes anatomical structures as **homologous** (features in different species that point to a common ancestor), **analogous** (structures that have superficial similarities because of similar functions, but do not point to a common ancestor), and **vestigial** (structures that have no modern function, indicating that different species diverged and evolved). Through the study of **comparative embryology**, homologous structures that do not appear in mature organisms may be found between different species in their embryological development.

There have been two basic **theories of evolution: Lamarck's and Darwin's**. Lamarck's theory proposed that an organism can change its structure through use or disuse and that acquired traits can be inherited; this theory has been disproved.

Darwin's theory of **natural selection** is the basis of all evolutionary theory. His theory has four basic points:

1. Each species produces more offspring than can survive.
2. The individual organisms that make up a larger population are born with certain variations.
3. The overabundance of offspring creates competition for survival among individual organisms (**survival of the fittest**).
4. Variations are passed down from parent to offspring.

Points 2 and 4 form the genetic basis for evolution.

New species develop from two types of evolution: divergent and convergent. **Divergent evolution**, also known as **speciation**, is the divergence of a new species from a previous form of that species. There are two main ways in which speciation may occur: **allopatric speciation** (resulting from geographical isolation so that species cannot interbreed) and **adaptive radiation** (creation of several new species from a single parent species). **Convergent evolution** is a process whereby different species develop similar traits from inhabiting similar environments, facing similar selection pressures, and/or use parts of their bodies for similar functions. This type of evolution is only superficial. It can never result in two species being able to interbreed.

Skill 39.3 Understanding of how tectonic forces have shaped landforms over time.

Data obtained from many sources led scientists to develop the theory of plate tectonics. This theory is the most current model that explains not only the movement of the continents, but also the changes in the earth's crust caused by internal forces.

Plates are rigid blocks of earth's crust and upper mantle. These rigid solid blocks make up the lithosphere. The earth's lithosphere is broken into nine large sections and several small ones. These moving slabs are called plates. The major plates are named after the continents they are "transporting."

The plates float on and move with a layer of hot, plastic-like rock in the upper mantle. Geologists believe that the heat currents circulating within the mantle cause this plastic zone of rock to slowly flow, carrying along the overlying crustal plates.

Movement of these crustal plates creates areas where the plates diverge as well as areas where the plates converge. A major area of divergence is located in the Mid-Atlantic. Currents of hot mantle rock rise and separate at this point of divergence creating new oceanic crust at the rate of 2 to 10 centimeters per year. Convergence is when the oceanic crust collides with either another oceanic plate or a continental plate. The oceanic crust sinks forming an enormous trench and generating volcanic activity. Convergence also includes continent to continent plate collisions. When two plates slide past one another a transform fault is created.

These movements produce many major features of the earth's surface, such as mountain ranges, volcanoes, and earthquake zones. Most of these features are located at plate boundaries, where the plates interact by spreading apart, pressing together, or sliding past each other. These movements are very slow, averaging only a few centimeters a year.

Boundaries form between spreading plates where the crust is forced apart in a process called rifting. Rifting generally occurs at mid-ocean ridges. Rifting can also take place within a continent, splitting the continent into smaller landmasses that drift away from each other, thereby forming an ocean basin between them. The Red Sea is a product of rifting. As the seafloor spreading takes place, new material is added to the inner edges of the separating plates. In this way the plates grow larger, and the ocean basin widens. This is the process that broke up the super continent Pangaea and created the Atlantic Ocean.

Boundaries between plates that are colliding are zones of intense crustal activity. When a plate of ocean crust collides with a plate of continental crust, the more dense oceanic plate slides under the lighter continental plate and plunges into the mantle. This process is called **subduction**, and the site where it takes place is called a subduction zone. A subduction zone is usually seen on the sea-floor as a deep depression called a trench.

The crustal movement which is identified by plates sliding sideways past each other produces a plate boundary characterized by major faults that are capable of unleashing powerful earth-quakes. The San Andreas Fault forms such a boundary between the Pacific Plate and the North American Plate.

Skill 39.4 The formation of fossils and the importance of the fossil record in explaining the earth's history.

A fossil is the remains or trace of an ancient organism that has been preserved naturally in the Earth's crust. Sedimentary rocks usually are rich sources of fossil remains. Those fossils found in layers of sediment were embedded in the slowly forming sedimentary rock strata. The oldest fossils known are the traces of 3.5 billion year old bacteria found in sedimentary rocks. Few fossils are found in metamorphic rock and virtually none found in igneous rocks. The magma is so hot that any organism trapped in the magma is destroyed.

The fossil remains of a woolly mammoth embedded in ice were found by a group of Russian explorers. However, the best-preserved animal remains have been discovered in natural tar pits. When an animal accidentally fell into the tar, it became trapped sinking to the bottom. Preserved bones of the saber-toothed cat have been found in tar pits.

Prehistoric insects have been found trapped in ancient amber or fossil resin that was excreted by some extinct species of pine trees.

Fossil molds are the hollow spaces in a rock previously occupied by bones or shells. A fossil cast is a fossil mold that fills with sediments or minerals that later hardens forming a cast.

Fossil tracks are the imprints in hardened mud left behind by birds or animals.

DOMAIN V. SCIENCE LEARNING, INSTRUCTION, AND ASSESSMENT

COMPETENCY 040   **THEORETICAL AND PRACTICAL KNOWLEDGE ABOUT TEACHING SCIENCE AND HOW STUDENTS LEARN SCIENCE**

Skill 40.1   **Understand how the developmental characteristics, prior knowledge, and experience, and attitudes of students influence science learning**

Learning styles refers to the ways in which individuals learn best. Physical settings, instructional arrangements, materials available, techniques, and individual preferences are all factors in the teacher's choice of instructional strategies and materials. Information about the student's preference can be done through a direct interview or a Likert-style checklist where the student rates his preferences.

Physical Settings

A. **Noise**: Students vary in the degree of quiet that they need and the amount of background noise or talking that they can tolerate without getting distracted or frustrated.

B. **Temperature and Lighting**: Students also vary in their preference for lighter or darker areas of the room, tolerance for coolness or heat, and ability to see the chalkboard, screen, or other areas of the room.

C. **Physical Factors**: This refers to the student's need for workspace and preference for type of work area, such as desk, table, or learning center. Proximity factors such as closeness to other students, the teacher or high traffic areas such as doorways or pencil sharpeners, may help the student to feel secure and stay on task, or may serve as distractions, depending on the individual.

Instructional Arrangements

Some students work well in large groups; others prefer small groups or one-to-one instruction with the teacher, aide or volunteer. Instructional arrangements also involve peer-tutoring situations with the student as tutor or tutee. The teacher also needs to consider how well the student works independently with seatwork.

Instructional Techniques

Consideration of the following factors will affect the teacher's choice of instructional techniques, as well as selecting optimal times to schedule various types of assignments. Some of these factors are listed below:

- How much time the student needs to complete assignments
- Time of day the student performs best
- How student functions under timed conditions
- How much teacher demonstration and attention is needed for the task
- The student's willingness to approach new tasks
- Student's willingness to give up
- Student's preference for verbal or written instruction
- Student's frustration tolerance when faced with difficulty
- Number of prompts, cues, and attention needed for the student to maintain expected behavior

Material and Textbook Preferences

Students vary in their ability to respond and learn with different techniques of lesson delivery. They likewise vary in their preference and ability to learn with different types of materials. Depending on the student's preference and success, the teacher can choose from among these types of instructional materials:

- Self-correcting materials
- Worksheets with or without visual cues
- Worksheets with a reduced number of items or lots of writing space
- Manipulative materials
- Flash cards, commercial or student-prepared
- Computers
- Commercial materials
- Teacher-made materials
- Games, board or card
- Student-made instructional materials

Learning Styles
Students also display preferences for certain learning styles and these differences are also factors in the teacher's choice of presentation and materials.

A) **Visual:** Students who are visual may enjoy working with and remember best from books, films, pictures, pictures, modeling, overheads, demonstrating and writing.

B) **Auditory:** Students who are auditory may enjoy working with and remember best from hearing records of tapes, auditory directions, listening to people, radio, read-aloud stories, and lectures.

Kinesthetic: Indicators include learning through writing, experiments, operating machines, such as typewriters or calculators, motor activities and games, and taking pictures.

Skill 40.2   Select and adapt science curricula, content, instructional materials, and activities to meet the interests, knowledge, understanding, abilities, experiences, and needs of all students, including English Language Learners

Teaching students of diverse backgrounds is very challenging and must be handled very carefully. It is important that the teacher remain politically correct, when handling students from diverse cultures. . Moreover, from a humanitarian point of view, the teacher needs to be compassionate and empathetic, since it is a challenge to settle in a different country and call it home.

The lesson plans must reflect the teachers' understanding and respect towards the diverse students.

One way a teacher can show respect for diverse cultures is to incorporate the different cultural practices into the lessons and connect it to science. For example, students can study the contributions to science made by the Latino scientists, African American scientists, Native American scientists, and Asian scientists etc. In February, we can study about famous scientists of African American decent. The same principle applies to other cultures. When this is done, the students feel very happy and appreciate the effort and thought of their teacher, who took time to recognize their heritage.

Decorating the classroom using ethnic material would be interesting and also creates an atmosphere of being at home. Incorporating the cultural and linking them to science needs a little bit of time and ingenuity, which will go a long way in establishing good relationships with students and their families.

Skill 40.3 Understanding how to use situations from students' daily lives to develop instructional materials that investigate how science can be used to make informed decisions

Before the teacher begins instruction, he or she should choose activities that are at the appropriate level of student difficulty, are meaningful, and relevant. Because biology is the study of living things, we can easily apply the knowledge of biology to daily life and personal decision-making. For example, biology greatly influences the health decisions humans make everyday. What foods to eat, when and how to exercise, and personal hygiene are just three of the many decisions we make everyday that are based on our knowledge of biology. Other areas of daily life where biology affects decision-making are parenting, interpersonal relationships, family planning, and consumer spending.

Skill 40.4 Understanding common misconceptions in science and effective ways to address these misconceptions

There are many common misconceptions about science. The following are a few scientific misconceptions that are or have been common in the past:

* The Earth is the center of the solar system.
* The Earth is the largest object in the solar system
* Rain comes from the holes in the clouds
* Acquired characters can be inherited
* The eye receives upright images
* Energy is a thing
* Heat is not energy

Some strategies to uncover and dispel misconceptions include:

1. Planning appropriate activities, so that the students will see for themselves where there are misconceptions.

2. Web search is a very useful tool to dispel misconceptions. Students need to be guided in how to look for answers on the Internet, and if necessary the teacher should explain scientific literature to help the students understand it.

3. Science journals are a great source of information. Recent research is highly beneficial for the senior science students.

4. Critical thinking and reasoning are two important skills that the students should be encouraged to use to discover facts – for example, that heat is a form of energy. Here, the students have to be challenged to use their critical thinking skills to reason that heat can cause change – for example, causing water to boil – and so it is not a thing but a form of energy, since only energy can cause change.

Skill 40.5 Understanding the rationale for the use of active learning and inquiry processes for students

The purpose of practice is to help the student move through the acquisition of learning a skill (initial learning), to maintenance (remembering how to do the skill), to generalization (applying the skill to new or different situations),

During guided or semi-independent practice, the teacher should provide specific directions and model the procedure on the practice materials while the student follows along. Gradually, the teacher prompts, and modeling will decrease as the student becomes more proficient. The teacher should apply positive and corrective feedback at this stage.

During independent practice, the teacher's role is to monitor the students and provide individual attention and modeling as necessary. The student should be encouraged to "think aloud" so the teacher can monitor what strategies and problem-solving skills are being used to answer questions. Again, positive and/or corrective feedback with praise should be used for achievement.

Transfer of learning occurs when experience with one task influences performance on another task. Positive transfer occurs when the required responses are about the same and the stimuli are similar, such as moving from baseball, handball, to racquetball, or field hockey to soccer. Negative transfer occurs when the stimuli remain similar, but the required responses change, such as shifting from soccer to football, tennis to racquetball, and boxing to sports karate. Instructional procedures should stress the similar features between the activities and the dimensions that are transferable. Specific information should emphasize when stimuli in the old and new situations are the same as or similar, and when responses used in the old situation apply to the new.

To facilitate learning, instructional objectives should be arranged in order according to their patterns of similarity. Objectives involving similar responses should be closely sequenced; thus, the possibility for positive transfer is stressed. Likewise, learning objectives that involve different responses should be programmed within instructional procedures in the most appropriate way possible. For example, students should have little difficulty transferring handwriting instruction to writing in other areas; however, there might be some negative transfer when moving from manuscript to cursive writing. By using transitional methods and focusing upon the similarities between manuscript and cursive writing, negative transfer can be reduced.

Generalization is the occurrence of a learned behavior in the presence of a stimulus other than the one that produced the initial response (e.g. novel stimulus). It is the expansion of a student's performance beyond conditions initially anticipated. Students must be able to generalize what is learned to other settings.

Generalization may be enhanced by the following:

1. Use many examples in teaching to deepen application of learned skills.
2. Use consistency in initial teaching situations, and later introduce variety in format, procedure, and use of examples.
3. Have the same information presented by different teachers, in different settings, and under varying conditions.
4. Include a continuous reinforcement schedule at first, later changing to delayed, and intermittent schedules as instruction progresses.
5. Teach students to record instances of generalization and to reward themselves at that time.
6. Associate naturally occurring stimuli when possible.

Skill 40.6 Understanding questioning strategies designed to elicit higher-level thinking and how to use them to move students from concrete to more abstract understanding

Inquiry learning provides opportunities for students to experience and acquire thought processes through which they can gather information about the world. This requires a higher level of interaction among the learner, the teacher, the area of study, available resources, and the learning environment. Students become actively involved in the learning process as they :

1. Act upon their curiosity and interests
2. Develop questions that are relevant
3. Think their way through controversies or dilemmas
4. Analyze problems
5. Develop, clarify, and test hypotheses
6. Draw conclusions
7. Find possible solutions

The most important element in inquiry-based learning is questioning. Students must ask relevant questions and develop ways to search for answers and generate explanations. High order thinking is encouraged.

Here are some **inquiry strategies**:

1. Deductive inquiry: The main goal of this strategy is moving the student from a generalized principle to specific instances. The process of testing general assumptions, applying them, and exploring the relationships between specific elements is stressed. The teacher coordinates the information and presents important principles, themes, or hypotheses. Students are actively engaged in testing generalizations, gathering information, and applying it to specific examples.

2. Inductive inquiry: The information-seeking process of the inductive inquiry method helps students to establish facts, determine relevant questions, and develop ways to pursue these questions and build explanations. Students are encouraged to develop and support their own hypotheses. Through inductive inquiry, students experience the thought processes which require them to move from specific facts and observations to inferences.

3. Interactive instruction: This strategy relies heavily on discussion and sharing among participants. Students develop social skills, learning from teacher and peers. They also learn organizational skills. Examples are debates, brainstorming, discussion, laboratory groups, etc.

4. Direct instruction strategy: This is highly teacher-oriented and is among the most commonly used strategies. It is effective for providing information or developing step-by-step skills. Examples are lecture, demonstrations, explicit teaching, etc.

5. Indirect instruction: This is mainly student-centered. Direct and indirect instruction strategies can compliment each other. Indirect instruction seeks a high level of student involvement such as observing, investigating, drawing inferences from data, or forming hypotheses. In this strategy, the role of the teacher shifts from that of teacher/lecturer to that of facilitator, supporter, and resource person.

Examples are problem solving, inquiry, concept formation, etc.

6. <u>Independent study</u>: Independent study refers to the range of instructional methods that are purposely provided to foster the development of individual student initiative, self reliance, and self improvement. Examples are research projects, homework, etc.

The above mentioned strategies promote higher-level thinking skills such as problem solving, synthesizing (hypothesizing), designing (identifying the problem), analyzing (analyzing data in an experiment), and connecting (logical thinking).

Skill 40.7 Understanding the importance of planning activities that are inclusive and accommodate the needs of all students

The term diversity is defined as the presence of a wide range of variation in the qualities or attributes under discussion.

In the human context, particularly in a social context, the term diversity refers to the presence in one population of a variety of cultures, ethnic groups, languages, physical features, socio-economic backgrounds, religious faiths, sexuality, gender identity and neurology.

At the international level, diversity refers to the existence of many peoples contributing their unique experiences to humanity's culture.

In a class, there may be students from different ethnicities, cultures, nationalities etc. The teacher is responsible for recognizing the diversity of the students, respecting their cultures planning lessons , keeping in mind the various students who are diverse and for some of them, their first language is not English.

Skill 40.8 Understanding how to sequence learning activities in a way that allows students to build upon their prior knowledge and challenges them to expand their understanding of science

Subject matter should be presented in a fashion that helps students <u>organize, understand,</u> and <u>remember</u> important information. Advance organizers and other instructional devices can help students to:

- Connect information to what is already known
- Make abstract ideas more concrete
- Capture students' interest in the material
- Help students to organize the information and visualize the relationships.

Organizers can be visual aids such as diagrams, tables, charts, guides or verbal cues that alert students to the nature and content of the lesson. Organizers may be used:

- **Before the lesson** to alert the student the student to the main point of the lesson, establish a rationale for learning, and activate background information.
- **During the lesson** to help students organize information, keep focused on important points, and aid comprehension.
- **At the close of the lesson** to summarize and remember important points.

Examples of organizers include:

- Question and graphic-oriented study guide.
- Concept diagramming: students brainstorm a concept and organize information into three lists (always present, sometimes present, and never present).
- Semantic feature analysis: students construct a table with examples of the concept in one column and important features or characteristics in the other column opposite.
- Semantic webbing: The concept is placed in the middle of the chart or chalkboard and relevant information is placed around it. Lines show the relationships.
- Memory (mnemonic) devices. Diagrams charts and tables

COMPETENCY 041 UNDERSTANDING THE PROCESS OF SCIENTIFIC INQUIRY AND ITS ROLE IN SCIENCE INSTRUCTION

Skill 41.1 Plan and implement instruction that provides opportunities for all students to engage in non-experimental and experimental inquiry investigations.

Individual teaching should be the method for exceptional students and those who need more attention than the regular student. A few minutes of explaining the lesson or the task on hand will be very helpful. In the case of pair share or collaborative pairs, a small assignment could be given and a time frame set, at the end of which students will share as a class what they have learned. This is very good if an exceptional student and a higher functioning student are paired. A small group is very productive since there are many things involved in that situation, such as sharing information, waiting for one's turn, listening to other's ideas, views, and suggestions, and taking responsibility for doing a job in the group (writing/presenting/drawing etc.) which also teach basic manners. Teaching as a class involves traditional and modern methods such as lecture, lecture/demonstration, pause and lecture, etc. The same applies for experimental, field and non experimental work.

Today's learning, especially science, is largely inquiry-based. Sometimes it becomes part of teaching to encourage the students to ask questions. Sufficient time must be given to students to ask these questions.

As a teacher, one must be a good manager of not only the classroom but also of time, resources, and space. The teacher needs to plan how much time should be given to exceptional students, higher functioning students, students that are performing on level, and students that present behavioral concerns. The exceptional and the students with behavioral concerns must get more of the teacher's time. Next will be the regular and last the bright students, since they are a few steps ahead of the rest. If they finish work quickly, however, bright students need to be engaged, so some extra work must be available. In terms of space the same things apply. Resources must be shared equally as far as possible, since everybody has the right to have equal opportunity. However, there must be modification of resources suitable for the exceptional students, if required.

One thing is most important - a teacher must use logic and be able to think laterally since all the answers are not in books. The best teaching is part original thinking and part innovation and ingenuity.

Skill 41.2 Inquiry-based instruction on questions and issues relevant to students and uses strategies to assist students with generating, refining, and focusing scientific questions and hypotheses.

Scientific questions are very important because they are the basis for learning. Students need to be encouraged, provoked and challenged to ask questions. The questions need not necessarily make sense at the beginning, but as the time goes by, these questions begin to make lot of sense.

The first and foremost thing is to encourage the students to ask **questions**. They need to learn to frame questions.

There are a few ways in which the students can be encouraged to ask questions:
1. Brainstorming the topic under study
2. Discussing it in the class and inviting students to ask questions
3. Letting students discuss in small groups and come up with questions - this is extremely useful to students who are introverts and shy by nature.

There can be other ways as well besides those mentioned above. The teacher must realize that questioning is an important tool in teaching and it can be an effective one in learning as well. It must be mentioned here that not all students are curious and inquisitive and that not all parents encourage their children to ask questions. In such cases the teacher needs to show lot of patience in encouraging the students to be inquisitive and curious. This takes time and with time, this could be achieved to a large extent.

The next step in this process of teaching students to question is, **refining** questions. By now the students have learned to ask questions. These questions may not be completely relevant to the topic under discussion, but still the students have a set of questions. It is the responsibility of the teacher to take these questions and to convert them to "How" and "Why" type of open ended questions. Many times the students may end up asking questions such as: Who landed on the moon? These sorts of questions are not really knowledge generating questions. They are not thought provoking questions. The teacher can modify this question to " what did the missions to moon accomplish? With this type of question, a lot of discussion will be generated. Who landed on the moon first, as well as the weather of the moon, moon rock samples, etc.

The next step is **focusing**. The questions need to be focused on the topic under discussion or investigation. Focusing is absolutely important because it is very easy to be carried away and to be side tracked. The students need to be made aware of being able to focus on a topic, understand and not to deviating from it, however tempting it may be.

The last step in this is **testing scientific questions and hypotheses.** All questions can not be tested; some questions can answered by doing research. Like the question that was cited above regarding the moon, can not be tested, but on the other hand, can be answered by research. A wealth of information could be discovered and most of the questions will be answered.

However, some questions can be tested and answers could be found. For example, "which fertilizer is best for rose cuttings? This kind of question will be best answered by experimentation.

Skill 41.3 **Instruct students in the safe and proper use of a variety of grade-appropriate tools, equipment, resources, technology, and techniques to access, gather, store, retrieve, organize, and analyze data.**

Some of the most common laboratory techniques are: dissections, preserving, staining and mounting microscopic specimens, and preparing laboratory solutions.

1. Dissections

Animals that are not obtained from recognized sources should not be used. Decaying animals or those of unknown origin may harbor pathogens and/or parasites. Specimens should be rinsed before handling. Latex gloves are desirable, if gloves are not available, students with sores or scratches should be excused from the activity. Formaldehyde is a carcinogenic and should be avoided or disposed of according to district regulations. Students objecting to dissections for moral reasons should be given an alternative assignment.

Live specimens - No dissections may be performed on living mammalian vertebrates or birds, however lower order life and invertebrates may be used. Biological experiments may be done with all animals except mammalian vertebrates or birds. No physiological harm may result to the animal. All animals housed and cared for in the school must be handled in a safe and humane manner. Animals are not to remain on school premises during extended vacations unless adequate care is provided. Many state laws state that any instructor who intentionally refuses to comply with the laws may be suspended or dismissed. Interactive dissections are available online or from software companies for those students who object to performing dissections. There should be no penalty for those students who refuse to physically perform a dissection.

2. Staining

Specimens have to be stained because they are mostly transparent (except plant cells which are green) under the microscope and are difficult to be seen under microscope against a white background. The stains add color to the picture, making the image much easier to see. The stains actually work by fixing themselves to various structures on or in the cell. The exact structure determines the staining process used.

It is amazing to know that the variety of stains available are numerous, and are a vital tool to determine what the cellular components are made of. Starch, protein and even nucleic acids can be brought out using special stains.

Some common stains used in the laboratories are methylene blue, chlorazol black, lignin pink, gentian violet, etc.

3. Mounting of specimens

In order to observe microscopic specimens or minute parts, mounting them on a microscope slide is essential. There are two different ways of mounting. One kind of procedure is adapted for keeping mounted slides for a long time to be used again. The second type of procedure is for temporary slides. We will discus about temporary mounting since 12th Grade students are mostly concerned with the temporary mounting. Their work does not require permanent mounting.

Water is a very common mounting medium in High school laboratories since it is cheap and best suited for temporary mounting. However, one problem with water mounting is the fact that water evaporates.

Glycerin is also used for mounting. One advantage with glycerin is that it is non-toxic and is stable for years. It provides good contrast to the specimens under microscopic examination. The only problem with glycerin as a medium is that it supports mold formation.

3a. Care of microscopes

Light microscopes are commonly used in high school laboratory experiments. Total magnification is determined by multiplying the ocular (usually 10X) and the objective (usually 10X on low, 40X on high) lenses. A few steps should be followed to properly care for this equipment.

- Clean all lenses with lens paper only.
- Carry microscopes with two hands, one on the arm and one on the base.
- Always begin focusing on low power, then switch to high power.
- Store microscopes with the low power objective down.
- Always use a cover slip when viewing wet mount slides.
- Bring the objective down to its lowest position then focus moving up to avoid the slide from breaking or scratching.

4. Preparation of laboratory solutions

This is a critical skill needed for any experimental success. The procedure for making solutions must be followed to get maximum accuracy.

i) weigh out the required amount of each solute
ii) dissolve the solute in less than the total desired volume (about 75%)
iii) add enough solvent to get the desired volume

1. Weight/volume:
This is usually expressed as mg/ml for small amounts of chemicals and other specialized biological solutions. e.g. 100 mg/ml ampicillin = 100 mg. of ampicillin dissolved in 1 ml of water.

2. Molarity: moles of solute dissolved/ liter of solution

Mole = 6.02 times 10^23 atoms = Avagadro's number
Mole = gram formula weight (FW) or gram molecular weight (MW)

* These values are usually found on the labels or in Periodic Table.
e.g. Na2SO4

2 sodium atoms - 2 times 22.99g = 45.98 g
1 sulfur atom   - 1 times 32.06g = 32.06 g
4 oxygen atoms – 4 times 16.00g = 64.00 g
  Total   = 142.04g

1M = 1 mole/liter, 1 mM = 1 millimole/liter, 1 uM = 1 umole/liter

* How much sodium is needed to make 1L of 1M solution?
Formula weight of sodium sulfate = 142.04g
Dissolve 142.04g of sodium sulfate in about 750mL of water, dissolve sodium sulfate thoroughly and make up the volume to 1 liter (L)

Skill 41.4   Knows how to guide students in making systematic observations and measurements

The starting point for any science is systematic observation. Systematic observation is observing and recording the occurrence of certain specific (naturally occurring) behaviors.

There are four descriptive observation methods:

1. Naturalistic observation: observers record occurrence of naturally occurring behavior.
2. Systematic observation: observers record the occurrence of certain specific (naturally occurring) behaviors.
3. Case study: gather detailed information about one individual.
4. Archival research: use existing behavior to establish occurrence of behavior.
With each type of approach, there are potential problems and limitations.

We discuss here systematic observation.

Systematic observation emphasizes gathering quantitative data on certain specific behaviors. The researcher is interested in a limited set of behaviors. This allows them to study and test specific hypotheses.

The first step is to develop a coding system. The coding system is a description of behaviors and how they will be recorded. The key idea is to delimit the range of behaviors that are observed. The operational definitions of each behavior that will be recorded are defined, and occurrence of each behavior and its duration is recorded.

For example, consider the recording of animal behavior. Potential problems that can occur during the observation include:

1. Remaining vigilant: Following an animal in its natural habitat is difficult, especially when human presence is required. Recording devices (audio, video) are used to deal with this problem.

2. Reactivity: Humans and animals often change their behavior when they are being observed. The observer must take steps to remain unobtrusive or become a participant observer.

3. Reliability: Ensuring that the coding of behavior is accurate (two or more observers are used consistently and their results compared). This is known as inter-rater or inter-observer reliability.

4. Sampling: Setting up a schedule of observation intervals, using multiple observations over a range of time. This measures the behavior of an animal over a period of time and is considered to be reliable.

It is the responsibility of the teacher to introduce the process of systematic observation to students in a meaningful way. It is also the responsibility of the teacher to instruct students on how to take correct scientific measurements.

Measurements in science

In science the system of measurements is called the SI or Standard International system of units, or Systeme Internacionale in French. This method is a refined version of the Metric system used in France. This method is unique in that all the measurements are in multiples of tens, which makes it very easy to do basic mathematical operations such as multiplication and division.

SI prefixes:

kilo - k - 1000
hecto - h - 100
deka - da - 10
deci - d - 0.1 (1/10)
centi - c - 0.01 (1/100)
milli - m - 0.001 (1/1000)

Length: The SI units of length are millimeter, centimeter, meter and kilometer.
10 millimeters make 1 centimeter
100 centimeters make 1 meter
1000 meters make 1 kilometer

Mass and weight:
The SI units of mass and weight are milligrams and kilograms
1000 milligrams make 1 kilogram

Volume:
The SI units of volume are milliliters and liters.
There are 1000 milliliters in a liter.

Time:
This is the same in any system of measurement.

Temperature:
The SI scale for measuring temperature is Kelvin (K). Scientists also use Celsius for temperature measurement.

Skill 41.5 Promoting the use of critical-thinking skills, logical reasoning, and scientific problem solving to reach conclusions based on evidence.

It is imperative that teachers are able to teach students to use the skills necessary to evaluate scientific principles and ideas, as well as lead students to their own discoveries. This promotes a better understanding of science and its topics of study. Critical-thinking skills are a necessity for a student to adequately understand all processes of science. For example, students must be able to apply the principles of sciences to the question being considered. It requires a higher level thinking in students, rather than just a repeat of terms or facts found in the text. A good example of a higher level thinking or critical-thinking question would be as follows.

If five appliances are all in place in a parallel circuit and all separately connected to the voltage source, and there is a microwave plugged in at the beginning of the circuit, will all the appliances still operate? Why or why not?

Asking such questions force student to think about the answer rather than just define a vocabulary term. Using logical reasoning skills are equally as important for students' understanding. Logical reasoning skills involve comprehending what the scientific possibilities are and how the possibilities may occur. For example, when studying earthquakes and natural disasters, students should be able to reason that fault lines which normally have some movement, and have not moved in the past ten years, would be expected within the next few years to produce major movements resulting in an earthquake.

Finally, scientific problem solving will lead students to a greater understanding of scientific tools and processes. Scientific problem solving involves inquiry, assessment, lastly solving or drawing conclusions. This can be observed in each step of The Scientific Method. Students should be given the opportunity to test problem solving during a display of their scientific knowledge. An excellent opportunity for this display is during a science fair or an engineering fair. Students should build, develop, and display their findings.

Skill 41.6 How to teach students to develop, analyze, and evaluate different explanations for a given scientific investigation.

Testing of scientific questions can be carried out by the using scientific method. This is a process consisting of a series of steps, designed to solve a problem. This method is designed in such a way that most of our bias / prejudices are eliminated.

This method consists of the following steps:

1. Question / problem: any investigation, big or small has a beginning as a question / problem. It is important to begin with a well-defined problem. The problem needs to be stated in simple and clear language. Any body who reads the problem should be able to understand it.

2. Gathering information: It is very important that relevant information is gathered for a better understanding of the problem.

3. Forming a hypothesis: Hypothesis is otherwise known as an educated guess. It is based on information and knowledge and that is the reason why it is called an educated guess.
Again, hypothesis needs to be simple and put in clear language for any body to understand.

4. The fourth step is **designing an experiment**. This involves identifying control / standard, the constants and the variables. There has to be a control /standard for any experiment to compare the results with at the end of the experiment. Some things have to be kept constant through out the experiment. The more constants an experiment has, the better the experiment. Variables are of two types - the first one, independent variables that the experimenter changes, and the dependent variable, which is the factor that is measured in an experiment. Independent variables are the ones that are tested in an experiment. These should not be more than four in an ideal experiment. If there are more independent variables, the experiment gets complicated and tedious usually resulting in the experimenter losing interest. There should be only one dependent variable. More than one dependent veritable complicates an experiment at the High School level. Students need to be educated to do quality science experiments and get good results than doing complicated and lengthy experiments. All experiments must be repeated twice, which means there are three sets of data, since reliability of results is absolutely important science experiments.

5. Analyzing data: The data collected must well presented by graphing and in tables and analyzed for any patterns that may exist. The numbers are important, but the students must be trained to recognize patterns and trends, which are very important.

6. Conclusion: Drawing conclusions is very important because those are the answers for the question the students started with. The conclusion tells us whether the hypothesis is proved or disproved. If the hypothesis is disproved after experimentation, the hypothesis needs to be modified.

The teacher must emphasize one important thing - data must be recorded truthfully and that calls for honesty and integrity.

Skill 41.7 How to teach students to demonstrate an understanding of potential sources or error in inquiry-based investigation.

Students should be taught that there are many ways in which errors could creep in measurements. Errors in measurements could occur because of:

1. Improper use of instruments used for measuring – weighing etc.
2. Parallax error – not positioning the eyes during reading of measurements
3. Not using same instruments and methods of measurement during an experiment
4. Not using the same source of materials, resulting in the content of a certain compound used for experimentation

Besides these mentioned above, there could be other possible sources of error as well. When erroneous results are used for interpreting data, the conclusions are not reliable. An experiment is valid only when all the constants (time, place, method of measurement, etc.) are strictly controlled. Students should be aware of this when conducting investigations.

Skill 41.8 How to teach students to demonstrate an understanding of how to communicate and defend the results of an inquiry-based investigation.

Conclusions must be communicated by clearly describing the information using accurate data, visual presentation and other appropriate media such as a power point presentation. Examples of visual presentations are graphs (bar/line/pie), tables/charts, diagrams, and artwork. Modern technology must be used whenever necessary. The method of communication must be suitable to the audience. Written communication is as important as oral communication. The scientist's strongest ally is a solid set of reproducible data.

COMPETENCY 042 VARIED AND APPROPRIATE ASSESSMENTS AND ASSESSMENT PRACTICES TO MONITOR SCIENCE LEARNING IN LABORATORY, FIELD, AND CLASSROOM SETTINGS.

Skill 42.1 The relationship among science curriculum, assessment, and instruction and bases instruction on information gathered through assessment of students' strengths and needs.

All children enrolled in our educational system will experience testing throughout their schooling, whether it is preschool sensory screenings, teacher-made quizzes, or annual standardized assessments. Assessment is continuous, and occurs on a regular basis. There are a variety of assessment instruments: standardized, criterion-referenced, curriculum-based, and teacher-made. Teachers in the field should possess sufficient knowledge to be able to determine quantitative dimensions such as the validity and reliability of tests, to recognize sound test content, and to choose appropriate tests for specific purposes. Continuous assessment allows the teacher to direct, and sometimes redirect, resources appropriately.

Skill 42.2 The importance of monitoring and assessing students' understanding of science concepts and skills on an ongoing basis.

Much of science is based on past research. Because the concepts build upon one another it is vital that the student understands all previous concepts. For instance, a child who does not sufficiently grasp the difference between chemical and physical properties will not be unable to make the leap towards understanding chemical reactions, nor would s/he then understand the law of conservation of matter. For this reason, assessment should occur regularly. Where a student or students is/are not proficient, the subject matter should be covered again to enhance understanding.

Skill 42.3   The importance of carefully selecting or designing formative and summative assessments for the specific decisions they are intended to inform.

Formative assessment is the sum of all activities undertaken by teachers and by their students that provide information. That information can be considered feedback, and the teacher should modify the teaching and learning activities in response.

Teachers are responsible for assessment. They are required to give a report on each student's progress to both parents and administrators, as well as above and beyond regular administrators (often state mandates require a report as well). These reports inform parents, other teachers, officials, and also serve accountability purposes. In addition, other teachers are informed in regards to placement and teaching skills necessary for students to be successful. Teachers must take on multiple roles and must use formative assessment to help support and enhance student learning. The teacher ultimately decides the future for most students and must make summative judgments about a student's achievement at a specific point in time for purposes of placement, grading, accountability, and informing parents and future teachers about student performance. Teachers have special skills and observe students on a daily basis therefore giving them the power to use their perspective and knowledge to make recommendations for students in the future. Teachers are able to observe and assess over a period of time. This assessment may conflict with other performances and a teacher may be able to make better recommendations for students. Teachers must also remember that assessment-based judgments must be adjusted and reexamined over long periods of time to insure that conclusions are accurate.

Summative assessments are typically used to evaluate instructional programs and services. Typically these take place at the end of each school year. The overall goal of summative assessments is to make a judgment of student competency after an instructional phase is complete, in an attempt to see what has been learned or mastered by the students. Summative evaluations are used to determine if students have mastered specific competencies and to identify instructional areas that need additional attention. Some formative and summative assessments that are common in K-12 schools are listed below:

Formative Assessments	Summative Assessments
Anecdotal records	Final Exams
Quizzes and essays	State wide tests
Diagnostic tests	National tests
Lab reports	Entrance exams (SAT & ACT)

Skill 42.4 Select or design and administer a variety of appropriate assessment methods (e.g., performance assessment, self-assessment, formal/informal, formative/summative) to monitor students understanding and progress.

Some assessment methods can be both formal and informal tools. For example, observation may incorporate structured observation instruments as well as other informal observation procedures, including professional judgment. When evaluating a child's developmental level, a professional may use a formal adaptive rating scale while simultaneously using professional judgment to assess the child's motivation and behavior during the evaluation process.

Curriculum-Based Assessment—Assessment of an individual's performance of objectives of a curriculum, such as a reading, math, or science program. The individual's performance is measured in terms of what objectives were mastered. This type of testing could be verbal, written, or demonstration based. Its general structure may include such factors as how much time to complete, length of time needed to complete, and group or individual testing. The level of response may be multiple choice, essay, or recall of facts.

Momentary time sampling—This is a technique used for measuring behaviors of a group of individuals or several behaviors from the same individual. Time samples are usually brief, and may be conducted at fixed or variable intervals. The advantage of using variable intervals is increased reliability, since the students will not be able to predict when the time sample will be taken.

Multiple Baseline Design—This may be used to test the effectiveness of an intervention in a skill performance or to determine if the intervention accounted for the observed changes in a target behavior. First, the initial baseline data is collected, followed by the data during the intervention period. To get the second baseline, the intervention is removed for a period of time and data is collected again. The intervention is then reapplied, and data collected on the target behavior. An example of a multiple baseline design might be ignoring a child who calls out in class without raising his hand. Initially, the baseline could involve counting the number of times the child calls out before applying interventions. During the time the teacher ignores the child's call-outs, data is collected. For the second baseline, the teacher would resume the response to the child's call-outs in the way she did before ignoring. The child's call-outs would probably increase again, if ignoring actually accounted for the decrease. If the teacher reapplies the ignoring strategy, the child's call-outs would probably decrease again.

Group Tests And Individual Tests

The distinction between a group test and in individual test is that individual tests must be administered to only one person at a time, whereas group tests are administered to several people simultaneously, or can be administered individually. However, there are several other subtle differences.

When administering an individual test, the tester has the opportunity to observe the individual's responses and to determine how such things as problem solving are accomplished. Within limits, the tester is able to control the pace and tempo of the testing session, and to rephrase and probe responses in order to elicit the individual's best performance. If the child becomes tired, the examiner can break between sub tests or end the test; if he loses his place on the test, the tester can help him to regain it; if he dawdles or loses interest, the tester can encourage or redirect him. If the child lacks self-confidence, the examiner can reinforce his efforts. In short, individual tests allow the examiner to encourage best efforts, and to observe how a student uses his skills to answer questions. Thus, individual tests provide for the gathering of both quantitative and qualitative information. On the other hand, with a group test, the examiner may provide oral directions for younger children, but beyond the fourth grade, directions are usually written. The children write or mark their own responses, and the examiner monitors the progress of several students at the same time. He cannot rephrase questions, or probe or prompt responses. Even when a group test is administered to only one child, qualitative information is very difficult, if not impossible, to obtain.

The choice between group and individual testing should be primarily determined by purpose and efficiency. When testing for program evaluation, screening, and some types of program planning (such as tracking), group tests are appropriate. Individual tests could be used but are impractical in terms of time and expense. Special consideration may need to be given if there are any motivational, personalities, linguistic, or physically disabling factors that might impair the examinee's performance on group tests.

Skill 42.5 **Use of formal and informal assessments of student performance and products (e.g., projects, lab journals, rubrics, portfolios, student profiles, checklists) to evaluate student participation in and understanding of the inquiry process.**

It is vital for teachers to track students' performances through a variety of methods. Teachers should be able to asses students on a daily basis using informal assessments such as monitoring during work time, class discussions, and note taking. Often these assessments are a great way to determine whether or not students are "on track" learning selected objectives. Generally teachers can assess students just by class discussion and often this is a great time to give participation points, especially to those students who have special needs and participate well in class but may struggle with alternative assignments. More formal assessments are necessary to ensure students fully understand selected objectives. Regular grading using selected performance skills is necessary: however teachers should develop personal ways of grading using a variety of assessment tools. For example, students could keep a "Science Journal" and track progress of ongoing assignments, projects, and labs. Another great tool for observing and evaluating students is the use of rubrics and checklists. Student profiles and checklists are a great way to quickly determine if students are meeting selected objectives. These are useful during monitoring time for teachers. It is easy for a teacher to check off students who are meeting objectives using a checklist while monitoring students' work done in class. See example below:

Student Name	Objective #1	Objective #2	On task
Joe Student	X	X	X
Jon Student		X	X

Checklists provide an easy way for teachers to track students and quickly see who is behind or needs a lesson or objective reviewed.

Skill 42.6   The importance of sharing evaluation criteria and assessment results with students.

Effective teachers:
- offer students a safe and supportive learning environment, including clearly expressed and reasonable expectations for behavior;
- create learning environments that encourage self-advocacy and developmentally appropriate independence; and
- offer learning environments that promote active participation in independent or group activities.

Such an environment is an excellent foundation for building rapport and trust with students, and communicating a teacher's respect for and expectation that they take a measure of responsibility for their educational development. Ideally, mutual trust and respect will afford teachers opportunities to learn of and engage students' ideas, preferences and abilities.

Teacher behaviors that motivate students include:

- Maintain Success Expectations through teaching, goal setting, establishing connections between effort and outcome, and self-appraisal and reinforcement.
- Have a supply of intrinsic incentives such as rewards, appropriate competition between students, and the value of the academic activities.
- Focus on students' intrinsic motivation through adapting the tasks to students' interests, providing opportunities for active response, including a variety of tasks, providing rapid feedback, incorporating games into the lesson, allowing students the opportunity to make choices, create, and interact with peers.
- Stimulate students' learning by modeling positive expectations and attributions. Project enthusiasm and personalize abstract concepts. Students will be better motivated if they know what they will be learning about. The teacher should also model problem-solving and task-related thinking so students can see how the process is done.

For adolescents, motivation strategies are usually aimed at getting the student actively involved in the learning process. Since the adolescent has the opportunity to get involved in a wider range of activities outside the classroom (job, car, being with friends), stimulating motivation may be the focus even more than academics. Motivation may be achieved through extrinsic reinforcers or intrinsic reinforcers. This is accomplished by allowing the student a degree of choice in what is being taught or how it will be taught. The teacher will, if possible, obtain a commitment either through a verbal or written contract between the student and the teacher. Adolescents also respond to regular feedback, especially when that feedback shows that they are making progress. Motivation is a key component in learning.

Sample Test: Science

Directions: Read each item and select the correct response.

1. In an experiment, the scientist states that he believes a change in the color of a liquid is due to a change of pH. This is an example of _____ .

A. observing.

B. inferring.

C. measuring.

D. classifying.

2. **When is a hypothesis formed?**

A. Before the data is collected.

B. After the data is collected.

C. After the data is analyzed.

D. Concurrent with graphing the data.

3. **Who determines the laws regarding the use of safety glasses in the classroom?**

A. The state.

B. The school site.

C. The Federal government.

D. The district level.

4. **If one inch equals 2.54 cm how many mm in 1.5 feet? (APPROXIMATELY)**

A. 18 mm.

B. 1800 mm.

C. 460 mm.

D. 4,600 mm.

5. **Which of the following instruments measures wind speed?**

A. A barometer.

B. An anemometer.

C. A wind sock.

D. A weather vane.

6. **Sonar works by _____ .**

A. Timing how long it takes for sound to reach a certain speed.

B. Bouncing sound waves between two metal plates.

C. Bouncing sound waves off of an underwater object and timing how long it takes for the sound to return.

D. Evaluating the motion and amplitude of sound.

7. The measure of the pull of the earth's gravity on an object is called _____ .

A. mass number.

B. atomic number.

C. mass.

D. weight.

8. Which reaction below is a decomposition reaction?

A. HCl + NaOH → NaCl + H_2O

B. C + O_2 → CO_2

C. $2H_2O$ → $2H_2$ + O_2

D. $CuSO_4$ + Fe → $FeSO_4$ + Cu

9. The Law of Conservation of Energy states that _____ .

A. there must be the same number of products and reactants in any chemical equation.

B. objects always fall toward large masses such as planets.

C. energy is neither created nor destroyed, but may change form.

D. lights must be turned off when not in use, by state regulation.

10. Which parts of an atom are located inside the nucleus?

A. Electrons and neutrons.

B. Protons and neutrons.

C. Protons only.

D. Neutrons only.

11. The elements in the modern Periodic Table are arranged _____ .

A. in numerical order by atomic number.

B. randomly.

C. in alphabetical order by chemical symbol.

D. in numerical order by atomic mass.

12. Carbon bonds with hydrogen by _____ .

A. ionic bonding.

B. non-polar covalent bonding.

C. polar covalent bonding.

D. strong nuclear force.

13. Vinegar is an example of a _____.

A. strong acid.

B. strong base.

C. weak acid.

D. weak base.

14. Which of the following is not a nucleotide?

A. adenine.

B. alanine.

C. cytosine.

D. guanine.

15. When measuring the volume of water in a graduated cylinder, where does one read the measurement?

A. At the highest point of the liquid.

B. At the bottom of the meniscus.

C. At the closest mark to the top of the liquid.

D. At the top of the plastic safety ring.

16. A duck's webbed feet are examples of _____.

A. mimicry.

B. structural adaptation.

C. protective resemblance.

D. protective coloration.

17. What organelle contains the cell's stored food?

A. Vacuoles.

B. Golgi Apparatus.

C. Ribosomes.

D. Lysosomes.

18. The first stage of mitosis is called _____.

A. telophase.

B. anaphase.

C. prophase.

D. mitophase.

19. The Doppler Effect is associated most closely with which property of waves?

A. Amplitude.

B. Wavelength.

C. Frequency.

D. Intensity.

20. Viruses are responsible for many human diseases including all of the following *except* _____?

A. influenza.

B. A.I.D.S.

C. the common cold.

D. strep throat.

21. A series of experiments on pea plants formed by _____ showed that two invisible markers existed for each trait, and one marker dominated the other.

A. Pasteur.

B. Watson and Crick.

C. Mendel.

D. Mendeleyev.

22. Formaldehyde should not be used in school laboratories for the following reason:

A. it smells unpleasant.

B. it is a known carcinogen.

C. it is expensive to obtain.

D. it is explosive.

23. Amino acids are carried to the ribosome in protein synthesis by _____.

A. transfer RNA (tRNA).

B. messenger RNA (mRNA).

C. ribosomal RNA (rRNA).

D. transformation RNA (trRNA).

24. When designing a scientific experiment, a student considers all the factors that may influence the results. The goal of the process is to _____.

A. recognize and manipulate independent variables.

B. recognize and record independent variables.

C. recognize and manipulate dependent variables.

D. recognize and record dependent variables.

25. Since ancient times, people have been entranced with bird flight. What is the key to bird flight?

A. Bird wings are a particular shape and composition.

B. Birds flap their wings quickly enough to propel themselves.

C. Birds take advantage of tailwinds.

D. Birds take advantage of crosswinds.

26. Laboratory researchers have classified fungi as distinct from plants because the cell walls of fungi _____ .

A. contain chitin.

B. contain yeast.

C. are more solid.

D. are less solid.

27. In a fission reactor, "heavy water" is used to _____ .

A. terminate fission reactions.

B. slow down neutrons and moderate reactions.

C. rehydrate the chemicals.

D. initiate a chain reaction.

28. The transfer of heat by electromagnetic waves is called _____ .

A. conduction.

B. convection.

C. phase change.

D. radiation.

29. When heat is added to most solids, they expand. Why is this the case?

A. The molecules get bigger.

B. The faster molecular motion leads to greater distance between the molecules.

C. The molecules develop greater repelling electric forces.

D. The molecules form a more rigid structure.

30. The force of gravity on earth causes all bodies in free fall to _____ .

A. fall at the same speed.

B. accelerate at the same rate.

C. reach the same terminal Velocity.

D. move in the same direction.

31. Sound waves are produced by _____ .

A. pitch.

B. noise.

C. vibrations.

D. sonar.

32. Resistance is measured in units called _____ .

A. watts.

B. volts.

C. ohms.

D. current.

33. Sound can be transmitted in all of the following *except* _____ .

A. air.

B. water.

C. a diamond.

D. a vacuum.

34. As a train approaches, the sound of the whistle seems _____ .

A. higher, because it has a higher apparent frequency.

B. lower, because it has a lower apparent frequency.

C. higher, because it has a lower apparent frequency.

D. lower, because it has a higher apparent frequency.

35. The speed of light is different in different materials. This is responsible for _____ .

A. interference.

B. refraction.

C. reflection.

D. relativity.

36. A converging lens produces a real image _____.

A. always.

B . never.

C. when the object is within one focal length of the lens.

D. when the object is further than one focal length from the lens.

37. The electromagnetic radiation with the longest wave length is/are _____.

A. radio waves.

B. red light.

C. X-rays.

D. ultraviolet light.

38. Under a 440 power microscope, an object with diameter 0.1 millimeter appears to have a diameter of _____ .

A. 4.4 millimeters.

B. 44 millimeters.

C. 440 millimeters.

D. 4400 millimeters.

39. Separating blood into blood cells and plasma involves the process of _____.

A. electrophoresis.

B. spectrophotometry.

C. centrifugation.

D. chromatography.

40. Experiments may be done with any of the following animals except _____.

A. birds.

B. invertebrates.

C. lower order life.

D. frogs.

41. A student is designing a science experiment to test the effects of light and water on plant growth. You recommend that she _____.

A. manipulate the temperature also.

B. manipulate the water pH also.

C. determine the relationship between light and water unrelated to plant growth.

D. omit either water or light as a variable.

42. In a laboratory report, what is the abstract?

A. The abstract is a summary of the report, and is the first section of the report.

B. The abstract is a summary of the report, and is the last section of the report.

C. The abstract is predictions for future experiments, and is the first section of the report.

D. The abstract is predictions for future experiments, and is the last section of the report.

43. What is the scientific method?

A. It is the process of doing an experiment and writing a laboratory report.

B. It is the process of using open inquiry and repeatable results to establish theories.

C. It is the process of reinforcing scientific principles by confirming results.

D. It is the process of recording data and observations.

44. A student had four corn plants and was measuring photosynthetic rate (by measuring growth mass). Half of the plants were exposed to full (constant) sunlight, and the other half were kept in 50% (constant) sunlight. Identify the control in the experiment:

A. The control is a set of plants grown in full (constant) sunlight.

B. The control is a set of plants grown in 50% (constant) sunlight.

C. The control is a set of plants grown in the dark.

D. The control is a set of plants grown in a mixture of natural levels of sunlight.

45. In an experiment measuring the growth of bacteria at different temperatures, what is the independent variable?

A. Number of bacteria.

B. Growth rate of bacteria.

C. Temperature.

D. Light intensity.

46. A scientific law_____.

A. proves scientific accuracy.

B. may never be broken.

C. may be revised in light of new data.

D. is the result of one excellent experiment.

47. Which is the correct order of methodology?

1. collecting data
2. planning a controlled experiment
3. drawing a conclusion
4. hypothesizing a result
5. re-visiting a hypothesis to answer a question

A. 1,2,3,4,5

B. 4,2,1,3,5

C. 4,5,1,3,2

D. 1,3,4,5,2

48. Which is the most desirable tool to use to heat substances in a middle school laboratory?

A. Alcohol burner.

B. Freestanding gas burner.

C. Bunsen burner.

D. Hot plate.

49. Newton's Laws are taught in science classes because _____.

A. they are the correct analysis of inertia, gravity, and forces.

B. they are a close approximation to correct physics, for usual Earth conditions.

C. they accurately incorporate relativity into studies of forces.

D. Newton was a well-respected scientist in his time.

50. Which of the following statements is most accurate?

A. Mass is always constant; Weight may vary by location.

B. Mass and Weight are both always constant.

C. Weight is always constant; Mass may vary by location.

D. Mass and Weight may both vary by location.

51. Chemicals should be stored

A. in the principal's office.

B. in a dark room.

C. in an off-site research facility.

D. according to their reactivity with other substances.

52. Which of the following is the worst choice for a school laboratory activity?

A. A genetics experiment tracking the fur color of mice.

B. Dissection of a preserved fetal pig.

C. Measurement of goldfish respiration rate at different temperatures.

D. Pithing a frog to watch the circulatory system.

53. Who should be notified in the case of a serious chemical spill?

A. The custodian.

B. The fire department or the municipal authority.

C. The science department chair.

D. The School Board.

54. A scientist exposes mice to cigarette smoke, and notes that their lungs develop tumors. Mice that were not exposed to the smoke do not develop as many tumors. Which of the following conclusions may be drawn from these results?

I. Cigarette smoke causes lung tumors.

II. Cigarette smoke exposure has a positive correlation with lung tumors in mice.

III. Some mice are predisposed to develop lung tumors.

IV. Mice are often a good model for humans in scientific research.

A. I and II only.

B. II only.

C. I, II, and III only.

D. II and IV only.

55. In which situation would a science teacher be legally liable?

A. The teacher leaves the classroom for a telephone call and a student slips and injures him/herself.

B. A student removes his/her goggles and gets acid in his/her eye.

C. A faulty gas line in the classroom causes a fire.

D. A student cuts him/herself with a dissection scalpel.

56. Which of these is the best example of 'negligence'?

A. A teacher fails to give oral instructions to those with reading disabilities.

B. A teacher fails to exercise ordinary care to ensure safety in the classroom.

C. A teacher displays inability to supervise a large group of students.

D. A teacher reasonably anticipates that an event may occur, and plans accordingly.

57. Which item should always be used when handling glassware?

A. Tongs.

B. Safety goggles.

C. Gloves.

D. Buret stand.

58. Which of the following is *not* a necessary characteristic of living things?

A. Movement.

B. Reduction of local entropy.

C. Ability to cause change in local energy form.

D. Reproduction.

59. What are the most significant and prevalent elements in the biosphere?

A. Carbon, Hydrogen, Oxygen, Nitrogen, Phosphorus.

B. Carbon, Hydrogen, Sodium, Iron, Calcium.

C. Carbon, Oxygen, Sulfur, Manganese, Iron.

D. Carbon, Hydrogen, Oxygen, Nickel, Sodium, Nitrogen.

60. All of the following are units of energy *except* _____

A. joules.

B. calories.

C. watts.

D. ergs.

61. Identify the correct sequence of organization of living things from lower to higher order:

A. Cell, Organelle, Organ, Tissue, System, Organism.

B. Cell, Tissue, Organ, Organelle, System, Organism.

C. Organelle, Cell, Tissue, Organ, System, Organism.

D. Organelle, Tissue, Cell, Organ, System, Organism.

62. Which kingdom is comprised of organisms made of one cell with no nuclear membrane?

A. Monera.

B. Protista.

C. Fungi.

D. Algae.

63. Which of the following is found in the least abundance in organic molecules?

A. Phosphorus.

B. Potassium.

C. Carbon.

D. Oxygen.

64. The significance of a catalyst in a reaction is that it _____ .

A. lowers effective activation energy.

B. maintains precise pH levels.

C. keeps systems at equilibrium.

D. adjusts reaction speed.

65. Accepted procedures for preparing solutions should be made with _____.

A. alcohol.

B. hydrochloric acid.

C. distilled water.

D. tap water.

66. Enzymes speed up reactions by _____.

A. utilizing ATP.

B. lowering pH thereby allowing reaction speed to increase.

C. increasing volume of substrate.

D. lowering energy of activation.

67. When you step out of the shower, the floor feels colder on your feet than the bathmat. Which of the following is the correct explanation for this phenomenon?

A. The floor is colder than the bathmat.

B. Your feet have a chemical reaction with the floor, but not the bathmat.

C. Heat is conducted more easily into the floor.

D. Water is absorbed from your feet into the bathmat.

68. Which of the following is *not* considered ethical behavior for a scientist?

A. Using unpublished data and citing the source.

B. Publishing data before other scientists have had a chance to replicate results.

C. Collaborating with other scientists from different laboratories.

D. Publishing work with an incomplete list of citations.

69. The chemical equation for water formation is: $2H_2 + O_2 \rightarrow 2H_2O$. Which of the following is an *incorrect* interpretation of this equation?

A. Two moles of hydrogen gas and one mole of oxygen gas combine to make two moles of water.

B. Two grams of hydrogen gas and one gram of oxygen gas combine to make two grams of water.

C. Two molecules of hydrogen gas and one molecule of oxygen gas combine to make two molecules of water.

D. Four atoms of hydrogen (combined as a diatomic gas) and two atoms of oxygen (combined as a diatomic gas) combine to make two molecules of water.

70. Energy is measured with the same units as _____.

A. force.

B. momentum.

C. work.

D. power.

71. If the volume of a confined gas is increased, what happens to the pressure of the gas? You may assume that the gas behaves ideally, and that temperature and number of gas molecules remain constant.

A. The pressure increases.

B. The pressure decreases.

C. The pressure stays the same.

D. There is not enough information given to answer this question.

72. A product of anaerobic respiration in animals is _____.

A. carbon dioxide.

B. lactic acid.

C. oxygen.

D. sodium chloride

73. A Newton is fundamentally a measure of _____.

A. force.

B. momentum.

C. energy.

D. gravity.

74. Which change does *not* affect enzyme rate?

A. Increasing the temperature.

B. Adding more substrate.

C. Adjusting the pH.

D. Using a larger cell.

75. Which of the following types of rock are made from magma?

A. Fossils

B. Sedimentary

C. Metamorphic

D. Igneous

76. Which of the following is *not* an acceptable way for a student to acknowledge sources in a laboratory report?

A. The student tells his/her teacher what sources s/he used to write the report.

B. The student uses footnotes in the text, with sources cited, but not in correct MLA format.

C. The student uses endnotes in the text, with sources cited, in correct MLA format.

D. The student attaches a separate bibliography, noting each use of sources.

77. Animals with a notochord or backbone are in the phylum

A. arthropoda.

B. chordata.

C. mollusca.

D. mammalia.

78. Which of the following is a correct example of scientific 'evolution'?

A. Giraffes need to reach higher for leaves to eat, so their necks stretch. The giraffe babies are then born with longer necks. Eventually, there are more long-necked giraffes in the population.

B. Giraffes with longer necks are able to reach more leaves, so they eat more and have more babies than other giraffes. Eventually, there are more long-necked giraffes in the population.

C. Giraffes want to reach higher for leaves to eat, so they release enzymes into their bloodstream, which in turn causes fetal development of longer-necked giraffes. Eventually, there are more long-necked giraffes in the population.

D. Giraffes with long necks are more attractive to other giraffes, so they get the best mating partners and have more babies. Eventually, there are more long-necked giraffes in the population.

79. Which of the following is a correct definition of 'chemical equilibrium'?

A. Chemical equilibrium is when the forward and backward reaction rates are equal. The reaction may continue to proceed forward and backward.

B. Chemical equilibrium is when the forward and backward reaction rates are equal, and equal to zero. The reaction does not continue.

C. Chemical equilibrium is when there are equal quantities of reactants and products.

D. Chemical equilibrium is when acids and bases neutralize each other fully.

80. Which of the following data sets is properly represented by a bar graph?

A. Number of people choosing to buy cars, vs. color of car bought.

B. Number of people choosing to buy cars, vs. age of car customer.

C. Number of people choosing to buy cars, vs. distance from car lot to customer home.

D. Number of people choosing to buy cars, vs. time since last car purchase.

81. In a science experiment, a student needs to dispense very small measured amounts of liquid into a well-mixed solution. Which of the following is the best choice of equipment to use?

A. Buret with buret stand, Stir-plate, Stirring rod, Beaker.

B. Buret with Buret stand, Stir-plate, Beaker.

C. Volumetric Flask, Dropper, Graduated Cylinder, Stirring Rod.

D. Beaker, Graduated Cylinder, Stir-plate.

82. A laboratory balance is most appropriately used to measure the mass of which of the following?

A. Seven paper clips.

B. Three oranges.

C. Two hundred cells.

D. One student's elbow.

83. All of the following are measured in units of length, *except* for

A. perimeter.

B. distance.

C. radius.

D. area.

84. What is specific gravity?

A. The mass of an object.

B. The ratio of the density of a substance to the density of water.

C. Density.

D. The pull of the earth's gravity on an object.

85. What is the most accurate description of the Water Cycle?

A. Rain comes from clouds, filling the ocean. The water then evaporates and becomes clouds again.

B. Water circulates from rivers into groundwater and back, while water vapor circulates in the atmosphere.

C. Water is conserved except for chemical or nuclear reactions, and any drop of water could circulate through clouds, rain, ground-water, and surface-water.

D. Weather systems cause chemical reactions to break water into its atoms.

86. The scientific name *Canis familiaris* refers to the animal's _____.

A. kingdom and phylum.

B. genus and species.

C. class and species.

D. type and family.

87. Members of the same animal species _____.

A. look identical.

B. never adapt differently.

C. are able to reproduce with one another.

D. are found in the same location.

88. Which of the following is/are true about scientists?

I. Scientists usually work alone.
II. Scientists usually work with other scientists.
III. Scientists achieve more prestige from new discoveries than from replicating established results.
IV. Scientists keep records of their own work, but do not publish it for outside review.

I and IV only.

II only.

II and III only.

D. I and IV only.

89. What is necessary for ion diffusion to occur spontaneously?

A. Carrier proteins.

B. Energy from an outside source.

C. A concentration gradient.

D. Cell flagellae.

90. All of the following are considered Newton's Laws *except*

A. An object in motion will continue in motion unless acted upon by an outside force.

B. For every action force, there is an equal and opposite reaction force.

C. Nature abhors a vacuum.

D. Mass can be considered the ratio of force to acceleration.

91. A cup of hot liquid and a cup of cold liquid are both sitting in a room at comfortable room temperature and humidity. Both cups are thin plastic. Which of the following is a true statement?

A. There will be fog on the outside of the hot liquid cup, and also fog on the outside of the cold liquid cup.

B. There will be fog on the outside of the hot liquid cup, but not on the cold liquid cup.

C. There will be fog on the outside of the cold liquid cup, but not on the hot liquid cup.

D. There will not be fog on the outside of either cup.

92. A ball rolls down a smooth hill. You may ignore air resistance. Which of the following is a true statement?

A. The ball has more energy at the start of its descent than just before it hits the bottom of the hill, because it is higher up at the beginning.

B. The ball has less energy at the start of its descent than just before it hits the bottom of the hill, because it is moving more quickly at the end.

C. The ball has the same energy throughout its descent, because positional energy is converted to energy of motion.

D. The ball has the same energy throughout its descent, because a single object (such as a ball) cannot gain or lose energy.

93. A long silver bar has a temperature of 50 degrees Celsius at one end and 0 degrees Celsius at the other end. The bar will reach thermal equilibrium (barring outside influence) by the process of heat _____.

A. conduction.

B. convection.

C. radiation.

D. phase change.

94. _____ are cracks in the plates of the earth's crust, along which the plates move.

A. Faults.

B. Ridges.

C. Earthquakes.

D. Volcanoes.

95. Fossils are usually found in _____ rock.

A. igneous.

B. sedimentary.

C. metamorphic.

D. cumulus.

96. Which of the following is *not* a common type of acid in 'acid rain' or acidified surface water?

A. Nitric acid.

B. Sulfuric acid.

C. Carbonic acid.

D. Hydrofluoric acid.

97. Which of the following is *not* true about phase change in matter?

A. Solid water and liquid ice can coexist at water's freezing point.

B. At 7° C, water is always in liquid phase.

C. Matter changes phase when enough energy is gained or lost.

D. Different phases of matter are characterized by differences in molecular motion.

98. Which of the following is the longest (largest) unit of geological time?

A. Solar Year.

B. Epoch.

C. Period.

D. Era.

99. Extensive use of antibacterial soap has been found to increase the virulence of certain infections in hospitals. Which of the following might be an explanation for this phenomenon?

A. Antibacterial soaps do not kill viruses.

B. Antibacterial soaps do not incorporate the same antibiotics used as medicine.

C. Antibacterial soaps kill a lot of bacteria, and only the hardiest ones survive to reproduce.

D. Antibacterial soaps can be very drying to the skin.

100. Which of the following is a correct explanation for astronaut 'weightlessness'?

A. Astronauts continue to feel the pull of gravity in space, but they are so far from planets that the force is small.

B. Astronauts continue to feel the pull of gravity in space, but spacecraft have such powerful engines that those forces dominate, reducing effective weight.

C. Astronauts do not feel the pull of gravity in space, because space is a vacuum.

D. Astronauts do not feel the pull of gravity in space, because black hole forces dominate the force field, reducing their masses.

101. The theory of 'sea floor spreading' explains _____.

A. the shapes of the continents.

B. how continents collide.

C. how continents move apart.

D. how continents sink to become part of the ocean floor.

102. Which of the following animals are most likely to live in a tropical rain forest?

A. Reindeer.

B. Monkeys.

C. Puffins.

D. Bears.

103. Which of the following is *not* a type of volcano?

A. Shield Volcanoes.

B. Composite Volcanoes.

C. Stratus Volcanoes.

D. Cinder Cone Volcanoes.

104. Which of the following is *not* a property of metalloids?

A. Metalloids are solids at standard temperature and pressure.

B. Metalloids can conduct electricity to a limited extent.

C. Metalloids are found in groups 13 through 17.

D. Metalloids all favor ionic bonding.

105. Which of these is a true statement about loamy soil?

A. Loamy soil is gritty and porous.

B. Loamy soil is smooth and a good barrier to water.

C. Loamy soil is hostile to microorganisms.

D. Loamy soil is velvety and clumpy.

106. Lithification refers to the process by which unconsolidated sediments are transformed into _____.

A. metamorphic rocks.

B. sedimentary rocks.

C. igneous rocks.

D. lithium oxide.

107. Igneous rocks can be classified according to which of the following?

A. Texture.

B. Composition.

C. Formation process.

D. All of the above.

108. Which of the following is the most accurate definition of a non-renewable resource?

A. A nonrenewable resource is never replaced once used.

B. A nonrenewable resource is replaced on a timescale that is very long relative to human life-spans.

C. A nonrenewable resource is a resource that can only be manufactured by humans.

D. A nonrenewable resource is a species that has already become extinct.

109. The theory of 'continental drift' is supported by which of the following?

A. The way the shapes of South America and Europe fit together.

B. The way the shapes of Europe and Asia fit together.

C. The way the shapes of South America and Africa fit together.

D. The way the shapes of North America and Antarctica fit together.

110. When water falls to a cave floor and evaporates, it may deposit calcium carbonate. This process leads to the formation of which of the following?

A. Stalactites.

B. Stalagmites.

C. Fault lines.

D. Sedimentary rocks.

111. A child has type O blood. Her father has type A blood, and her mother has type B blood. What are the genotypes of the father and mother, respectively?

A. AO and BO.

B. AA and AB.

C. OO and BO.

D. AO and BB.

112. Which of the following is the best definition for 'meteorite'?

A. A meteorite is a mineral composed of mica and feldspar.

B. A meteorite is material from outer space, that has struck the earth's surface.

C. A meteorite is an element that has properties of both metals and nonmetals.

D. A meteorite is a very small unit of length measurement.

113. A white flower is crossed with a red flower. Which of the following is a sign of incomplete dominance?

A. Pink flowers.

B. Red flowers.

C. White flowers.

D. No flowers.

114. What is the source for most of the United States' drinking water?

A. Desalinated ocean water.

B. Surface water (lakes, streams, mountain runoff).

C. Rainfall into municipal reservoirs.

D. Groundwater.

115. Which is the correct sequence of insect development?

A. Egg, pupa, larva, adult.

B. Egg, larva, pupa, adult.

C. Egg, adult, larva, pupa.

D. Pupa, egg, larva, adult.

116. A wrasse (fish) cleans the teeth of other fish by eating away plaque. This is an example of _____ between the fish.

A. parasitism.

B. symbiosis (mutualism).

C. competition.

D. predation.

117. What is the main obstacle to using nuclear fusion for obtaining electricity?

A. Nuclear fusion produces much more pollution than nuclear fission.

B. There is no obstacle; most power plants us nuclear fusion today.

C. Nuclear fusion requires very high temperature and activation energy.

D. The fuel for nuclear fusion is extremely expensive.

118. Which of the following is a true statement about radiation exposure and air travel?

A. Air travel exposes humans to radiation, but the level is not significant for most people.

B. Air travel exposes humans to so much radiation that it is recommended as a method of cancer treatment.

C. Air travel does not expose humans to radiation.

D. Air travel may or may not expose humans to radiation, but it has not yet been determined.

119. Which process(es) result(s) in a haploid chromosome number?

A. Mitosis.

B. Meiosis.

C. Both mitosis and meiosis.

D. Neither mitosis nor meiosis.

120. Which of the following is *not* a member of Kingdom Fungi?

A. Mold.

B. Blue-green algae.

C. Mildew.

D. Mushrooms.

121. Which of the following organisms use spores to reproduce?

A. Fish.

B. Flowering plants.

C. Conifers.

D. Ferns.

122. What is the main difference between the 'condensation hypothesis' and the 'tidal hypothesis' for the origin of the solar system?

A. The tidal hypothesis can be tested, but the condensation hypothesis cannot.

B. The tidal hypothesis proposes a near collision of two stars pulling on each other, but the condensation hypothesis proposes condensation of rotating clouds of dust and gas.

C. The tidal hypothesis explains how tides began on planets such as Earth, but the condensation hypothesis explains how water vapor became liquid on Earth.

D. The tidal hypothesis is based on Aristotelian physics, but the condensation hypothesis is based on Newtonian mechanics.

123. Which of the following units is *not* a measure of distance?

A. AU (astronomical unit).

B. Light year.

C. Parsec.

D. Lunar year.

124. The salinity of ocean water is closest to _____ .

A. 0.035 %

B. 0.35 %

C. 3.5 %

D. 35 %

125. Which of the following will not change in a chemical reaction?

A. Number of moles of products.

B. Atomic number of one of the reactants.

C. Mass (in grams) of one of the reactants.

D. Rate of reaction.

Answer Key: Science

1. B
2. A
3. A
4. C
5. B
6. C
7. D
8. C
9. C
10. B
11. A
12. C
13. C
14. B
15. B
16. B
17. A
18. C
19. C
20. D
21. C
22. B
23. A
24. A
25. A
26. A
27. B
28. D
29. B
30. B
31. C
32. C
33. D
34. A
35. B
36. D
37. A
38. B
39. C
40. A
41. D
42. A
43. B
44. A
45. C
46. C
47. B
48. D
49. B
50. A
51. D
52. D
53. B
54. B
55. A
56. B
57. B
58. A
59. A
60. C
61. C
62. A
63. B
64. A
65. C
66. D
67. C
68. D
69. B
70. C
71. B
72. B
73. A
74. D
75. D
76. A
77. B
78. B
79. A
80. A
81. B
82. A
83. D
84. B
85. C
86. B
87. C
88. C
89. C
90. C
91. C
92. C
93. A
94. A
95. B
96. D
97. B
98. D
99. C
100. A
101. C
102. B
103. C
104. D
105. D
106. B
107. D
108. B
109. C
110. B
111. A
112. B
113. A
114. D
115. B
116. B
117. C
118. A
119. B
120. B
121. D
122. B
123. D
124. C
125. B

Rationales with Sample Questions: Science

1. After an experiment, the scientist states that s/he believes a change in color is due to a change in pH. This is an example of

A. observing.

B. inferring.

C. measuring.

D. classifying.

B. Inferring.

To answer this question, note that the scientist has observed a change in color, and has then made a guess as to its reason. This is an example of inferring. The scientist has not measured or classified in this case. Although s/he has observed [the color change], the explanation of this observation is **inferring (B)**.

2. When is a hypothesis formed?

A. Before the data is taken.

B. After the data is taken.

C. After the data is analyzed.

D. While the data is being graphed.

A. Before the data is taken.

A hypothesis is an educated guess, made before undertaking an experiment. The hypothesis is then evaluated based on the observed data. Therefore, the hypothesis must be formed before the data is taken, not during or after the experiment. This is consistent only with **answer (A)**.

3. Who determines the laws regarding the use of safety glasses in the classroom?

A. The state government.

B. The school site.

C. The federal government.

D. The local district.

A. The state government.

Health and safety regulations are set by the state government, and apply to all school districts. Federal regulations may accompany specific federal grants, and local districts or school sites may enact local guidelines that are stricter than the state standards. All schools, however, must abide by safety precautions as set by state government. This is consistent only with **answer (A)**.

4. If one inch equals 2.54 centimeters, how many millimeters are in 1.5 feet? (Approximately)

A. 18

B. 1800

C. 460

D. 4600

C. 460

To solve this problem, note that if one inch is 2.54 centimeters, then 1.5 feet (which is 18 inches), must be (18)(2.54) centimeters, i.e. approximately 46 centimeters. Because there are ten millimeters in a centimeter, this is approximately 460 millimeters:

(1.5 ft) (12 in/ft) (2.54 cm/in) (10 mm/cm) = (1.5) (12) (2.54) (10) mm = 457.2 mm

This is consistent only with **answer (C)**.

5. Which of the following instruments measures wind speed?

A. Barometer.

B. Anemometer.

C. Thermometer.

D. Weather Vane.

B. Anemometer.

An anemometer is a device to measure wind speed, while a barometer measures pressure, a thermometer measures temperature, and a weather vane indicates wind direction. This is consistent only with **answer (B)**.

If you chose "barometer," here is an old physics joke to console you:

A physics teacher asks a student the following question:
"Suppose you want to find out the height of a building, and the only tool you have is a barometer. How could you find out the height?"
(The teacher hopes that the student will remember that pressure is inversely proportional to height, and will measure the pressure at the top of the building and then use the data to calculate the height of the building.)
"Well," says the student, "I could tie a string to the barometer and lower it from the top of the building, and then measure the amount of string required."
"You could," answers the teacher, "but try to think of a method that uses your physics knowledge from our class."
"All right," replies the student, "I could drop the barometer from the roof and measure the time it takes to fall, and then use free-fall equations to calculate the height from which it fell."
"Yes," says the teacher, "but what about using the barometer per se?"
"Oh," answers the student, "I could find the building superintendent, and offer to exchange the barometer for a set of blueprints, and look up the height!"

6. Sonar works by _____

A. timing how long it takes sound to reach a certain speed.

B. bouncing sound waves between two metal plates.

C. bouncing sound waves off an object and timing how long it takes for the sound to return.

D. evaluating the motion and amplitude of sound.

C. Bouncing sound waves off an object and timing how long it takes for the sound to return.

Sonar is used to measure distances. Sound waves are sent out, and the time is measured for the sound to hit an obstacle and bounce back. By using the known speed of sound, observers (or machines) can calculate the distance to the obstacle. This is consistent only with **answer (C)**.

7. The measure of the pull of Earth's gravity on an object is called

A. mass number.

B. atomic number.

C. mass.

D. weight.

D. Weight.

To answer this question, recall that mass number is the total number of protons and neutrons in an atom, atomic number is the number of protons in an atom, and mass is the amount of matter in an object. The only remaining **choice is (D)**, weight, which is correct because weight is the force of gravity on an object.

8. Which reaction below is a decomposition reaction?

A. $HCl + NaOH \rightarrow NaCl + H_2O$

B. $C + O_2 \rightarrow CO_2$

C. $2H_2O \rightarrow 2H_2 + O_2$

D. $CuSO_4 + Fe \rightarrow FeSO_4 + Cu$

C. $2H_2O \rightarrow 2H_2 + O_2$

To answer this question, recall that a decomposition reaction is one in which there are fewer reactants (on the left) than products (on the right). This is consistent only with **answer (C).** Meanwhile, note that answer (A) shows a double-replacement reaction (in which two sets of ions switch bonds), answer (B) shows a synthesis reaction (in which there are fewer products than reactants), and answer (D) shows a single-replacement reaction (in which one substance replaces another in its bond, but the other does not get a new bond).

9. The Law of Conservation of Energy states that

A. there must be the same number of products and reactants in any chemical equation.

B. objects always fall toward large masses such as planets.

C. energy is neither created nor destroyed, but may change form.

D. lights must be turned off when not in use, by state regulation.

C. Energy is neither created nor destroyed, but may change form.

Answer (C) is a summary of the Law of Conservation of Energy (for non-nuclear reactions). In other words, energy can be transformed into various forms such as kinetic, potential, electric, or heat energy, but the total amount of energy remains constant. Answer (A) is untrue, as demonstrated by many synthesis and decomposition reactions. Answers (B) and (D) may be sensible, but they are not relevant in this case. Therefore, the **answer is (C).**

10. Which parts of an atom are located inside the nucleus?

A. Protons and Electrons.

B. Protons and Neutrons.

C. Protons only.

D. Neutrons only.

B. Protons and Neutrons.

Protons and neutrons are located in the nucleus, while electrons move around outside the nucleus. This is consistent only with **answer (B)**.

11. The elements in the modern Periodic Table are arranged

A. in numerical order by atomic number.

B. randomly.

C. in alphabetical order by chemical symbol.

D. in numerical order by atomic mass.

A. In numerical order by atomic number.

Although the first periodic tables were arranged by atomic mass, the modern table is arranged by atomic number, i.e. the number of protons in each element. (This allows the element list to be complete and unique.) The elements are not arranged either randomly or in alphabetical order. The answer to this question is **therefore (A)**.

12. Carbon bonds with hydrogen by

A. ionic bonding.

B. non-polar covalent bonding.

C. polar covalent bonding.

D. strong nuclear force.

C. Polar covalent bonding.

Each carbon atom contains four valence electrons, while each hydrogen atom contains one valence electron. A carbon atom can bond with one or more hydrogen atoms, such that two electrons are shared in each bond. This is covalent bonding, because the electrons are shared. (In ionic bonding, atoms must gain or lose electrons to form ions. The ions are then electrically attracted in oppositely-charged pairs.) Covalent bonds are always polar when between two non-identical atoms, so this bond must be polar. ("Polar" means that the electrons are shared unequally, forming a pair of partial charges, i.e. poles.) In any case, the strong nuclear force is not relevant to this problem. The answer to this question is **therefore (C)**.

13. Vinegar is an example of a _____

A. strong acid.

B. strong base.

C. weak acid.

D. weak base.

C. Weak acid.

The main ingredient in vinegar is acetic acid, a weak acid. Vinegar is a useful acid in science classes, because it makes a frothy reaction with bases such as baking soda (e.g. in the quintessential volcano model). Vinegar is not a strong acid, such as hydrochloric acid, because it does not dissociate as fully or cause as much corrosion. It is not a base. Therefore, the **answer is (C)**.

14. Which of the following is not a nucleotide?

A. Adenine.

B. Alanine.

C. Cytosine.

D. Guanine.

B. Alanine.

Alanine is an amino acid. Adenine, cytosine, guanine, thymine, and uracil are nucleotides. The correct **answer is (B).**

15. When measuring the volume of water in a graduated cylinder, where does one read the measurement?

A. At the highest point of the liquid.

B. At the bottom of the meniscus curve.

C. At the closest mark to the top of the liquid.

D. At the top of the plastic safety ring.

B. At the bottom of the meniscus curve.

To measure water in glass, you must look at the top surface at eye-level, and ascertain the location of the bottom of the meniscus (the curved surface at the top of the water). The meniscus forms because water molecules adhere to the sides of the glass, which is a slightly stronger force than their cohesion to each other. This leads to a U-shaped top of the liquid column, the bottom of which gives the most accurate volume measurement. (Other liquids have different forces, e.g. mercury in glass, which has a convex meniscus.) This is consistent only with **answer (B).**

16. A duck's webbed feet are examples of

A. mimicry.

B. structural adaptation.

C. protective resemblance.

D. protective coloration.

B. Structural adaptation.

Ducks (and other aquatic birds) have webbed feet, which makes them more efficient swimmers. This is most likely due to evolutionary patterns where webbed-footed-birds were more successful at feeding and reproducing, and eventually became the majority of aquatic birds. Because the structure of the duck adapted to its environment over generations, this is termed 'structural adaptation'. Mimicry, protective resemblance, and protective coloration refer to other evolutionary mechanisms for survival. The answer to this question is **therefore (B)**.

17. What cell organelle contains the cell's stored food?

A. Vacuoles.

B. Golgi Apparatus.

C. Ribosomes.

D. Lysosomes.

A. Vacuoles.

In a cell, the sub-parts are called organelles. Of these, the vacuoles hold stored food (and water and pigments). The Golgi Apparatus sorts molecules from other parts of the cell; the ribosomes are sites of protein synthesis; the lysosomes contain digestive enzymes. This is consistent only with **answer (A)**.

18. The first stage of mitosis is called _____

A. telophase.

B. anaphase.

C. prophase.

D. mitophase.

C. Prophase.

In mitosis, the division of somatic cells, prophase is the stage where the cell enters mitosis. The four stages of mitosis, in order, are: prophase, metaphase, anaphase, and telophase. ("Mitophase" is not one of the steps.) During prophase, the cell begins the nonstop process of division. Its chromatin condenses, its nucleolus disappears, the nuclear membrane breaks apart, mitotic spindles form, its cytoskeleton breaks down, and centrioles push the spindles apart. Note that interphase, the stage where chromatin is loose, chromosomes are replicated, and cell metabolism is occurring, is technically not a stage of mitosis; it is a precursor to cell division.

19. The Doppler Effect is associated most closely with which property of waves?

A. Amplitude.

B. Wavelength.

C. Frequency.

D. Intensity.

C. Frequency.

The Doppler Effect accounts for an apparent increase in frequency when a wave source moves toward a wave receiver or apparent decrease in frequency when a wave source moves away from a wave receiver. (Note that the receiver could also be moving toward or away from the source.) As the wave fronts are released, motion toward the receiver mimics more frequent wave fronts, while motion away from the receiver mimics less frequent wave fronts. Meanwhile, the amplitude, wavelength, and intensity of the wave are not as relevant to this process (although moving closer to a wave source makes it seem more intense). The **answer to this question is therefore (C)**.

20. Viruses are responsible for many human diseases including all of the following *except*

A. influenza.

B. A.I.D.S.

C. the common cold.

D. strep throat.

D. Strep throat.

Influenza, A.I.D.S., and the "common cold" (rhinovirus infection), are all caused by viruses. (This is the reason that doctors should not be pressured to prescribe antibiotics for colds or 'flu—i.e. they will not be effective since the infections are not bacterial.) Strep throat (properly called 'streptococcal throat' and caused by streptococcus bacteria) is not a virus, but a bacterial infection. Thus, the **answer is (D)**.

21. A series of experiments on pea plants formed by _____ showed that two invisible markers existed for each trait, and one marker dominated the other.

A. Pasteur.

B. Watson and Crick.

C. Mendel.

D. Mendeleev.

C. Mendel.

Gregor Mendel was a nineteenth-century Austrian botanist, who derived "laws" governing inherited traits. His work led to the understanding of dominant and recessive traits, carried by biological markers. Mendel cross-bred different kinds of pea plants with varying features and observed the resulting new plants. He showed that genetic characteristics are not passed identically from one generation to the next. (Pasteur, Watson, Crick, and Mendeleev were other scientists with different specialties.) This is consistent only with **answer (C)**.

22. Formaldehyde should not be used in school laboratories for the following reason:

A. it smells unpleasant.

B. it is a known carcinogen.

C. it is expensive to obtain.

D. it is an explosive.

B. It is a known carcinogen.

Formaldehyde is a known carcinogen, so it is too dangerous for use in schools. In general, teachers should not use carcinogens in school laboratories. Although formaldehyde also smells unpleasant, a smell alone is not a definitive marker of danger. For example, many people find the smell of vinegar to be unpleasant, but vinegar is considered a very safe classroom/laboratory chemical. Furthermore, some odorless materials are toxic. Formaldehyde is neither particularly expensive nor explosive. Thus, the **answer is (B)**.

23. Amino acids are carried to the ribosome in protein synthesis by:

A. transfer RNA (tRNA).

B. messenger RNA (mRNA).

C. ribosomal RNA (rRNA).

D. transformation RNA (trRNA).

A. Transfer RNA (tRNA).

The job of tRNA is to carry and position amino acids to/on the ribosomes. mRNA copies DNA code and brings it to the ribosomes; rRNA is in the ribosome itself. There is no such thing as trRNA. Thus, the **answer is (A)**.

24. When designing a scientific experiment, a student considers all the factors that may influence the results. The process goal is to _____

A. recognize and manipulate independent variables.

B. recognize and record independent variables.

C. recognize and manipulate dependent variables.

D. recognize and record dependent variables.

A. Recognize and manipulate independent variables.

When a student designs a scientific experiment, s/he must decide what to measure, and what independent variables will play a role in the experiment. S/he must determine how to manipulate these independent variables to refine his/her procedure and to prepare for meaningful observations. Although s/he will eventually record dependent variables (D), this does not take place during the experimental design phase. Although the student will likely recognize and record the independent variables (B), this is not the process goal, but a helpful step in manipulating the variables. It is unlikely that the student will manipulate dependent variables directly in his/her experiment (C), or the data would be suspect. Thus, the **answer is (A)**.

25. Since ancient times, people have been entranced with bird flight. What is the key to bird flight?

A. Bird wings are a particular shape and composition.

B. Birds flap their wings quickly enough to propel themselves.

C. Birds take advantage of tailwinds.

D. Birds take advantage of crosswinds.

A. Bird wings are a particular shape and composition.

Bird wings are shaped for wide area, and their bones are very light. This creates a large surface-area-to-mass ratio, enabling birds to glide in air. Birds do flap their wings and float on winds, but none of these is the main reason for their flight ability. Thus, the **answer is (A)**.

26. Laboratory researchers have classified fungi as distinct from plants because the cell walls of fungi

A. contain chitin.

B. contain yeast.

C. are more solid.

D. are less solid.

A. Contain chitin.

Kingdom Fungi consists of organisms that are eukaryotic, multicellular, absorptive consumers. They have a chitin cell wall, which is the only universally present feature in fungi that is never present in plants. Thus, the **answer is (A)**.

27. In a fission reactor, "heavy water" is used to _____

A. terminate fission reactions.

B. slow down neutrons and moderate reactions.

C. rehydrate the chemicals.

D. initiate a chain reaction.

B. Slow down neutrons and moderate reactions.

"Heavy water" is used in a nuclear [fission] reactor to slow down neutrons, controlling and moderating the nuclear reactions. It does not terminate the reaction, and it does not initiate the reaction. Also, although the reactor takes advantage of water's other properties (e.g. high specific heat for cooling), the water does not "rehydrate" the chemicals. Therefore, the **answer is (B)**.

28. The transfer of heat by electromagnetic waves is called _____

A. conduction.

B. convection.

C. phase change.

D. radiation.

D. Radiation.

Heat transfer via electromagnetic waves (which can occur even in a vacuum) is called radiation. (Heat can also be transferred by direct contact (conduction), by fluid current (convection), and by matter changing phase, but these are not relevant here.) The answer to this question is **therefore (D)**.

29. When heat is added to most solids, they expand. Why is this the case?

A. The molecules get bigger.

B. The faster molecular motion leads to greater distance between the molecules.

C. The molecules develop greater repelling electric forces.

D. The molecules form a more rigid structure.

B. The faster molecular motion leads to greater distance between the molecules.

The atomic theory of matter states that matter is made up of tiny, rapidly moving particles. These particles move more quickly when warmer, because temperature is a measure of average kinetic energy of the particles. Warmer molecules therefore move further away from each other, with enough energy to separate from each other more often and for greater distances. The individual molecules do not get bigger, by conservation of mass, eliminating answer (A). The molecules do not develop greater repelling electric forces, eliminating answer (C). Occasionally, molecules form a more rigid structure when becoming colder and freezing (such as water)—but this gives rise to the exceptions to heat expansion, so it is not relevant here, eliminating answer (D). Therefore, the **answer is (B)**.

30. The force of gravity on earth causes all bodies in free fall to _____

A. fall at the same speed.

B. accelerate at the same rate.

C. reach the same terminal velocity.

D. move in the same direction.

B. Accelerate at the same rate.

Gravity causes approximately the same acceleration on all falling bodies close to earth's surface. (It is only "approximately" because there are very small variations in the strength of earth's gravitational field.) More massive bodies continue to accelerate at this rate for longer, before their air resistance is great enough to cause terminal velocity, so answers (A) and (C) are eliminated. Bodies on different parts of the planet move in different directions (always toward the center of mass of earth), so answer (D) is eliminated. Thus, the **answer is (B)**.

31. Sound waves are produced by _____

A. pitch.

B. noise.

C. vibrations.

D. sonar.

C. Vibrations.

Sound waves are produced by a vibrating body. The vibrating object moves forward and compresses the air in front of it, then reverses direction so that the pressure on the air is lessened and expansion of the air molecules occurs. The vibrating air molecules move back and forth parallel to the direction of motion of the wave as they pass the energy from adjacent air molecules closer to the source to air molecules farther away from the source. Therefore, the **answer is (C)**.

32. Resistance is measured in units called

A. watts.

B. volts.

C. ohms.

D. current.

C. Ohms.

A watt is a unit of energy. Potential difference is measured in a unit called the volt. Current is the number of electrons per second that flow past a point in a circuit. An ohm is the unit for resistance. The correct **answer is (C)**.

33. Sound can be transmitted in all of the following *except*

A. air.

B. water.

C. diamond.

D. a vacuum.

D. A vacuum.

Sound, a longitudinal wave, is transmitted by vibrations of molecules. Therefore, it can be transmitted through any gas, liquid, or solid. However, it cannot be transmitted through a vacuum, because there are no particles present to vibrate and bump into their adjacent particles to transmit the waves. This is consistent only with **answer (D)**. (It is interesting also to note that sound is actually faster in solids and liquids than in air.)

TEACHER CERTIFICATION STUDY GUIDE

34. As a train approaches, the whistle sounds

A. higher, because it has a higher apparent frequency.

B. lower, because it has a lower apparent frequency.

C. higher, because it has a lower apparent frequency.

D. lower, because it has a higher apparent frequency.

A. Higher, because it has a higher apparent frequency.

By the Doppler effect, when a source of sound is moving toward an observer, the wave fronts are released closer together, i.e. with a greater apparent frequency. Higher frequency sounds are higher in pitch. This is consistent only with **answer (A)**.

35. The speed of light is different in different materials. This is responsible for _____

A. interference.

B. refraction.

C. reflection.

D. relativity.

B. Refraction.

Refraction (B) is the bending of light because it hits a material at an angle wherein it has a different speed. (This is analogous to a cart rolling on a smooth road. If it hits a rough patch at an angle, the wheel on the rough patch slows down first, leading to a change in direction.) Interference (A) is when light waves interfere with each other to form brighter or dimmer patterns; reflection (C) is when light bounces off a surface; relativity (D) is a general topic related to light speed and its implications, but not specifically indicated here. Therefore, the **answer is (B)**.

36. A converging lens produces a real image _____

A. always.

B. never.

C. when the object is within one focal length of the lens.

D. when the object is further than one focal length from the lens.

D. When the object is further than one focal length from the lens.

A converging lens produces a real image whenever the object is far enough from the lens (outside one focal length) so that the rays of light from the object can hit the lens and be focused into a real image on the other side of the lens. When the object is closer than one focal length from the lens, rays of light do not converge on the other side; they diverge. This means that only a virtual image can be formed, i.e. the theoretical place where those diverging rays would have converged if they had originated behind the object. Thus, the correct **answer is (D)**.

37. The electromagnetic radiation with the longest wave length is _____

A. radio waves.

B. red light.

C. X-rays.

D. ultraviolet light.

A. Radio waves.

As one can see on a diagram of the electromagnetic spectrum, radio waves have longer wave lengths (and smaller frequencies) than visible light, which in turn has longer wave lengths than ultraviolet or X-ray radiation. If you did not remember this sequence, you might recall that wave length is inversely proportional to frequency, and that radio waves are considered much less harmful (less energetic, i.e. lower frequency) than ultraviolet or X-ray radiation. The correct answer is **therefore (A)**.

38. **Under a 440 power microscope, an object with diameter 0.1 millimeter appears to have diameter** _____

A. 4.4 millimeters.

B. 44 millimeters.

C. 440 millimeters.

D. 4400 millimeters.

B. 44 millimeters.

To answer this question, recall that to calculate a new length, you multiply the original length by the magnification power of the instrument. Therefore, the 0.1 millimeter diameter is multiplied by 440. This equals 44, so the image appears to be 44 millimeters in diameter. You could also reason that since a 440 power microscope is considered a "high power" microscope, you would expect a 0.1 millimeter object to appear a few centimeters long. Therefore, the correct **answer is (B)**.

39. **To separate blood into blood cells and plasma involves the process of**

A. electrophoresis.

B. spectrophotometry.

C. centrifugation.

D. chromatography.

C. Centrifugation.

Electrophoresis uses electrical charges of molecules to separate them according to their size. Spectrophotometry uses percent light absorbance to measure a color change, thus giving qualitative data a quantitative value. Chromatography uses the principles of capillarity to separate substances. Centrifugation involves spinning substances at a high speed. The more dense part of a solution will settle to the bottom of the test tube, where the lighter material will stay on top. The **answer is (C)**.

40. Experiments may be done with any of the following animals except

A. birds.

B. invertebrates.

C. lower order life.

D. frogs.

A. Birds.

No dissections may be performed on living mammalian vertebrates or birds. Lower order life and invertebrates may be used. Biological experiments may be done with all animals except mammalian vertebrates or birds. Therefore the **answer is (A)**.

41. For her first project of the year, a student is designing a science experiment to test the effects of light and water on plant growth. You should recommend that she _____

A. manipulate the temperature also.

B. manipulate the water pH also.

C. determine the relationship between light and water unrelated to plant growth.

D. omit either water or light as a variable.

D. Omit either water or light as a variable.

As a science teacher for middle-school-aged kids, it is important to reinforce the idea of 'constant' vs. 'variable' in science experiments. At this level, it is wisest to have only one variable examined in each science experiment. (Later, students can hold different variables constant while investigating others.) Therefore it is counterproductive to add in other variables (answers (A) or (B)). It is also irrelevant to determine the light-water interactions aside from plant growth (C). So the only possible **answer is (D)**.

42. In a laboratory report, what is the abstract?

A. The abstract is a summary of the report, and is the first section of the report.

B. The abstract is a summary of the report, and is the last section of the report.

C. The abstract is predictions for future experiments, and is the first section of the report.

D. The abstract is predictions for future experiments, and is the last section of the report.

A. The abstract is a summary of the report, and is the first section of the report.

In a laboratory report, the abstract is the section that summarizes the entire report (often containing one representative sentence from each section). It appears at the very beginning of the report, even before the introduction, often on its own page (instead of a title page). This format is consistent with articles in scientific journals. Therefore, the **answer is (A).**

43. What is the scientific method?

A. It is the process of doing an experiment and writing a laboratory report.

B. It is the process of using open inquiry and repeatable results to establish theories.

C. It is the process of reinforcing scientific principles by confirming results.

D. It is the process of recording data and observations.

B. It is the process of using open inquiry and repeatable results to establish theories.

Scientific research often includes elements from answers (A), (C), and (D), but the basic underlying principle of the scientific method is that people ask questions and do repeatable experiments to answer those questions and develop informed theories of why and how things happen. Therefore, the best **answer is (B).**

TEACHER CERTIFICATION STUDY GUIDE

44. Identify the control in the following experiment: A student had four corn plants and was measuring photosynthetic rate (by measuring growth mass). Half of the plants were exposed to full (constant) sunlight, and the other half were kept in 50% (constant) sunlight.

A. The control is a set of plants grown in full (constant) sunlight.

B. The control is a set of plants grown in 50% (constant) sunlight.

C. The control is a set of plants grown in the dark.

D. The control is a set of plants grown in a mixture of natural levels of sunlight.

A. The control is a set of plants grown in full (constant) sunlight.

In this experiment, the goal was to measure how two different amounts of sunlight affected plant growth. The control in any experiment is the 'base case,' or the usual situation without a change in variable. Because the control must be studied alongside the variable, answers (C) and (D) are omitted (because they were not in the experiment). The **better answer of (A) and (B) is (A)**, because usually plants are assumed to have the best growth and their usual growing circumstances in full sunlight. This is particularly true for crops like the corn plants in this question.

45. In an experiment measuring the growth of bacteria at different temperatures, what is the independent variable?

A. Number of bacteria.

B. Growth rate of bacteria.

C. Temperature.

D. Light intensity.

C. Temperature.

To answer this question, recall that the independent variable in an experiment is the entity that is changed by the scientist, in order to observe the effects (the dependent variable(s)). In this experiment, temperature is changed in order to measure growth of bacteria, so **(C) is the answer**. Note that answer (A) is the dependent variable, and neither (B) nor (D) is directly relevant to the question.

TEACHER CERTIFICATION STUDY GUIDE

46. A scientific law _____

A. proves scientific accuracy.

B. may never be broken.

C. may be revised in light of new data.

D. is the result of one excellent experiment.

C. May be revised in light of new data.

A scientific law is the same as a scientific theory, except that it has lasted for longer, and has been supported by more extensive data. Therefore, such a law may be revised in light of new data, and may be broken by that new data. Furthermore, a scientific law is always the result of many experiments, and never 'proves' anything but rather is implied or supported by various results. Therefore, the **answer must be (C)**.

47. Which is the correct order of methodology?

**1. collecting data
2. planning a controlled experiment
3. drawing a conclusion
4. hypothesizing a result
5. re-visiting a hypothesis to answer a question**

A. 1,2,3,4,5

B. 4,2,1,3,5

C. 4,5,1,3,2

D. 1,3,4,5,2

B. 4,2,1,3,5

The correct methodology for the scientific method is first to make a meaningful hypothesis (educated guess), then plan and execute a controlled experiment to test that hypothesis. Using the data collected in that experiment, the scientist then draws conclusions and attempts to answer the original question related to the hypothesis. This is consistent only with **answer (B)**.

48. Which is the most desirable tool to use to heat substances in a middle school laboratory?

A. Alcohol burner.

B. Freestanding gas burner.

C. Bunsen burner.

D. Hot plate.

D. Hot plate.

Due to safety considerations, the use of open flame should be minimized, so a hot plate is the best choice. Any kind of burner may be used with proper precautions, but it is difficult to maintain a completely safe middle school environment. Therefore, the best **answer is (D)**.

49. Newton's Laws are taught in science classes because _____.

A. they are the correct analysis of inertia, gravity, and forces.

B. they are a close approximation to correct physics, for usual Earth conditions.

C. they accurately incorporate Relativity into studies of forces.

D. Newton was a well-respected scientist in his time.

They are a close approximation to correct physics, for usual Earth conditions.

Although Newton's Laws are often taught as fully correct for inertia, gravity, and forces, it is important to realize that Einstein's work (and that of others) has indicated that Newton's Laws are reliable only at speeds much lower than that of light. This is reasonable, though, for most middle- and high-school applications. At speeds close to the speed of light, Relativity considerations must be used. Therefore, the only correct **answer is (B)**.

TEACHER CERTIFICATION STUDY GUIDE

50. Which of the following is most accurate?

A. Mass is always constant; Weight may vary by location.

B. Mass and Weight are both always constant.

C. Weight is always constant; Mass may vary by location.

D. Mass and Weight may both vary by location.

A. Mass is always constant; Weight may vary by location.

When considering situations exclusive of nuclear reactions, mass is constant (mass, the amount of matter in a system, is conserved). Weight, on the other hand, is the force of gravity on an object, which is subject to change due to changes in the gravitational field and/or the location of the object. Thus, the **best answer is (A)**.

51. Chemicals should be stored _____

A. in the principal's office.

B. in a dark room.

C. in an off-site research facility.

D. according to their reactivity with other substances.

D. According to their reactivity with other substances.

Chemicals should be stored with other chemicals of similar properties (e.g. acids with other acids), to reduce the potential for either hazardous reactions in the store-room, or mistakes in reagent use. Certainly, chemicals should not be stored in anyone's office, and the light intensity of the room is not very important because light-sensitive chemicals are usually stored in dark containers. In fact, good lighting is desirable in a store-room, so that labels can be read easily. Chemicals may be stored off-site, but that makes their use inconvenient. Therefore, the best **answer is (D)**.

52. Which of the following is the worst choice for a school laboratory activity?

A. A genetics experiment tracking the fur color of mice.

B. Dissection of a preserved fetal pig.

C. Measurement of goldfish respiration rate at different temperatures.

D. Pithing a frog to watch the circulatory system.

D. Pithing a frog to watch the circulatory system.

While any use of animals (alive or dead) must be done with care to respect ethics and laws, it is possible to perform choices (A), (B), or (C) with due care. (Note that students will need significant assistance and maturity to perform these experiments.) However, modern practice precludes pithing animals (causing partial brain death while allowing some systems to function), as inhumane. Therefore, the answer to this **question is (D)**.

53. Who should be notified in the case of a serious chemical spill?

A. The custodian.

B. The fire department or other municipal authority.

C. The science department chair.

D. The School Board.

B. The fire department or other municipal authority.

Although the custodian may help to clean up laboratory messes, and the science department chair should be involved in discussions of ways to avoid spills, a serious chemical spill may require action by the fire department or other trained emergency personnel. It is best to be safe by notifying them in case of a serious chemical accident. Therefore, the **best answer is (B)**.

54. A scientist exposes mice to cigarette smoke, and notes that their lungs develop tumors. Mice that were not exposed to the smoke do not develop as many tumors. Which of the following conclusions may be drawn from these results?

I. Cigarette smoke causes lung tumors.
II. Cigarette smoke exposure has a positive correlation with lung tumors in mice.
III. Some mice are predisposed to develop lung tumors.
IV. Mice are often a good model for humans in scientific research.

A. I and II only.

B. II only.

C. I , II, and III only.

D. II and IV only.

B. II only.

Although cigarette smoke has been found to cause lung tumors (and many other problems), this particular experiment shows only that there is a positive correlation between smoke exposure and tumor development in these mice. It may be true that some mice are more likely to develop tumors than others, which is why a control group of identical mice should have been used for comparison. Mice are often used to model human reactions, but this is as much due to their low financial and emotional cost as it is due to their being a "good model" for humans. Therefore, the **answer must be (B)**.

TEACHER CERTIFICATION STUDY GUIDE

55. In which situation would a science teacher be legally liable?

A. The teacher leaves the classroom for a telephone call and a student slips and injures him/herself.

B. A student removes his/her goggles and gets acid in his/her eye.

C. A faulty gas line in the classroom causes a fire.

D. A student cuts him/herself with a dissection scalpel.

The teacher leaves the classroom for a telephone call and a student slips and injures him/herself.

Teachers are required to exercise a "reasonable duty of care" for their students. Accidents may happen (e.g. (D)), or students may make poor decisions (e.g. (B)), or facilities may break down (e.g. (C)). However, the teacher has the responsibility to be present and to do his/her best to create a safe and effective learning environment. Therefore, the **answer is (A)**.

56. Which of these is the best example of 'negligence'?

A. A teacher fails to give oral instructions to those with reading disabilities.

B. A teacher fails to exercise ordinary care to ensure safety in the classroom.

C. A teacher displays inability to supervise a large group of students.

D. A teacher reasonably anticipates that an event may occur, and plans accordingly.

A teacher fails to exercise ordinary care to ensure safety in the classroom.

'Negligence' is the failure to "exercise ordinary care" to ensure an appropriate and safe classroom environment. It is best for a teacher to meet all special requirements for disabled students, and to be good at supervising large groups. However, if a teacher can prove that s/he has done a reasonable job to ensure a safe and effective learning environment, then it is unlikely that she/he would be found negligent. Therefore, **the answer is (B)**.

57. Which item should always be used when handling glassware?

A. Tongs.

B. Safety goggles.

C. Gloves.

D. Buret stand.

B. Safety goggles.

Safety goggles are the single most important piece of safety equipment in the laboratory, and should be used any time a scientist is using glassware, heat, or chemicals. Other equipment (e.g. tongs, gloves, or even a buret stand) has its place for various applications. However, the most important is safety goggles. Therefore, the **answer is (B)**.

58. Which of the following is *not* a necessary characteristic of living things?

A. Movement.

B. Reduction of local entropy.

C. Ability to cause local energy form changes.

D. Reproduction.

A. Movement.

There are many definitions of "life," but in all cases, a living organism reduces local entropy, changes chemical energy into other forms, and reproduces. Not all living things move, however, so the correct **answer is (A)**.

59. What are the most significant and prevalent elements in the biosphere?

A. Carbon, Hydrogen, Oxygen, Nitrogen, Phosphorus.

B. Carbon, Hydrogen, Sodium, Iron, Calcium.

C. Carbon, Oxygen, Sulfur, Manganese, Iron.

D. Carbon, Hydrogen, Oxygen, Nickel, Sodium, Nitrogen.

A. Carbon, Hydrogen, Oxygen, Nitrogen, Phosphorus.

Organic matter (and life as we know it) is based on Carbon atoms, bonded to Hydrogen and Oxygen. Nitrogen and Phosphorus are the next most significant elements, followed by Sulfur and then trace nutrients such as Iron, Sodium, Calcium, and others. Therefore, the **answer is (A)**. If you know that the formula for any carbohydrate contains Carbon, Hydrogen, and Oxygen, that will help you narrow the choices to (A) and (D) in any case.

60. All of the following measure energy *except* for _____

A. joules.

B. calories.

C. watts.

D. ergs.

C. Watts.

Energy units must be dimensionally equivalent to (force)x(length), which equals (mass)x(length squared)/(time squared). Joules, Calories, and Ergs are all metric measures of energy. Joules are the SI units of energy, while Calories are used to allow water to have a Specific Heat of one unit. Ergs are used in the 'cgs' (centimeter-gram-second) system, for smaller quantities. Watts, however, are units of power, i.e. Joules per Second. Therefore, the **answer is (C)**.

61. Identify the correct sequence of organization of living things from lower to higher order:

A. Cell, Organelle, Organ, Tissue, System, Organism.

B. Cell, Tissue, Organ, Organelle, System, Organism.

C. Organelle, Cell, Tissue, Organ, System, Organism.

D. Organelle, Tissue, Cell, Organ, System, Organism.

C. Organelle, Cell, Tissue, Organ, System, Organism.

Organelles are parts of the cell; cells make up tissue, which makes up organs. Organs work together in systems (e.g. the respiratory system), and the organism is the living thing as a whole. Therefore, the **answer must be (C)**.

62. Which kingdom is comprised of organisms made of one cell with no nuclear membrane?

A. Monera.

B. Protista.

C. Fungi.

D. Algae.

A. Monera.

To answer this question, first note that algae are not a kingdom of their own. Some algae are in monera, the kingdom that consists of unicellular prokaryotes with no true nucleus. Protista and fungi are both eukaryotic, with true nuclei, and are sometimes multi-cellular. Therefore, the **answer is (A)**.

63. Which of the following is found in the least abundance in organic molecules?

A. Phosphorus.

B. Potassium.

C. Carbon.

D. Oxygen.

B. Potassium.

Organic molecules consist mainly of Carbon, Hydrogen, and Oxygen, with significant amounts of Nitrogen, Phosphorus, and often Sulfur. Other elements, such as Potassium, are present in much smaller quantities. Therefore, the **answer is (B)**. If you were not aware of this ranking, you might have been able to eliminate Carbon and Oxygen because of their prevalence, in any case.

64. Catalysts assist reactions by _____

A. lowering effective activation energy.

B. maintaining precise pH levels.

C. keeping systems at equilibrium.

D. adjusting reaction speed.

A. Lowering effective activation energy.

Chemical reactions can be enhanced or accelerated by catalysts, which are present both with reactants and with products. They induce the formation of activated complexes, thereby lowering the effective activation energy—so that less energy is necessary for the reaction to begin. Although this often makes reactions faster, answer (D) is not as good a choice as the more generally applicable **answer (A)**, which is correct.

65. Accepted procedures for preparing solutions should be made with

A. alcohol.

B. hydrochloric acid.

C. distilled water.

D. tap water.

C. Distilled water.

Alcohol and hydrochloric acid should never be used to make solutions unless instructed to do so. All solutions should be made with distilled water as tap water contains dissolved particles which may affect the results of an experiment. The correct **answer is (C)**.

66. Enzymes speed up reactions by _____

A. utilizing ATP.

B. lowering pH, allowing reaction speed to increase.

C. increasing volume of substrate.

D. lowering energy of activation.

D. Lowering energy of activation.

Because enzymes are catalysts, they work the same way—they cause the formation of activated chemical complexes, which require a lower activation energy. Therefore, the **answer is (D)**. ATP is an energy source for cells, and pH or volume changes may or may not affect reaction rate, so these answers can be eliminated.

TEACHER CERTIFICATION STUDY GUIDE

67. When you step out of the shower, the floor feels colder on your feet than the bathmat. Which of the following is the correct explanation for this phenomenon?

A. The floor is colder than the bathmat.

B. Your feet have a chemical reaction with the floor, but not the bathmat.

C. Heat is conducted more easily into the floor.

D. Water is absorbed from your feet into the bathmat.

C. Heat is conducted more easily into the floor.

When you step out of the shower and onto a surface, the surface is most likely at room temperature, regardless of its composition (eliminating answer (A)). Your feet feel cold when heat is transferred from them to the surface, which happens more easily on a hard floor than a soft bathmat. This is because of differences in specific heat (the energy required to change temperature, which varies by material). Therefore, the **answer must be (C)**, i.e. heat is conducted more easily into the floor from your feet.

68. Which of the following is *not* considered ethical behavior for a scientist?

A. Using unpublished data and citing the source.

B. Publishing data before other scientists have had a chance to replicate results.

C. Collaborating with other scientists from different laboratories.

D. Publishing work with an incomplete list of citations.

D. Publishing work with an incomplete list of citations.

One of the most important ethical principles for scientists is to cite all sources of data and analysis when publishing work. It is reasonable to use unpublished data (A), as long as the source is cited. Most science is published before other scientists replicate it (B), and frequently scientists collaborate with each other, in the same or different laboratories (C). These are all ethical choices. However, publishing work without the appropriate citations, is unethical. Therefore, the **answer is (D)**.

TEACHER CERTIFICATION STUDY GUIDE

69. The chemical equation for water formation is: $2H_2 + O_2 \rightarrow 2H_2O$. Which of the following is an *incorrect* interpretation of this equation?

A. Two moles of hydrogen gas and one mole of oxygen gas combine to make two moles of water.

B. Two grams of hydrogen gas and one gram of oxygen gas combine to make two grams of water.

C. Two molecules of hydrogen gas and one molecule of oxygen gas combine to make two molecules of water.

D. Four atoms of hydrogen (combined as a diatomic gas) and two atoms of oxygen (combined as a diatomic gas) combine to make two molecules of water.

B. Two grams of hydrogen gas and one gram of oxygen gas combine to make two grams of water.

In any chemical equation, the coefficients indicate the relative proportions of molecules (or atoms), or of moles of molecules. They do not refer to mass, because chemicals combine in repeatable combinations of molar ratio (i.e. number of moles), but vary in mass per mole of material. Therefore, the answer must be the only choice that does not refer to numbers of particles, i.e. **answer (B)**, which refers to grams, a unit of mass.

70. Energy is measured with the same units as _____

A. force.

B. momentum.

C. work.

D. power.

C. Work.

In SI units, energy is measured in Joules, i.e. (mass)(length squared)/(time squared). This is the same unit as is used for work. You can verify this by calculating that since work is force times distance, the units work out to be the same. Force is measured in Newtons in SI; momentum is measured in (mass)(length)/(time); power is measured in Watts (which equal Joules/second). Therefore, the **answer must be (C)**.

71. If the volume of a confined gas is increased, what happens to the pressure of the gas? You may assume that the gas behaves ideally, and that temperature and number of gas molecules remain constant.

A. The pressure increases.

B. The pressure decreases.

C. The pressure stays the same.

D. There is not enough information given to answer this question.

B. The pressure decreases.

Because we are told that the gas behaves ideally, you may assume that it follows the Ideal Gas Law, i.e. $PV = nRT$. This means that an increase in volume must be associated with a decrease in pressure (i.e. higher T means lower P), because we are also given that all the components of the right side of the equation remain constant. Therefore, the **answer must be (B)**.

72. A product of anaerobic respiration in animals is _____

A. carbon dioxide.

B. lactic acid.

C. oxygen.

D. sodium chloride.

B. Lactic acid.

In animals, anaerobic respiration (i.e. respiration without the presence of oxygen) generates lactic acid as a byproduct. (Note that some anaerobic bacteria generate carbon dioxide from respiration of methane, and animals generate carbon dioxide in aerobic respiration.) Oxygen is not normally a by-product of respiration, though it is a product of photosynthesis, and sodium chloride is not strictly relevant in this question. Therefore, the **answer must be (B)**. By the way, lactic acid is believed to cause muscle soreness after anaerobic weight-lifting.

73. A Newton is fundamentally a measure of _____.

A. force.

B. momentum.

C. energy.

D. gravity.

A. Force.

In SI units, force is measured in Newtons. Momentum and energy each have different units, without equivalent dimensions. A Newton is one (kilogram)(meter)/(second squared), while momentum is measured in (kilgram)(meter)/(second) and energy, in Joules, is (kilogram)(meter squared)/(second squared). Although "gravity" can be interpreted as the force of gravity, i.e. measured in Newtons, fundamentally it is not required. Therefore, the **answer is (A)**.

74. Which change does *not* affect enzyme rate?

A. Increase the temperature.

B. Add more substrate.

C. Adjust the pH.

D. Use a larger cell.

D. Use a larger cell.

Temperature, chemical amounts, and pH can all affect enzyme rate. However, the chemical reactions take place on a small enough scale that the overall cell size is not relevant. Therefore, the **answer is (D)**.

75. Which of the following types of rock are made from magma?

A. Fossils.

B. Sedimentary.

C. Metamorphic.

D. Igneous.

D. Igneous.

Few fossils are found in metamorphic rock and virtually none found in igneous rocks. Igneous rocks are formed from magma and magma is so hot that any organisms trapped by it are destroyed. Metamorphic rocks are formed by high temperatures and great pressures. When fluid sediments are transformed into solid sedimentary rocks, the process is known as lithification. The **answer is (D)**.

76. Which of the following is *not* an acceptable way for a student to acknowledge sources in a laboratory report?

A. The student tells his/her teacher what sources s/he used to write the report.

B. The student uses footnotes in the text, with sources cited, but not in correct MLA format.

C. The student uses endnotes in the text, with sources cited, in correct MLA format.

D. The student attaches a separate bibliography, noting each use of sources.

A. The student tells his/her teacher what sources s/he used to write the report.

It may seem obvious, but students are often unaware that scientists need to cite all sources used. For the young adolescent, it is not always necessary to use official MLA format (though this should be taught at some point). Students may properly cite references in many ways, but these references must be in writing, with the original assignment. Therefore, the **answer is (A)**.

77. Animals with a notochord or a backbone are in the phylum

A. arthropoda.

B. chordata.

C. mollusca.

D. mammalia.

B. Chordata.

The phylum arthropoda contains spiders and insects and phylum mollusca contain snails and squid. Mammalia is a class in the phylum chordata. The **answer is (B)**.

78. Which of the following is a correct explanation for scientific 'evolution'?

A. Giraffes need to reach higher for leaves to eat, so their necks stretch. The giraffe babies are then born with longer necks. Eventually, there are more long-necked giraffes in the population.

B. Giraffes with longer necks are able to reach more leaves, so they eat more and have more babies than other giraffes. Eventually, there are more long-necked giraffes in the population.

C. Giraffes want to reach higher for leaves to eat, so they release enzymes into their bloodstream, which in turn causes fetal development of longer-necked giraffes. Eventually, there are more long-necked giraffes in the population.

D. Giraffes with long necks are more attractive to other giraffes, so they get the best mating partners and have more babies. Eventually, there are more long-necked giraffes in the population.

B. Giraffes with longer necks are able to reach more leaves, so they eat more and have more babies than other giraffes. Eventually, there are more long-necked giraffes in the population.

Although evolution is often misunderstood, it occurs via natural selection. Organisms with a life/reproductive advantage will produce more offspring. Over many generations, this changes the proportions of the population. In any case, it is impossible for a stretched neck (A) or a fervent desire (C) to result in a biologically mutated baby. Although there are traits that are naturally selected because of mate attractiveness and fitness (D), this is not the primary situation here, **so answer (B) is the best choice**.

79. Which of the following is a correct definition for 'chemical equilibrium'?

A. Chemical equilibrium is when the forward and backward reaction rates are equal. The reaction may continue to proceed forward and backward.

B. Chemical equilibrium is when the forward and backward reaction rates are equal, and equal to zero. The reaction does not continue.

C. Chemical equilibrium is when there are equal quantities of reactants and products.

D. Chemical equilibrium is when acids and bases neutralize each other fully.

A. Chemical equilibrium is when the forward and backward reaction rates are equal. The reaction may continue to proceed forward and backward.

Chemical equilibrium is defined as when the quantities of reactants and products are at a 'steady state' and are no longer shifting, but the reaction may still proceed forward and backward. The rate of forward reaction must equal the rate of backward reaction. Note that there may or may not be equal amounts of chemicals, and that this is not restricted to a completed reaction or to an acid-base reaction. Therefore, the **answer is (A)**.

80. Which of the following data sets is properly represented by a bar graph?

A. Number of people choosing to buy cars, vs. Color of car bought.

B. Number of people choosing to buy cars, vs. Age of car customer.

C. Number of people choosing to buy cars, vs. Distance from car lot to customer home.

D. Number of people choosing to buy cars, vs. Time since last car purchase.

A. Number of people choosing to buy cars, vs. Color of car bought.

A bar graph should be used only for data sets in which the independent variable is non-continuous (discrete), e.g. gender, color, etc. Any continuous independent variable (age, distance, time, etc.) should yield a scatter-plot when the dependent variable is plotted. Therefore, the **answer must be (A)**.

81. In a science experiment, a student needs to dispense very small measured amounts of liquid into a well-mixed solution. Which of the following is the \best choice for his/her equipment to use?

A. Buret with Buret Stand, Stir-plate, Stirring Rod, Beaker.

B. Buret with Buret Stand, Stir-plate, Beaker.

C. Volumetric Flask, Dropper, Graduated Cylinder, Stirring Rod.

D. Beaker, Graduated Cylinder, Stir-plate.

B. Buret with Buret Stand, Stir-plate, Beaker.

The most accurate and convenient way to dispense small measured amounts of liquid in the laboratory is with a buret, on a buret stand. To keep a solution well-mixed, a magnetic stir-plate is the most sensible choice, and the solution will usually be mixed in a beaker. Although other combinations of materials could be used for this experiment, **choice (B)** is thus the simplest and best.

82. A laboratory balance is most appropriately used to measure the mass of which of the following?

A. Seven paper clips.

B. Three oranges.

C. Two hundred cells.

D. One student's elbow.

A. Seven paper clips.

Usually, laboratory/classroom balances can measure masses between approximately 0.01 gram and 1 kilogram. Therefore, answer (B) is too heavy and answer (C) is too light. Answer (D) is silly, but it is a reminder to instruct students not to lean on the balances or put their things near them. **Answer (A)**, which is likely to have a mass of a few grams, is correct in this case.

83. All of the following are measured in units of length, *except* for:

A. Perimeter.

B. Distance.

C. Radius.

D. Area.

D. Area.

Perimeter is the distance around a shape; distance is equivalent to length; radius is the distance from the center (e.g. in a circle) to the edge. Area, however, is the squared-length-units measure of the size of a two-dimensional surface. Therefore, **the answer is (D)**.

84. What is specific gravity?

A. The mass of an object.

B. The ratio of the density of a substance to the density of water.

C. Density.

D. The pull of the earth's gravity on an object.

B. The ratio of the density of a substance to the density of water.

Mass is a measure of the amount of matter in an object. Density is the mass of a substance contained per unit of volume. Weight is the measure of the earth's pull of gravity on an object. The only option here is the ratio of the density of a substance to the density of water, **answer (B)**.

85. What is the most accurate description of the Water Cycle?

A. Rain comes from clouds, filling the ocean. The water then evaporates and becomes clouds again.

B. Water circulates from rivers into groundwater and back, while water vapor circulates in the atmosphere.

C. Water is conserved except for chemical or nuclear reactions, and any drop of water could circulate through clouds, rain, ground-water, and surface-water.

D. Weather systems cause chemical reactions to break water into its atoms.

C. Water is conserved except for chemical or nuclear reactions, and any drop of water could circulate through clouds, rain, ground-water, and surface-water.

All natural chemical cycles, including the Water Cycle, depend on the principle of Conservation of Mass. (For water, unlike for elements such as Nitrogen, chemical reactions may cause sources or sinks of water molecules.) Any drop of water may circulate through the hydrologic system, ending up in a cloud, as rain, or as surface- or ground-water. Although answers (A) and (B) describe parts of the water cycle, the most comprehensive and correct **answer is (C)**.

86. The scientific name *Canis familiaris* refers to the animal's _____.

A. kingdom and phylum.

B. genus and species.

C. class and species.

D. type and family.

B. Genus and species.

To answer this question, you must be aware that genus and species are the most specific way to identify an organism, and that usually the genus is capitalized and the species, immediately following, is not. Furthermore, it helps to recall that 'Canis' is the genus for dogs, or canines. Therefore, the **answer must be (B)**. If you did not remember these details, you might recall that there is no such kingdom as 'Canis,' and that there isn't a category 'type' in official taxonomy. This could eliminate answers (A) and (D).

87. Members of the same animal species _____

A. look identical.

B. never adapt differently.

C. are able to reproduce with one another.

D. are found in the same location.

C. Are able to reproduce with one another.

Although members of the same animal species may look alike (A), adapt alike (B), or be found near one another (D), the only requirement is that they be able to reproduce with one another. This ability to reproduce within the group is considered the hallmark of a species. Therefore, the **answer is (C)**.

88. Which of the following is/are true about scientists?

I. Scientists usually work alone.
II. Scientists usually work with other scientists.
III. Scientists achieve more prestige from new discoveries than from replicating established results.
IV. Scientists keep records of their own work, but do not publish it for outside review.

A. I and IV only.

B. II only.

C. II and III only.

D. III and IV only.

C. II and III only.

In the scientific community, scientists nearly always work in teams, both within their institutions and across several institutions. This eliminates (I) and requires (II), leaving only **answer choices (B) and (C)**. Scientists do achieve greater prestige from new discoveries, so the answer must be (C). Note that scientists must publish their work in peer-reviewed journals, eliminating (IV) in any case.

89. What is necessary for ion diffusion to occur spontaneously?

A. Carrier proteins.

B. Energy from an outside source.

C. A concentration gradient.

D. Cell flagellae.

C. A concentration gradient.

Spontaneous diffusion occurs when random motion leads particles to increase entropy by equalizing concentrations. Particles tend to move into places of lower concentration. Therefore, a concentration gradient is required, and the **answer is (C)**. No proteins (A), outside energy (B), or flagellae (D) are required for this process.

90. All of the following are considered Newton's Laws *except* for:

A. An object in motion will continue in motion unless acted upon by an outside force.

B. For every action force, there is an equal and opposite reaction force.

C. Nature abhors a vacuum.

D. Mass can be considered the ratio of force to acceleration.

C. Nature abhors a vacuum.

Newton's Laws include his law of inertia (an object in motion (or at rest) will stay in motion (or at rest) until acted upon by an outside force) (A), his law that (Force)=(Mass)(Acceleration) (D), and his equal and opposite reaction force law (B). Therefore, the **answer to this question is (C)**, because "Nature abhors a vacuum" is not one of these.

91. A cup of hot liquid and a cup of cold liquid are both sitting in a room at comfortable room temperature and humidity. Both cups are thin plastic. Which of the following is a true statement?

A. There will be fog on the outside of the hot liquid cup, and also fog on the outside of the cold liquid cup.

B. There will be fog on the outside of the hot liquid cup, but not on the cold liquid cup.

C. There will be fog on the outside of the cold liquid cup, but not on the hot liquid cup.

D. There will not be fog on the outside of either cup.

C. There will be fog on the outside of the cold liquid cup, but not on the hot liquid cup.

Fog forms on the outside of a cup when the contents of the cup are colder than the surrounding air, and the cup material is not a perfect insulator. This happens because the air surrounding the cup is cooled to a lower temperature than the ambient room, so it has a lower saturation point for water vapor. Although the humidity had been reasonable in the warmer air, when that air circulates near the colder region and cools, water condenses onto the cup's outside surface. This phenomenon is also visible when someone takes a hot shower, and the mirror gets foggy. The mirror surface is cooler than the ambient air, and provides a surface for water condensation. Furthermore, the same phenomenon is why defrosters on car windows send heat to the windows—the warmer window does not permit as much condensation. Therefore, the correct **answer is (C)**.

92. A ball rolls down a smooth hill. You may ignore air resistance. Which of the following is a true statement?

A. The ball has more energy at the start of its descent than just before it hits the bottom of the hill, because it is higher up at the beginning.

B. The ball has less energy at the start of its descent than just before it hits the bottom of the hill, because it is moving more quickly at the end.

C. The ball has the same energy throughout its descent, because positional energy is converted to energy of motion.

D. The ball has the same energy throughout its descent, because a single object (such as a ball) cannot gain or lose energy.

The ball has the same energy throughout its descent, because positional energy is converted to energy of motion.

The principle of Conservation of Energy states that (except in cases of nuclear reaction, when energy may be created or destroyed by conversion to mass), "Energy is neither created nor destroyed, but may be transformed." Answers (A) and (B) give you a hint in this question—it is true that the ball has more Potential Energy when it is higher, and that it has more Kinetic Energy when it is moving quickly at the bottom of its descent. However, the total sum of all kinds of energy in the ball remains constant, if we neglect 'losses' to heat/friction. Note that a single object can and does gain or lose energy when the energy is transferred to or from a different object. Conservation of Energy applies to systems, not to individual objects unless they are isolated. Therefore, the **answer must be (C)**.

93. A long silver bar has a temperature of 50 degrees Celsius at one end and 0 degrees Celsius at the other end. The bar will reach thermal equilibrium (barring outside influence) by the process of heat _____.

A. conduction.

B. convection.

C. radiation.

D. phase change.

A. conduction.

Heat conduction is the process of heat transfer via solid contact. The molecules in a warmer region vibrate more rapidly, jostling neighboring molecules and accelerating them. This is the dominant heat transfer process in a solid with no outside influences. Recall, also, that convection is heat transfer by way of fluid currents; radiation is heat transfer via electromagnetic waves; phase change can account for heat transfer in the form of shifts in matter phase. The answer to this question must **therefore be (A)**.

94. _____ are cracks in the plates of the earth's crust, along which the plates move.

A. Faults

B. Ridges

C. Earthquakes

D. Volcanoes

A. Faults.

Faults are cracks in the earth's crust, and when the earth moves, an earthquake results. Faults may lead to mismatched edges of ground, forming ridges, and ground shape may also be determined by volcanoes. The answer to this question must **therefore be (A).**

95. Fossils are usually found in _____ rock.

A. igneous.

B. sedimentary.

C. metamorphic.

D. cumulus.

B. Sedimentary

Fossils are formed by layers of dirt and sand settling around organisms, hardening, and taking an imprint of the organisms. When the organism decays, the hardened imprint is left behind. This is most likely to happen in rocks that form from layers of settling dirt and sand, i.e. sedimentary rock. Note that igneous rock is formed from molten rock from volcanoes (lava), while metamorphic rock can be formed from any rock under very high temperature and pressure changes. 'Cumulus' is a descriptor for clouds, not rocks. The best answer is **therefore (B)**.

96. Which of the following is *not* a common type of acid in 'acid rain' or acidified surface water?

A. Nitric acid.

B. Sulfuric acid.

C. Carbonic acid.

D. Hydrofluoric acid.

D. Hydrofluoric acid.

Acid rain forms predominantly from pollutant oxides in the air (usually nitrogen-based NO_x or sulfur-based SO_x), which become hydrated into their acids (nitric or sulfuric acid). Because of increased levels of carbon dioxide pollution, carbonic acid is also common in acidified surface water environments. Hydrofluoric acid can be found, but it is much less common. In general, carbon, nitrogen, and sulfur are much more prevalent in the environment than fluorine. Therefore, the **answer is (D)**.

TEACHER CERTIFICATION STUDY GUIDE

97. Which of the following is *not* true about phase change in matter?

A. Solid water and liquid ice can coexist at water's freezing point.

B. At 7 degrees Celsius, water is always in liquid phase.

C. Matter changes phase when enough energy is gained or lost.

D. Different phases of matter are characterized by differences in molecular motion.

B. At 7 degrees Celsius, water is always in liquid phase.

According to the molecular theory of matter, molecular motion determines the 'phase' of the matter, and the energy in the matter determines the speed of molecular motion. Solids have vibrating molecules that are in fixed relative positions; liquids have faster molecular motion than their solid forms, and the molecules may move more freely but must still be in contact with one another; gases have even more energy and more molecular motion. (Other phases, such as plasma, are yet more energetic.) At the 'freezing point' or 'boiling point' of a substance, both relevant phases may be present. For instance, water at zero degrees Celsius may be composed of some liquid and some solid, or all liquid, or all solid. Pressure changes, in addition to temperature changes, can cause phase changes. For example, nitrogen can be liquefied under high pressure, even though its boiling temperature is very low. Therefore, the **correct answer must be (B)**. Water may be a liquid at that temperature, but it may also be a solid, depending on ambient pressure.

98. Which of the following is the longest (largest) unit of geological time?

A. Solar Year.

B. Epoch.

C. Period.

D. Era.

D. Era.

Geological time is measured by many units, but the longest unit listed here (and indeed the longest used to describe the biological development of the planet) is the Era. Eras are subdivided into Periods, which are further divided into Epochs. Therefore, the **answer is (D)**.

99. **Extensive use of antibacterial soap has been found to increase the virulence of certain infections in hospitals. Which of the following might be an explanation for this phenomenon?**

A. Antibacterial soaps do not kill viruses.

B. Antibacterial soaps do not incorporate the same antibiotics used as medicine.

C. Antibacterial soaps kill a lot of bacteria, and only the hardiest ones survive to reproduce.

D. Antibacterial soaps can be very drying to the skin.

C. Antibacterial soaps kill a lot of bacteria, and only the hardiest ones survive to reproduce.

All of the answer choices in this question are true statements, but the question specifically asks for a cause of increased disease virulence in hospitals. This phenomenon is due to natural selection. The bacteria that can survive contact with antibacterial soap are the strongest ones, and without other bacteria competing for resources, they have more opportunity to flourish. This problem has led to several antibiotic-resistant bacterial diseases in hospitals nation-wide. Therefore, the **answer is (C)**. However, note that answers (A) and (D) may be additional problems with over-reliance on antibacterial products.

100. Which of the following is a correct explanation for astronaut 'weightlessness'?

A. Astronauts continue to feel the pull of gravity in space, but they are so far from planets that the force is small.

B. Astronauts continue to feel the pull of gravity in space, but spacecraft have such powerful engines that those forces dominate, reducing effective weight.

C. Astronauts do not feel the pull of gravity in space, because space is a vacuum.

D. Astronauts do not feel the pull of gravity in space, because black hole forces dominate the force field, reducing their masses.

A. Astronauts continue to feel the pull of gravity in space, but they are so far from planets that the force is small.

Gravity acts over tremendous distances in space (theoretically, infinite distance, though certainly at least as far as any astronaut has traveled). However, gravitational force is inversely proportional to distance squared from a massive body. This means that when an astronaut is in space, s/he is far enough from the center of mass of any planet that the gravitational force is very small, and s/he feels 'weightless'. Space is mostly empty (i.e. vacuum), and there are some black holes, and spacecraft do have powerful engines. However, none of these has the effect attributed to it in the incorrect answer choices (B), (C), or (D). The answer to this question must **therefore be (A).**

101. The theory of 'sea floor spreading' explains _____

A. the shapes of the continents.

B. how continents collide.

C. how continents move apart.

D. how continents sink to become part of the ocean floor.

C. How continents move apart.

In the theory of 'sea floor spreading', the movement of the ocean floor causes continents to spread apart from one another. This occurs because crust plates split apart, and new material is added to the plate edges. This process pulls the continents apart, or may create new separations, and is believed to have caused the formation of the Atlantic Ocean. The **answer is (C).**

102. Which of the following animals are most likely to live in a tropical rain forest?

A. Reindeer.

B. Monkeys.

C. Puffins.

D. Bears.

B. Monkeys.

The tropical rain forest biome is hot and humid, and is very fertile—it is thought to contain almost half of the world's species. Reindeer (A), puffins (C), and bears (D), however, are usually found in much colder climates. There are several species of monkeys that thrive in hot, humid climates, so **answer (B) is correct.**

103. Which of the following is *not* a type of volcano?

A. Shield Volcanoes.

B. Composite Volcanoes.

C. Stratus Volcanoes.

D. Cinder Cone Volcanoes.

C. Stratus Volcanoes.

There are three types of volcanoes. Shield volcanoes (A) are associated with non-violent eruptions and repeated lava flow over time. Composite volcanoes (B) are built from both lava flow and layers of ash and cinders. Cinder cone volcanoes (D) are associated with violent eruptions, such that lava is thrown into the air and becomes ash or cinder before falling and accumulating. **'Stratus' (C)** is a type of cloud, not volcano, so it is the correct answer to this question.

104. Which of the following is *not* a property of metalloids?

A. Metalloids are solids at standard temperature and pressure.

B. Metalloids can conduct electricity to a limited extent.

C. Metalloids are found in groups 13 through 17.

D. Metalloids all favor ionic bonding.

D. Metalloids all favor ionic bonding.

Metalloids are substances that have characteristics of both metals and nonmetals, including limited conduction of electricity and solid phase at standard temperature and pressure. Metalloids are found in a 'stair-step' pattern from Boron in group 13 through Astatine in group 17. Some metalloids, e.g. Silicon, favor covalent bonding. Others, e.g. Astatine, can bond ionically. Therefore, **the answer is (D).** Recall that metals/nonmetals/metalloids are not strictly defined by Periodic Table group, so their bonding is unlikely to be consistent with one another.

105. Which of these is a true statement about loamy soil?

A. Loamy soil is gritty and porous.

B. Loamy soil is smooth and a good barrier to water.

C. Loamy soil is hostile to microorganisms.

D. Loamy soil is velvety and clumpy.

D. Loamy soil is velvety and clumpy.

The three classes of soil by texture are: Sandy (gritty and porous), Clay (smooth, greasy, and most impervious to water), and Loamy (velvety, clumpy, and able to hold water and let water flow through). In addition, loamy soils are often the most fertile soils. Therefore, the **answer must be (D)**.

106. Lithification refers to the process by which unconsolidated sediments are transformed into _____

A. metamorphic rocks.

B. sedimentary rocks.

C. igneous rocks.

D. lithium oxide.

B. Sedimentary rocks.

Lithification is the process of sediments coming together to form rocks, i.e. sedimentary rock formation. Metamorphic and igneous rocks are formed via other processes (heat and pressure or volcano, respectively). Lithium oxide shares a word root with 'lithification' but is otherwise unrelated to this question. Therefore, the **answer must be (B)**.

107. Igneous rocks can be classified according to which of the following?

A. Texture.

B. Composition.

C. Formation process.

D. All of the above.

D. All of the above.

Igneous rocks, which form from the crystallization of molten lava, are classified according to many of their characteristics, including texture, composition, and how they were formed. Therefore, **the answer is (D).**

108. Which of the following is the most accurate definition of a nonrenewable resource?

A. A nonrenewable resource is never replaced once used.

B. A nonrenewable resource is replaced on a timescale that is very long relative to human life-spans.

C. A nonrenewable resource is a resource that can only be manufactured by humans.

D. A nonrenewable resource is a species that has already become extinct.

B. A nonrenewable resource is replaced on a timescale that is very long relative to human life-spans.

Renewable resources are those that are renewed, or replaced, in time for humans to use more of them. Examples include fast-growing plants, animals, or oxygen gas. (Note that while sunlight is often considered a renewable resource, it is actually a nonrenewable but extremely abundant resource.) Nonrenewable resources are those that renew themselves only on very long timescales, usually geologic timescales. Examples include minerals, metals, or fossil fuels. Therefore, the **correct answer is (B)**.

109. The theory of 'continental drift' is supported by which of the following?

A. The way the shapes of South America and Europe fit together.

B. The way the shapes of Europe and Asia fit together.

C. The way the shapes of South America and Africa fit together.

D. The way the shapes of North America and Antarctica fit together.

C. The way the shapes of South America and Africa fit together.

The theory of 'continental drift' states that many years ago, there was one land mass on the earth ('pangea'). This land mass broke apart via earth crust motion, and the continents drifted apart as separate pieces. This is supported by the shapes of South America and Africa, which seem to fit together like puzzle pieces if you look at a globe. Note that answer choices (A), (B), and (D) give either land masses that do not fit together, or those that are still attached to each other. Therefore, the **answer must be (C)**.

110. When water falls to a cave floor and evaporates, it may deposit calcium carbonate. This process leads to the formation of which of the following?

A. Stalactites.

B. Stalagmites.

C. Fault lines.

10. Sedimentary rocks.

B. Stalagmites.

To answer this question, recall the trick to remember the kinds of crystals formed in caves. Stalactites have a 'T' in them, because they form hanging from the ceiling (resembling a 'T'). Stalagmites have an 'M' in them, because they make bumps on the floor (resembling an 'M'). Note that fault lines and sedimentary rocks are irrelevant to this question. Therefore, **the answer must be (B)**.

111. A child has type O blood. Her father has type A blood, and her mother has type B blood. What are the genotypes of the father and mother, respectively?

A. AO and BO.

B. AA and AB.

C. OO and BO.

D. AO and BB.

A. AO and BO.

Because O blood is recessive, the child must have inherited two O's—one from each of her parents. Since her father has type A blood, his genotype must be AO; likewise her mother's blood must be BO. Therefore, only **answer (A)** can be correct.

112. Which of the following is the best definition for 'meteorite'?

A. A meteorite is a mineral composed of mica and feldspar.

B. A meteorite is material from outer space, that has struck the earth's surface.

C. A meteorite is an element that has properties of both metals and nonmetals.

D. A meteorite is a very small unit of length measurement.

B. A meteorite is material from outer space, that has struck the earth's surface.

Meteoroids are pieces of matter in space, composed of particles of rock and metal. If a meteoroid travels through the earth's atmosphere, friction causes burning and a "shooting star"—i.e. a meteor. If the meteor strikes the earth's surface, it is known as a meteorite. Note that although the suffix –ite often means a mineral, answer (A) is incorrect. Answer (C) refers to a 'metalloid' rather than a 'meteorite', and answer (D) is simply a misleading pun on 'meter'. Therefore, the **answer is (B)**.

113. A white flower is crossed with a red flower. Which of the following is a sign of incomplete dominance?

A. Pink flowers.

B. Red flowers.

C. White flowers.

D. No flowers.

A. Pink flowers.

Incomplete dominance means that neither the red nor the white gene is strong enough to suppress the other. Therefore both are expressed, leading in this case to the formation of pink flowers. Therefore, the **answer is (A)**.

114. What is the source for most of the United States' drinking water?

A. Desalinated ocean water.

B. Surface water (lakes, streams, mountain runoff).

C. Rainfall into municipal reservoirs.

D. Groundwater.

D. Groundwater.

Groundwater currently provides drinking water for 53% of the population of the United States. (Although groundwater is often less polluted than surface water, it can be contaminated and it is very hard to clean once it is polluted. If too much groundwater is used from one area, then the ground may sink or shift, or local salt water may intrude from ocean boundaries.) The other answer choices can be used for drinking water, but they are not the most widely used. Therefore, **the answer is (D)**.

115. Which is the correct sequence of insect development?

A. Egg, pupa, larva, adult.

B. Egg, larva, pupa, adult.

C. Egg, adult, larva, pupa.

D. Pupa, egg, larva, adult.

B. Egg, larva, pupa, adult.

An insect begins as an egg, hatches into a larva (e.g. caterpillar), forms a pupa (e.g. cocoon), and emerges as an adult (e.g. moth). Therefore, the **answer is (B)**.

116. A wrasse (fish) cleans the teeth of other fish by eating away plaque. This is an example of _____ between the fish.

A. parasitism.

B. symbiosis (mutualism).

C. competition.

D. predation.

B. Symbiosis (mutualism).

When both species benefit from their interaction in their habitat, this is called 'symbiosis', or 'mutualism'. In this example, the wrasse benefits from having a source of food, and the other fish benefit by having healthier teeth. Note that 'parasitism' is when one species benefits at the expense of the other, 'competition' is when two species compete with one another for the same habitat or food, and 'predation' is when one species feeds on another. Therefore, the **answer is (B)**.

117. What is the main obstacle to using nuclear fusion for obtaining electricity?

A. Nuclear fusion produces much more pollution than nuclear fission.

B. There is no obstacle; most power plants us nuclear fusion today.

C. Nuclear fusion requires very high temperature and activation energy.

D. The fuel for nuclear fusion is extremely expensive.

C. Nuclear fusion requires very high temperature and activation energy.

Nuclear fission is the usual process for power generation in nuclear power plants. This is carried out by splitting nuclei to release energy. The sun's energy is generated by nuclear fusion, i.e. combination of smaller nuclei into a larger nucleus. Fusion creates much less radioactive waste, but it requires extremely high temperature and activation energy, so it is not yet feasible for electricity generation. Therefore, the **answer is (C)**.

118. Which of the following is a true statement about radiation exposure and air travel?

A. Air travel exposes humans to radiation, but the level is not significant for most people.

B. Air travel exposes humans to so much radiation that it is recommended as a method of cancer treatment.

C. Air travel does not expose humans to radiation.

D. Air travel may or may not expose humans to radiation, but it has not yet been determined.

A. Air travel exposes humans to radiation, but the level is not significant for most people.

Humans are exposed to background radiation from the ground and in the atmosphere, but these levels are not considered hazardous under most circumstances, and these levels have been studied extensively. Air travel does create more exposure to atmospheric radiation, though this is much less than people usually experience through dental X-rays or other medical treatment. People whose jobs or lifestyles include a great deal of air flight may be at increased risk for certain cancers from excessive radiation exposure. Therefore, the **answer is (A)**.

119. Which process(es) result(s) in a haploid chromosome number?

A. Mitosis.

B. Meiosis.

C. Both mitosis and meiosis.

D. Neither mitosis nor meiosis.

B. Meiosis.

Meiosis is the division of sex cells. The resulting chromosome number is half the number of parent cells, i.e. a 'haploid chromosome number'. Mitosis, however, is the division of other cells, in which the chromosome number is the same as the parent cell chromosome number. Therefore, the **answer is (B)**.

120. Which of the following is *not* a member of Kingdom Fungi?

A. Mold.

B. Blue-green algae.

C. Mildew.

D. Mushrooms.

B. Blue-green Algae.

Mold (A), mildew (C), and mushrooms (D) are all types of fungus. Blue-green algae, however, is in Kingdom Monera. Therefore, the **answer is (B)**.

121. Which of the following organisms use spores to reproduce?

A. Fish.

B. Flowering plants.

C. Conifers.

D. Ferns.

D. Ferns.

Ferns, in Division Pterophyta, reproduce with spores and flagellated sperm. Flowering plants reproduce via seeds, and conifers reproduce via seeds protected in cones (e.g. pinecone). Fish, of course, reproduce sexually. Therefore, the **answer is (D)**.

122. What is the main difference between the 'condensation hypothesis' and the 'tidal hypothesis' for the origin of the solar system?

A. The tidal hypothesis can be tested, but the condensation hypothesis cannot.

B. The tidal hypothesis proposes a near collision of two stars pulling on each other, but the condensation hypothesis proposes condensation of rotating clouds of dust and gas.

C. The tidal hypothesis explains how tides began on planets such as Earth, but the condensation hypothesis explains how water vapor became liquid on Earth.

D. The tidal hypothesis is based on Aristotelian physics, but the condensation hypothesis is based on Newtonian mechanics.

B. The tidal hypothesis proposes a near collision of two stars pulling on each other, but the condensation hypothesis proposes condensation of rotating clouds of dust and gas.

Most scientists believe the 'condensation hypothesis,' i.e. that the solar system began when rotating clouds of dust and gas condensed into the sun and planets. A minority opinion is the 'tidal hypothesis,' i.e. that the sun almost collided with a large star. The large star's gravitational field would have then pulled gases out of the sun; these gases are thought to have begun to orbit the sun and condense into planets. Because both of these hypotheses deal with ancient, unrepeatable events, neither can be tested, eliminating answer (A). Note that both 'tidal' and 'condensation' have additional meanings in physics, but those are not relevant here, eliminating answer (C). Both hypotheses are based on best guesses using modern physics, eliminating answer (D). Therefore, the **answer is (B)**.

TEACHER CERTIFICATION STUDY GUIDE

123. Which of the following units is *not* a measure of distance?

A. AU (astronomical unit).

B. Light year.

C. Parsec.

D. Lunar year.

D. Lunar year.

Although the terminology is sometimes confusing, it is important to remember that a 'light year' (B) refers to the distance that light can travel in a year. Astronomical Units (AU) (A) also measure distance, and one AU is the distance between the sun and the earth. Parsecs (C) also measure distance, and are used in astronomical measurement- they are very large, and are usually used to measure interstellar distances. A lunar year, or any other kind of year for a planet or moon, is the *time* measure of that body's orbit. Therefore, the answer to this **question is (D)**.

124. The salinity of ocean water is closest to _____ .

A. 0.035 %

B. 0.35 %

C. 3.5 %

D. 35 %

C. 3.5 %

Salinity, or concentration of dissolved salt, can be measured in mass ratio (i.e. mass of salt divided by mass of sea water). For Earth's oceans, the salinity is approximately 3.5 %, or 35 parts per thousand. Note that answers (A) and (D) can be eliminated, because (A) is so dilute as to be hardly saline, while (D) is so concentrated that it would not support ocean life. Therefore, the **answer is (C)**.

125. Which of the following will not change in a chemical reaction?

A. Number of moles of products.

B. Atomic number of one of the reactants.

C. Mass (in grams) of one of the reactants.

D. Rate of reaction.

B. Atomic number of one of the reactants.

Atomic number, i.e. the number of protons in a given element, is constant unless involved in a nuclear reaction. Meanwhile, the amounts (measured in moles (A) or in grams(C)) of reactants and products change over the course of a chemical reaction, and the rate of a chemical reaction (D) may change due to internal or external processes. Therefore, the **answer is (B)**.

XAMonline, INC. 21 Orient Ave. Melrose, MA 02176

Toll Free number 800-509-4128

TO ORDER Fax 781-662-9268 OR www.XAMonline.com

TEXAS EXAMINATION OF EDUCATOR STANDARDS- EXAMINATION FOR THE CERTIFICATION OF EDUCATORS - TEXES/EXCET - 2007

PO#   Store/School:

Address 1:

Address 2 (Ship to other):

City, State Zip

Credit card number _____-_____-_____-_____   expiration_____

EMAIL _____

PHONE   FAX

13# ISBN 2007	TITLE	Qty	Retail	Total
978-1-58197-925-1	ExCET ART SAMPLE TEST (ALL-LEVEL-SECONDARY) 005 006			
978-1-58197-949-7	TExES CHEMISTRY 8-12 140			
978-1-58197-938-1	TExES COMPUTER SCIENCE 141			
978-1-58197-933-6	TExES ENGLISH LANG-ARTS AND READING 4-8 117			
978-1-58197-935-0	TExES ENGLISH LANG-ARTS AND READING 8-12 131			
978-1-58197-926-8	ExCET FRENCH SAMPLE TEST (SECONDARY) 048			
978-1-58197-930-5	TExES GENERALIST 4-8 111			
978-1-58197-945-9	TExES GENERALIST EC-4 101			
978-1-58197-946-6	TExES SCIENCE 4-8 116			
978-1-58197-931-2	TExES SCIENCE 8-12 136			
978-1-58197-604-5	TExES LIFE SCIENCE 8-12 138			
978-1-58197-932-9	TExES MATHEMATICS 4-8 114-115			
978-1-58197-937-4	TExES MATHEMATICS 8-12 135			
978-1-58197-939-8	TExES MATHEMATICS-PHYSICS 8-12 143			
978-1-58197-948-0	TExES MATHEMATICS-SCIENCE 4-8 114			
978-1-58197-929-9	TExES PEDAGOGY AND PROFESSIONAL RESPONSIBILITIES 4-8 110			
978-1-58197-899-5	TExES PEDAGOGY AND PROFESSIONAL RESPONSIBILITIES 8-12 130			
978-1-58197-943-5	TExES PHYSICAL EDUCATON EC-12 158			
978-1-58197-928-2	TExES PRINCIPAL 068			
978-1-58197-941-1	TExES READING SPECIALIST 151			
978-1-58197-942-8	TExES SCHOOL COUNSELOR 152			
978-1-58197-940-4	TExES SCHOOL LIBRARIAN 150			
978-1-58197-934-3	TExES SOCIAL STUDIES 4-8 118			
978-1-58197-936-7	TExES SOCIAL STUDIES 8-12 132			
978-1-58197-927-5	ExCET SPANISH (SECONDARY) 047			
978-1-58197-944-2	TExES SPECIAL EDUCATION EC-12 161			
			SUBTOTAL	
	FOR PRODUCT PRICES GO TO WWW.XAMONLINE.COM		Ship	$8.25
			TOTAL	

www.ingramcontent.com/pod-product-compliance
Lightning Source LLC
Chambersburg PA
CBHW080531300426
44111CB00017B/2681